# AIR POLLUTION CONTROL TECHNOLOGIES AND POLICIES

# AIR POLLUTION CONTROL TECHNOLOGIES AND POLICIES

**Edited by**

**Dr. Kailash Malode**

Delve
PUBLISHING

www.delvepublishing.com

**Air Pollution Control Technologies and Policies**

*Dr. Kailash Malode*

**Delve Publishing**
4164 Lakeshore Road
Burlington, ON L7L 1A4
Canada
www.delvepublishing.com
Email: orders@arclereducation.com



ISBN: 978-1-77956-437-5 (Hardcover)

# ABOUT THE EDITOR

**Dr. Kailash Malode** (1980) is presently serving as Assistant Professor, Department of Soil Science and Agriculture Chemistry, college of Agriculture, Parul University, Vadodara, Gujarat. He obtained his B. Sc. (Ag.) in 2006 from Marathwada Agriculture University, Parbhani. M. Sc. (Ag.)2009 in Dr. Panjabrao Deshmukh Krishi Vidyapeeth,Akola and Ph. D. Soil Science And Agriculture Chemistry in 2014 from Vasantrao Naik Marathwada Krishi Vidhyapeeth, Parbhani. He started his career as Assistant professor in 2014. His field of specialization is soil fertility, Remote Sensing and Organic Farming. He has published 02 books and 15 research paper in National and international journals of reputed. He has participated many National, State Seminar and symposiums. He has written 02 practical manuals for under graduate course for the benefit of students besides several Radio talk, Lectures in farmer traning programme. Dr. Kailash has also written 21 popular articles in Shati Bharti, Shati pragati and other reputed Magazines. He has Best writer award on 2015 for Krishi dooth popular Magazines of Maharashtra and One Young scientist award in 2020.

# TABLE OF CONTENTS

# LIST OF FIGURES

# LIST OF TABLES

# ABSTRACT


This book, Air Pollution Control Technologies, and Policies, comprehensively explores air pollution, its impacts on human health and the environment, and the technologies and policies developed to control it. The text spans multiple scales of air pollution, from local to global, and provides an in-depth analysis of indoor and outdoor air quality. It details the health risks associated with air pollution, such as respiratory and cardiovascular diseases, and examines environmental effects, including acid rain and the greenhouse effect.

The book covers methods for monitoring ambient air quality and pollutant emissions, offering insight into the types of sensors, measurement techniques, and continuous emission monitoring systems in use today. Additionally, it provides practical strategies for controlling pollution from both stationary sources, such as factories, and mobile sources like vehicles. The final chapters discuss emission inventories and pollution prevention strategies, emphasizing regulatory frameworks, compliance, and sustainable practices.

Designed for environmental scientists, engineers, policymakers, and students, this resource offers valuable knowledge for addressing the global challenge of air pollution, balancing scientific understanding with practical approaches to pollution control.

# ABSTRACT 02

Air Pollution Control Technologies and Policies provides a detailed examination of air pollution, its sources, health effects, environmental impacts, and the various technologies and policies designed to control and mitigate its consequences. The book is structured across ten chapters, each focusing on key aspects of air pollution, from its sources and dispersion to its monitoring and prevention.

The initial chapters introduce air pollution, explaining its various scalesranging from local to global—and the types of pollutants involved. These include primary pollutants like sulfur dioxide and nitrogen oxides, and secondary pollutants such as ozone. The book describes how pollutants disperse in the atmosphere and contribute to global issues like climate change and ozone depletion. The Earth's atmosphere, its structure, composition, and energy balance are covered to provide context on how air pollution interacts with the environment. Subsequent chapters explore the health and environmental effects of air pollution. The text highlights the hazardous impact of pollutants on human health, including the respiratory and cardiovascular systems. It delves into the ways pollutants like particulate matter and toxic gases are linked to increased risks of diseases such as lung cancer, heart disease, and chronic respiratory disorders. Beyond human health, the environmental effects of air pollution are discussed, particularly acid rain, greenhouse gas emissions, and the depletion of the ozone layer. A critical part of air pollution management is monitoring. The book details techniques used to measure air quality and emissions from both industrial and natural sources. It covers tools for ambient air quality monitoring, including particulate matter sensors and gas analyzers. The role of continuous emission monitoring systems (CEMS) and the methodologies for collecting and analyzing emissions data are outlined, along with the regulatory frameworks that guide these efforts.

A significant focus is placed on the control technologies for stationary and mobile pollution sources. For stationary sources like factories and power plants, the book outlines technologies such as flue-gas desulfurization and catalytic reduction to manage particulate and gas emissions. The strategies for reducing emissions from mobile sources, including vehicles and off-road equipment, are also discussed, covering advancements in vehicle technology and fuel standards. The text emphasizes the importance of emissions inventories, which compile data on the sources and amounts of pollutants released into the atmosphere. These inventories are essential for regulatory development, air quality management, and compliance with environmental laws. The book concludes by addressing pollution prevention strategies, focusing on source reduction, recycling, and sustainable practices that minimize pollutant generation. Air Pollution Control Technologies and Policies serves as a crucial resource for environmental scientists, engineers, policymakers, and students. It provides a balanced understanding of the scientific, technical, and regulatory challenges of air pollution, offering practical solutions to combat one of the most pressing environmental issues of our time. Through detailed analysis and case studies, the book guides the reader toward effective air quality management and policy implementation, aiming to protect both human health and the environment.

# PREFACE

Air pollution and climate change have increasingly compromised human health and adversely affected plants and vegetation, posing severe threats to the entire ecosystem. The escalation of these environmental issues can be largely attributed to rising energy demands and the pervasive reliance on fossil fuels and nonrenewable energy sources. This dependence exacerbates air pollution, leading to the emission of a variety of harmful pollutants including gases such as sulfur dioxide ($SO_2$), nitrogen oxides ($NO_2$), ozone ($O_3$), and hydrocarbons, as well as particulate matter composed of various solid elements and dust particles. These pollutants inflict significant damage on plant life, reducing crop production and overall vegetation health.

The impact of air pollution on plants is both direct and indirect. Direct effects include the damage to plant tissues and the inhibition of physiological processes such as photosynthesis, respiration, and transpiration. For example, sulfur dioxide can enter plant leaves through stomata, causing chlorosis and reducing the plant's ability to photosynthesize. Nitrogen oxides can lead to the formation of ground-level ozone, which further harms plant tissues and impairs growth. Indirectly, air pollutants can alter soil chemistry and nutrient availability, affecting plant health and crop yields. The extent of these impacts depends on various factors, including the concentration of pollutants, the duration of exposure, and the age and species of plants. Young plants and crops are often more vulnerable to pollutant damage compared to mature vegetation.

The deteriorating quality of air and the environment necessitates a comprehensive and well-planned strategy to mitigate the adverse effects of air pollution. Addressing this challenge requires a thorough understanding of the causes, impacts, and control measures related to air pollution.

The book *Air Pollution Control Technologies and Policies* offers a detailed exploration of these aspects, providing valuable insights into the nature of pollutants affecting plants and vegetation, their sources, and the technologies adopted to control pollution levels. This book is structured into ten chapters, each addressing a critical dimension of air pollution control:

First chapter introduces the fundamental concepts of air pollution, including its sources, types, and the basic mechanisms through which pollutants impact the environment and human health. Second chapter examines the various ways air pollution affects both human health and the environment, highlighting the health risks posed by different pollutants and their impact on ecosystems.

Understanding how pollutants travel and disperse in the atmosphere is crucial for effective pollution management. Third chapter covers the atmospheric processes that influence pollutant distribution and concentration. While outdoor air pollution is a significant concern, indoor air quality also affects human health and plant life. Fourth chapter explores sources of indoor air pollution and their impact on indoor environments.

Accurate monitoring of air quality is essential for assessing pollution levels and informing control strategies. Fifth chapter discusses various techniques and tools used to measure ambient air quality. Sixth chapter covers the methodologies for measuring emissions from different sources, including industrial processes and transportation, to understand their contribution to overall air pollution.

An emissions inventory provides a comprehensive database of pollutant sources and their emission rates. Therefore, chapter seven explains how inventories are compiled and utilized for pollution management. Stationary sources such as factories and power plants are significant contributors to air pollution. Chapter eight details the technologies and strategies used to control emissions from these sources.

Vehicles and other mobile sources are major contributors to air pollution. Chapter nine discusses the methods and policies designed to reduce emissions from transportation. Prevention is often more effective than control. Chapter ten explores strategies and practices aimed at reducing the generation of pollutants and minimizing their impact on the environment.

This book is aimed at environmental scientists, engineers, and policymakers involved in air quality management. It is also a valuable resource for students and researchers in environmental science and engineering, providing a detailed overview of air pollution sources, effects, and control technologies. Additionally, the book serves as a practical guide for industry professionals tasked with implementing air pollution control measures. Its comprehensive coverage of monitoring, emission control, and policy development makes it an essential reference for anyone involved in designing, managing, or studying air pollution control strategies and regulations.

*Chapter* **1**

# Overview of Air Pollution

## Contents

# INTRODUCTION

Air pollution is the term used to describe the condition in which materials or the climate are harmed, or when humans and other living things are exposed to substances known as pollutants in the atmosphere. The natural characteristics of the atmosphere can also be changed by the introduction of chemical, physical, or biological agents into the indoor or outdoor environment. There are many distinct kinds of air pollutants, including particulates (both organic and inorganic), gases (such as ammonia, carbon monoxide, sulfur dioxide, nitrous oxides, methane, and chlorofluorocarbons), and biological molecules. Aside from harming humans, air pollution can also affect other living things like animals and crops. It can also degrade the built environment (such as acid rain) or the natural environment (such as ozone depletion, climate change, and habitat degradation). Both natural and human-caused factors can contribute to air pollution.

Air quality is closely related to the Earth's climate and ecosystems globally. Many of the contributors of air pollution are also sources of greenhouse emission i.e., burning of fossil fuel. Lung cancer, heart disease, stroke, chronic obstructive pulmonary disease (COPD), respiratory infections, and other pollution-related illnesses are all significantly increased by air pollution. An increasing body of research indicates that exposure to air pollution may be linked to lower IQs, cognitive decline, a higher risk of mental illnesses like depression, and poor perinatal health. Although the effects of poor air quality on human health are extensive, the respiratory and cardiovascular systems of the body are primarily affected. Individual responses to air pollution are influenced by the kind of pollutant they are exposed to, the level of exposure, their genetic makeup, and their current state of health.

The greatest environmental risk factor for illness and early death is air pollution, which also ranks as the fourth-highest risk factor for overall human health. Around 7 million people worldwide die prematurely from air pollution each year, translating into a global mean loss of life expectancy (LLE) of 2.9 years. Since at least 2015, there hasn't been a discernible shift in the total number of deaths from pollution-related causes. One of the leading causes of death for people worldwide, outdoor air pollution directly related to the burning of fossil fuels accounts for approximately 3.61 million deaths per year. An additional 2.1 million premature deaths and approximately 470,000 cases of fine particulate matter ($PM_{2.5}$) pollution are attributed to anthropogenic ozone each year. The extent of the air pollution crisis is significant: according to WHO estimates from 2018, 90% of people breathe air that is highly polluted. Despite the serious health effects, the way the issue is handled is mainly seen as careless or haphazard.

According to estimates from the World Bank, air pollution costs the global economy $5 trillion annually in lost labor and welfare losses (early deaths). The expenses associated with air pollution are typically an externality of the modern economy and the majority of human activity, though occasionally they can be covered by regulation, legislation, and monitoring. There are numerous approaches and technologies available to lessen air pollution. Despite the fact that most nations have laws pertaining to air pollution, the UNEP reports that 43% of nations do not have a legal definition of the term, 31% do not have standards for outdoor air quality, 49% limit their definition to outdoor pollution alone, and only 31% of nations have laws addressing pollution coming from outside their borders. National legislation pertaining to air quality, such as the US Clean Air Act of 1963 and the British Clean Air Act of 1956, have frequently proven to be very successful. International action on climate change has been less successful than some of these efforts, like the Montreal Protocol, which limited the release of dangerous chemicals that deplete the ozone layer, and the Helsinki Protocol, which reduced sulfur emissions in 1985.

## 1.1. SCALES OF AIR POLLUTION

There are levels of ambient air pollution, ranging from very local to very global. Five distinct scales are examined in this section: local, urban, continental, and global. The air pollutants themselves have a range of effects, from molecular (e.g., particles and gases) to planetary (e.g., greenhouse gas dispersion across the troposphere). The local scale covers a maximum of five kilometers on Earth's surface. The urban scale is roughly 50 kilometers in length. The range of the regional scale is 50–500 km. The range of continental scales is 500–1,000 km. Naturally, the global scale is international.

### 1.1.1. Local

A large number of relatively small emitters or one or more large emitters are typically the defining characteristics of local air pollution problems. The potential impact for a given release increases with a source's lower release height. One example of a gas that is released by motor vehicles and can accumulate near roads is carbon monoxide. Any ground-level source will produce the highest concentrations close to the source, usually decreasing with distance, such as the evaporation of volatile organic compounds from a waste treatment pond. A concentration gradient is the name given to this phenomenon. Large sources, like power plants or industrial sources, that release emissions high above the ground through stacks can also be a local source of pollution, particularly when weather patterns are erratic and allow high concentrations of part of the plume to fall to the ground. On top of one- or two-story buildings, there are numerous vents or relatively short stacks that release a lot of pollutants. These releases are typically caught in the tumultuous downwash downwind of the structure. This makes it possible to bring large concentrations to the surface. In this way, a wide range of pollutants, including odor-causing compounds and mixtures, can be released. The effects of unintentional releases are typically limited to the local level.

### 1.1.2. Urban

Urban air pollution issues typically fall into two categories. Primary pollutants, or those released straight from sources, are the first. The other is the creation of secondary pollutants, which result from the primary pollutants' chemical reactions. Both locally and urbanly focused individual sources can contribute to air pollution issues.

High concentrations are produced when the contributions from various sources combine to produce pollutants that are either relatively slowly reactive, like sulfur dioxide, or relatively nonreactive, like carbon monoxide and particulate matter. Motor vehicles are a major source of carbon monoxide, so hot spots with high concentrations can happen, particularly close to multilane intersections. Particularly high emissions come from vehicles that are idle. High buildings surrounding the intersection exacerbate the hot spots because they severely restrict the amount of air that can contain the pollution. High concentrations are the outcome of these factors working together.

The main issue facing cities as a result of secondary pollution formation is tropospheric ozone. Ozone is created by photochemical reactions between different types of hydrocarbons and oxides of nitrogen in many large urban areas. These processes are known as photochemical reactions because they are aided by the UV rays found in sunlight. Many urban areas do not meet the standards for air quality, or are in nonattainment for ozone. The Clean Air Act Amendments (CAAA) of 1990 acknowledge this as a significant issue and have categorized the different metropolitan areas as nonattainment based on the intensity of the issue in that particular area.

The National Ambient Air Quality Standards (NAAQS) attainment timelines are established by the CAAA for each classification. Cars and combustion processes release oxides of nitrogen, primarily nitric oxide (NO) but also nitrogen dioxide ($NO_2$). Hydrocarbons come from a wide range of sources. The reactivities of the different species vary greatly. Although it can be challenging to calculate the emissions of these chemical species from numerous sources as the foundation for pollution control programs, techniques are constantly improving.

### 1.1.3. Regional

Regional problems related to air pollution are caused by at least three different kinds of issues. The first is the combination of regional urban oxidant issues. Numerous large cities are close to one another and are still expanding. Some of the larger urban aggregations are referred to as megalopolises by urban geographers. Consequently, air from one metropolitan area flows into the next, carrying with it primary pollutants as well as secondary pollutants created by reactions. Then, the background from the first area is topped with the pollutants from the second area.

Relatively slow-reacting primary air pollutants that undergo reactions and transformations during long transit times are a different kind of problem. In environmental engineering, the correlation between chemical transformation, time, and spatial coverage is a common occurrence. Consequently, long transit times lead to regional transport distances for many transformation byproducts as well as the parent compounds. Long-distance transportation oxidizes the gas sulfur dioxide ($SO_2$), which is mostly released during the burning of fossil fuels, particularly coal and oil, to produce sulfur trioxide ($SO_3$).

$$2SO_2 + O_2 \rightarrow 2SO_3$$

Despite being a gas, $SO_2$ can be oxidized in both the gas and liquid phases in the troposphere. Sulfuric acid is created when water vapor and $SO_3$ combine.

$$SO_3 + H_2O \rightarrow H_2SO_4$$

Numerous substances react with sulfuric acid to produce sulfates. These particles are tiny—submicrometer. High-temperature combustion produces nitric oxide (NO), which is produced in internal combustion engine vehicles as well as stationary sources like power plants and industrial plants that produce process heat. In the atmosphere, NO is oxidized to nitrogen dioxide ($NO_2$), typically quite slowly but more quickly if ozone is present. Additionally, $NO_2$ undergoes additional reactions with other components to form nitrates, which are also finely divided particles.

Many processes can be used to remove the sulfates and nitrates that are present in the atmosphere as fine particulates, usually smaller than 1 μm. When the particles act as condensation nuclei and cause clouds to form, rain occurs. If the droplets get big enough to fall as raindrops, the particles are then precipitated. Rain is also a part of another mechanism called washout, but in this case, the raindrops fall through the air and collect the airborne particles. Acid rain is a result of both processes, which raises the acidity of lakes and streams by allowing sulfate and nitrate particles to enter them. Acid rain is therefore a problem that affects both regions and continents.

Visibility is a third category of regional issue that can be caused by particular plumes or by localized amounts of particulate matter that result in different degrees of haze. The fine sulfate and nitrate particulates that were just mentioned are mostly to blame for the decrease in visibility. In areas of natural beauty, where it is preferable to maintain picturesque vistas free of view-obstructing structures, this issue is concerning. One form of visibility impairment known as regional haze is brought on by the emissions of air pollutants from multiple sources over a large geographic area. For these types of areas, the CAAA offers extra protections; the strictest are known as required Federal Class I areas, which include 156 national parks and wilderness areas. The EPA released regulations on July 1, 1999, to address the regional haze in these areas. Safety can also be hampered by decreased visibility, particularly in aviation.

### 1.1.4. Continental

There isn't much of a distinction between what is regarded as regional and continental scale in a continent the size of Europe. Larger continents, however, would see a significant variation. On a continental scale, the possibility that a country's air pollution policies will affect its neighbors is possibly the biggest cause for concern. It has been suggested that Western Europe and Great Britain had an influence on the acid rain that affected Scandinavia. Japan has acknowledged that China and Korea are partly to blame for some of the country's air pollution issues, particularly in the west. Canada and the US have worked together for many years to investigate and resolve the issue of acid rain in North America. Numerous lakes in Scandinavia and North America have turned so acidic that fish cannot survive in them. According to a recent study on lakes in Norway, almost all of the lakes with a pH of 5.5 and above had fish, while more than 70% of the lakes with a pH of less than 4.5 had none at all. The issue is exacerbated by the low pH, which not only directly harms fish but also facilitates the release of potentially harmful metals like aluminum. In half of the high mountain lakes in the Adirondacks, acid rain has already completely destroyed the populations of many fish and plants in North America. In some lakes, the pH has gotten so low that acid-tolerant rough fish species and algal mats have taken the place of native plants and trout.

In response to the atmospheric acidity that is deposited on freshwater aquatic systems, the EPA proposed a 10–20 kg $SO_4^{2-}$ per hectare annual limit. When considering this limit in relation to current emissions, the state of Ohio alone emits $2.4 \times 10^6$ metric tons of gaseous sulfur dioxide ($SO_2$) annually. All of this would equal 360 kg per hectare annually if it were converted to $SO_4^{2-}$ and then deposited on the State of Ohio. Analogous computations regarding sulfur emissions in the northeastern United States reveal that the sulfur emission rate is four to five times higher than the deposition rate. Obviously, a large portion of the excess sulfur is oxidized into acidic chemical species, as the reactions above show, but what happens to the rest of it? For more than ten years, the United States and Canada have been debating this. Similarly, a large portion of the issue in Scandinavia is related to the use of tall stacks in continental Europe's lowland nations and Great Britain. For a while, British industries controlled air pollution by simply building taller stacks, which decreased the concentration of pollutants at ground level but released the same pollutants into the upper atmosphere. Although acid rain was experienced in other parts of Europe, the UK's air quality improved. The fact that addressing acid rain and climate change essentially means asking society to reduce combustion is one of the challenges. Furthermore, combustion is associated with advancement and development for many.

### 1.1.5. Global

It would be largely assumed that the radioactive leakage from the Chernobyl accident is a regional or continental issue. However, not long after the accident, higher than normal levels of radioactivity were found in the United States' Pacific Northwest. Similar to this, even thousands of miles away from their sources, Arctic mammals have been found to contain persistent organic pollutants like polychlorinated biphenyls (PCBs). These findings highlight the consequences of long-distance transportation. The release of chlorofluorocarbons used as propellants in air conditioners and spray cans, as well as their impact on the ozone layer high in the atmosphere, are particularly notable global air pollution issues.

Due to the introduction of radioactive debris into the stratosphere from atomic bomb tests in the Pacific, some understanding of the exchange processes between the troposphere and the stratosphere as well as between the Northern and Southern Hemispheres was gained in the late 1950s and early 1960s. The majority of the debris injection occurred in the stratosphere of the Northern Hemisphere. The stratosphere typically opposes vertical air exchange between layers and is fairly stable. It was discovered that there is a significant lag in the stratospheric exchange processes between the Northern and Southern Hemispheres. Within three years of the tests starting, radioactivity did, in fact, appear in the Southern Hemisphere, albeit at much lower levels than in the Northern Hemisphere. In a similar vein, the processes of exchange from the troposphere to the stratosphere move slowly. The tropopause, which separates the troposphere from the stratosphere, is occasionally breached by thunderstorm tops, which are the primary conduits for airborne particles from the troposphere into the stratosphere. Occasionally occurring holes in the tropopause also allow for some stratospheric air to be transferred downward. The ozone layer is located well above the troposphere, so it is anticipated that the upward transfer of chlorofluorocarbons to the ozone layer will occur very slowly. As a result, it took some time after the initial release of these gases for an impact to be noticed. In a similar vein, the restoration of the ozone layer has lagged behind the global cessation of use of these materials.

Climate change, which is a result of excessive emissions of radiant gases, also referred to as greenhouse gases, particularly methane ($CH_4$) and $CO_2$, which are not typically regarded as air pollutants, is another issue facing the entire world. However, some of this radiation is reradiated both downward and upward after being intercepted by the carbon dioxide in the atmosphere. The radiation that is directed downward prevents the earth from cooling too quickly. The earth's temperature is predicted to rise as long as the concentration of carbon dioxide keeps rising. The release of fine particulate matter into the atmosphere by volcanoes is one natural air pollution issue that has the potential to have an international impact. Around the world, the addition of this particulate matter has resulted in some breathtaking sunsets. The radiation balance can be altered if enough material is released. The earth's surface will warm less than usual during the day when incoming solar radiation is blocked. A volcanic mountain in the Pacific erupted in the middle of the 1800s, causing a mini-ice age. That year's summer was significantly colder than usual, and snow fell in New England in July.

Air pollution can have an impact on global phenomena like climate. The way that the climate reacts to various pollutants varies, just like most environmental systems. In actuality, certain pollutants cause the troposphere to cool, while others cause it to heat. Most scientists are currently very concerned about global warming. For instance, according to data from the International Panel of Climate Change (IPCC), greenhouse gas emissions worldwide rose by 70% between 1970 and 2004. While different models predict different effects of this increase on mean global temperatures and other climatic variables, many scientists and policy makers worry that failure to halt the trend could result in a significant environmental and societal disaster.

## 1.2. POLLUTANTS

A substance in the air that can negatively impact both people and the environment is known as an air pollutant. Aerosols, which are solid particles or liquid droplets dispersed and carried by a gas, are a common form of the substance (Hidy, 2012). The substance can also take the form of gases. A pollutant may be man-made or naturally occurring. There are two types of pollutants: primary and secondary. Primary pollutants are typically created by natural processes, like the eruption of volcanic ash. Sulfur dioxide emitted from factories and carbon monoxide gas from car exhausts are two more examples. Direct emissions of secondary pollutants do not occur. Instead, when primary pollutants interact or react, they form in the air. One well-known instance of a secondary pollutant is ground level ozone. Certain pollutants can be classified as primary or secondary depending on whether they originate from other primary pollutants or are released directly.

### 1.2.1. Primary Pollutants

Human activity releases a variety of pollutants into the atmosphere (Figure 1.1):

1. **Ammonia:** Primarily from agricultural waste emissions. $NH_3$ is the formula for the compound ammonia. It typically appears as a gas with a distinct, strong smell. Ammonia serves as a precursor to foods and fertilizers, greatly assisting in meeting the nutritional needs of terrestrial organisms. Many pharmaceuticals are also synthesized using ammonia, either directly or indirectly. Ammonia is dangerous and caustic even though it is widely used. Ammonia combines with sulfur and nitrogen oxides in the atmosphere to create secondary particles., ammonia reacts with oxides of nitrogen and sulfur to form secondary particles.
2. **Carbon Dioxide ($CO_2$):** Naturally occurring in the atmosphere, carbon dioxide is released by the human respiratory system and is necessary for plant life. At very high concentrations (usually 100 times normal atmospheric levels), it can be fatal. The World Health Organization does not include $CO_2$ in its Air Quality Guidelines or establish recommended targets for it, despite the fact that it acknowledges the gas as a climate pollutant. The worst climate pollutant has been dubbed $CO_2$ due to its function as a greenhouse gas. Such claims allude to the long-term effects of the gas on the atmosphere rather than the immediate effects on human health, food crops, and buildings. There are real-world implications to this nomenclature dispute, such as whether the U.S. It is believed that the Clean Air Act, which aims to enhance air quality, controls $CO_2$ emissions. The Inflation Reduction Act of 2022, which specifically amended the Clean Air Act to define carbon dioxide produced by the burning of fossil fuels as an air pollutant, resolved that issue in the United States. In pre-industrial times, $CO_2$ made up about 280 parts per million (ppm) of the Earth's atmosphere; today, billions of metric tons of $CO_2$ are released into the atmosphere each year

from the burning of fossil fuels. The Earth's atmosphere has been accumulating $CO_2$ faster. Since $CO_2$ is an asphyxiant, it cannot be generally categorized as toxic or harmful. In the UK, for example, there are workplace exposure limits of 5,000 ppm for long-term exposure and 15,000 ppm for short-term exposure. Massive volumes of $CO_2$ can also be released abruptly as a result of natural disasters like the limnic eruption at Lake Nyos.

3. **Carbon Monoxide (CO):** It is a harmful, odorless, and colorless gas. This gas is generated as a byproduct of fuel combustion, including natural gas, coal, or wood. Vehicle emissions account for the largest proportion of carbon monoxide released into the atmosphere. It contributes to the formation of a smog-like substance in the air, which has been associated with numerous lung ailments and disruptions to the ecosystem and animal populations.
4. **Chlorofluorocarbons (CFCs):** Released from products that are no longer in use; detrimental to the ozone layer. These are the gases released by freezers, air conditioners, aerosol sprays, and other appliances of a similar nature. After being released into the atmosphere, CFCs enter the stratosphere. This is where they interact with other gases, damaging the ozone layer. This allows UV rays to reach the surface of the Earth. Skin cancer, vision issues, and even damage to plants can arise from this.
5. **Nitrogen Oxides ($NO_x$):** These, in particular, are released from combustion at high temperatures and are also created by electric discharge during thunderstorms. They appear as a plume downwind of cities or as a dome of brown haze above. The chemical formula for nitrogen dioxide is $NO_2$. It's among the various nitrogen oxides. This reddish-brown, toxic gas is one of the most common air pollutants and has a distinct, sharp, and biting smell.
6. **Odors:** Like from sewage, trash, and industrial activities.
7. **Particulate Matter/Particles (PM):** These are microscopic solid or liquid particles suspended in a gas. They are also referred to as particulates, atmospheric particulate matter (APM), or fine particles. A mixture of gas and particles is called an aerosol. Particles can originate from volcanoes, dust storms, grassland and forest fires, living plants, and sea spray. Human activities include burning fossil fuels in power plants, cars, and other industrial processes can produce aerosols (Munsif, Zubair, Aziz, & Nadeem Zafar, 2021). Anthropogenic aerosols, or those produced by human activity, currently make up 10% of the atmosphere on average worldwide.

Elevations of particulate matter in the atmosphere are associated with health risks, including heart ailments, modified pulmonary function, and lung cancer. Particulates can be especially dangerous for people suffering from respiratory disorders like asthma (Balmes, Fine, & Sheppard, 1987). Particulates are linked to respiratory infections.

8. **Organic Pollutants:** These are persistent and can cling to particles. Organic compounds that are resistant to environmental degradation as a result of chemical, biological, or photolytic processes are known as persistent organic pollutants (POPs). They thus have been found to be able to endure in the environment, be long-range communicators, bioaccumulate in animal and human tissue, biomagnify in food chains, and represent a serious risk to both the ecosystem and human health (Singh, Kumar, Karmakar, Siddiqui, Mathur, Adnan, et al., 2021).
9. **Cardiopulmonary Disease:** It is associated with persistent free radicals that are linked to fine particles in the air (Kim, Jahan, Kabir, & Brown, 2013).
10. **Radioactive Pollutants:** Produced by natural processes like the radioactive decay of radon and by nuclear explosions, nuclear events, and war explosives.

**Figure 1.1.** *Before flue-gas desulfurization was installed, the emissions from this power plant in New Mexico contained excessive amounts of sulfur dioxide.*

*Source: By NPS – United States National Park Service (NPS). "Air Quality Core Slides: Sources of Air Pollution" (archived URL), Public Domain;* https://commons.wikimedia.org/w/index.php?curid=17444.

11. **Sulfur Oxides (SOx):** In particular, sulfur dioxide ($SO_2$), a chemical compound. Volcanoes and a number of industrial processes generate $SO_2$. Sulfur dioxide is produced when sulfur compounds found in coal and petroleum are burned. Acid rain is created when $SO_2$ is further oxidized, usually with the help of a catalyst like $NO_2$. This produces $H_2SO_4$. This is among the reasons why using these fuels as power sources has an adverse effect on the environment.
    - Toxic metals, particularly the compounds made of lead and mercury.
    - **Volatile Organic Compounds (VOC):** VOCs are air pollutants that can be found indoors and outdoors. They fall into one of two categories: non-methane (NMVOCs) or methane ($CH_4$). Methane is a highly effective greenhouse gas that intensifies global warming. Because they contribute to the formation of ozone and extend the atmospheric half-life of methane, other hydrocarbon volatile organic compounds (VOCs) are also important greenhouse gases. The impact varies according to the quality of the local air. Prolonged exposure to the aromatic NMVOCs benzene, toluene, and xylene may cause leukemia and raise the possibility of cancer. An additional hazardous substance that is frequently connected to industrial use is 1,3-butadiene.

### 1.2.2. Secondary Pollutants

Secondary pollutants include:

1. **Ground Level Ozone ($O_3$):** The combination of NOx and VOCs produces ozone. It makes up a sizeable portion of the troposphere. It is also a crucial component of the ozone layer, which is distributed throughout the stratosphere. Many of the chemical processes that take place in the atmosphere both during the day and at night are powered by photochemical processes and chemical reactions involving it. It is a pollutant that is created in large amounts by human activity, primarily the burning of fossil fuels, and it is a part of smog.
2. **Peroxyacetyl Nitrate ($C_2H_3NO_5$):** Similarly formed from $NO_x$ and VOCs.
3. **Photochemical Smog:** Chemicals and gaseous primary pollutants combine to form particles. One kind of pollution that exists in the atmosphere is smog. Smog is produced when a large amount of coal is burned in a particular area, releasing sulfur dioxide and smoke

into the atmosphere. Today's smog is mostly caused by industrial and automotive emissions that are exposed to UV light from the sun. These secondary pollutants then react with the primary emissions in the atmosphere to form photochemical smog.

### 1.2.3. Other Pollutants

Numerous other substances are categorized as dangerous air pollutants. A number of directives, such as the Air Framework Directive, 96/62/EC, on ambient air quality assessment and management, Directive 98/24/EC, on risks related to chemical agents at work, and Directive 2004/107/EC, covering heavy metals and polycyclic aromatic hydrocarbons in ambient air, regulate some of these in the United States under the Clean Air Act and in Europe under multiple directives.

## 1.3. THE EARTH'S ATMOSPHERE: STRUCTURE, COMPOSITION, AND ENERGY BALANCE

The gaseous layer surrounding the earth, known as the atmosphere, began to form more than 4 billion years ago. Volcanic eruptions released gases into the developing atmosphere during the solid earth's evolution. The gases released included water vapor ($H_2O$), carbon monoxide (CO), carbon dioxide ($CO_2$), hydrochloric acid (HCl), methane ($CH_4$), ammonia ($NH_3$), nitrogen ($N_2$), and sulfur gases, assuming the outgassing was comparable to that of modern volcanoes. Because there was no free oxygen, the atmosphere was decreasing. The earth's gravity could not have held on to the small masses of hydrogen and helium, so the majority of the gases that outgassed would have eventually escaped into space. Significant volatile contributions could have also come from the massive meteoritic bombardments that are known to have happened early in Earth's history.

As atmospheric water vapor condensed and began to rain, lakes and oceans were eventually created. The earliest creatures, which were presumably related to cyanobacteria, found habitat in the seas. These early organisms released oxygen into the atmosphere while sequestering carbon in sedimentary rocks. This resulted in the formation of our current oxidizing atmosphere, which is primarily made up of oxygen (about 28%) and nitrogen (about 71%). Together, carbon dioxide, argon, and water vapor make up a far smaller portion (about 1%). In trace amounts, the atmosphere also contains helium, neon, methane, and nitrous oxide, among other gases. Ozone is a crucial trace gas that the sun uses to absorb damaging UV rays.

### 1.3.1. Atmospheric Structure

The atmosphere of Earth reaches a distance of approximately 1,000 kilometers beyond which it merges with interplanetary space. However, the first 40 kilometers contain the majority of the atmosphere's mass (more than 99%). The primary sources of radiant energy in the atmosphere are the sun and the earth. While the majority of the radiation from the earth is infrared, the sun's radiation also extends into the visible and ultraviolet light spectrums.

The different types of radiation that impact each atmospheric layer determine how the vertical temperature profile of the atmosphere changes. This in turn is dependent on that layer's chemical makeup, which primarily consists of trace gases. The troposphere, stratosphere, mesosphere, and thermosphere are the four different layers that make up the atmosphere. The atmospheric layer that is closest to the earth's surface is called the troposphere. It is between 8 and 16 kilometers away from the surface of the earth. The layer's thickness varies by a few kilometers based on the season and latitude. In the summer and close to the equator, it is thicker; in the winter, it is thinner. Compared to the other layers, the troposphere holds the highest proportion of the atmosphere's mass. Additionally, it holds 99% of the atmosphere's total water vapor.

In close proximity to the earth's surface, the troposphere has a warm temperature of approximately 17°C. This is because other greenhouse gases, such as water vapor, absorb infrared radiation from the surface (e.g., carbon dioxide, nitrous oxide and methane) in the troposphere. The heating effect is highest close to the surface because the concentration of these gases decreases with altitude. The troposphere experiences a temperature drop of about 6.5°C per kilometer of altitude. Its upper boundary has an extremely low temperature (about -60°C). The troposphere is constantly experiencing a convective overturn of material due to hot air rising and cold air falling. In actuality, troposphere refers to a "region of mixing." All weather phenomena therefore take place in the troposphere. When water vapor from the earth's surface evaporates, it condenses in the troposphere's colder upper regions and returns to the surface as rain. When introduced into the troposphere, dust and other pollutants mix thoroughly and eventually get washed out by precipitation. Thus, the troposphere is self-cleaning.

The tropopause is a small area at the top of the troposphere. It successfully divides the stratosphere that is above and below the underlying troposphere. In the tropopause, the temperature is largely constant. Here, there are also strong winds that blow from the east, or the jet stream. The next major layer of the atmosphere is the stratosphere. This layer reaches up to about 50 kilometers above the surface of the earth from the tropopause, which is located at a height of about 12 km. The troposphere and stratosphere have very different temperature profiles. Up to about 25 kilometers, the temperature is fairly constant; after that, it progressively rises to the layer's upper boundary. Since there is very little water vapor in the stratosphere, it has little effect on how the layer's temperature is regulated. On the contrary, the observed temperature inversion is caused by ozone ($O_3$).

Between 20 and 30 kilometers up in the stratosphere is where the majority of the ozone in the atmosphere is found. Ultraviolet radiation (UV) from the sun is absorbed by this ozone layer and eventually released as heat in the stratosphere. The temperature rises as a result of this heat. For life to exist on Earth's surface, stratospheric ozone is crucial because it shields them from the majority of the sun's damaging UV rays. In the stratosphere, ozone is continuously created and destroyed by a natural cycle. The process of converting molecular oxygen into ozone necessitates the upper atmosphere's absorption of UVA radiation with high energy. Moderate and low energy radiation (UVB and UVC) are involved in the depletion of ozone through UV radiation absorption. At lower latitudes where UV radiation is most intense, the stratosphere is where the majority of ozone is created and destroyed.

Since ozone is so unstable, it reacts easily with nitrogen, hydrogen, bromine, and chlorine, among other airborne species. In actuality, this is how most ozone is destroyed. The natural ozone cycle has been significantly impacted by human use of chlorofluorocarbons (CFCs) in recent decades, as this has accelerated the rate of ozone depletion due to reactions with chlorine. There is minimal convective mixing of the gases due to the stratosphere's temperature rising with height. For this reason, the stratosphere is quite stable. Injectable particles, like volcanic ash, can remain in the air for a long time without falling back to the earth. The same holds true for toxins that come from people. The term "stratopause" refers to the abrupt drop in temperature that occurs at the upper edge of the stratosphere.

The mesosphere is the name given to the third layer of the earth's atmosphere. From the stratopause, which is located around 50 kilometers above the earth's surface, it rises to a height of about 85 kilometers. The mesosphere experiences temperature drops because there is very little ozone and water vapor present to produce heat. The stratosphere heats it from the bottom up. This area has extremely thin air, with a density of roughly 1/1,000 that of the surface. Lighter gases take over this layer as altitude rises, and the remaining gases in the outer reaches stratify according to molecular weight. The thermosphere, the fourth layer, reaches outward from roughly 85 kilometers to 600 kilometers. Its upper limit is not clearly defined. The thermosphere can reach temperatures of up to 1500°C or higher as one ascends in altitude. The last few oxygen molecules have absorbed a lot of solar radiation, which has caused the high temperatures. The amount of solar activity can have a significant impact on temperature variations. The ionosphere is another name for the lower thermosphere, which extends up to a distance of roughly 550 kilometers. Gas particles get ionized in this area due to the high temperatures. Because it reflects radio waves from the earth's surface, the ionosphere is crucial for long-distance radio communication. This area is also the site of the northern lights, an atmospheric visual phenomenon. The exosphere is the outermost part of the atmosphere. The last stage of the transition from the atmosphere to interplanetary space is known as the exosphere. It is roughly 1,000 kilometers long and mostly composed of hydrogen and helium. Here is where most satellites are operational. The primary energy source for atmospheric heating is solar radiation. When infrared radiation from the sun or reradiated from the earth's surface is absorbed by water vapor and other greenhouse gases in the troposphere, the atmosphere becomes warmer. Ocean water evaporation and heat transfer to the atmosphere are also caused by solar heat. The latitude of the earth affects the temperature of its surface. This results from the surface of the earth heating unevenly. Direct sunlight falls on the area close to the equator, while sunlight at higher latitudes is scattered and falls over a wider area at an angle. The earth's equatorial plane is tilted 23.5° with respect to its orbital plane around the sun, which causes variations in the angle at which sunlight strikes the higher latitudes throughout the year. The distinct seasons that the non-equatorial latitudes experience are caused by this variation.

### 1.3.2. Composition of the Atmosphere

The fractions of the major atmospheric components, $N_2$, $O_2$, and Ar, are remarkably uniform below about 100 km, with the exception of water vapor, whose atmospheric abundance varies from practically zero up to 4 percent. Diffusion takes over as the primary mode of transport at higher altitudes, where the lighter gases become comparatively more abundant. Furthermore, new species are formed by photochemical processes that would not allow them to exist in significant concentrations at the higher pressures found at lower elevations due to their high reactivities. There are three types of atmospheric gas abundances: major, minor, and trace. Because of its geochemical inertness, nitrogen is the most abundant component and has accumulated over time. A very small portion of nitrogen passes into the other phases due to biological activity and natural fixation by lightning. Denitrifying bacteria in marine sediments are thought to be the main pathway through which $N_2$ is released back into the atmosphere. Nearly all oxygen comes from biological sources and is recycled by sedimentary rocks, the hydrosphere, and the biosphere. $Ar^{40}$, a decay product of $K^{40}$ in the mantle and crust, makes up the majority of argon.

**Table 1.1.** *Major Components*

| | | |
|---|---|---|
| Nitrogen | $N_2$ | 78.08% |
| Oxygen | $O_2$ | 20.95% |
| Argon | Ar | 0.93% |

*Source:* https://chem.libretexts.org/Bookshelves/Environmental_Chemistry/Geochemistry_(Lower)/03%3A_The_Atmosphere/3.01%3A_Structure_and_Composition_of_the_Atmosphere.

Aside from water vapor, carbon dioxide is the most prevalent of the minor gases; further discussion of this will be provided below. Helium and neon are next in abundance. While neon and the other inert gases are primordial and most likely have existed in their current relative abundances since the formation of the earth, helium is a byproduct of radioactive elements in the earth. The amounts of two of the minor gases, carbon monoxide and ozone, change with the passage of time and place. An imbalance between the rates of formation and removal is implied by a variable abundance. When it comes to carbon monoxide, the main human source is likely localized variations in fuel consumption, especially in internal combustion engines. Only a small amount is produced by biological activity. It's unclear exactly what kind of carbon monoxide sink (removal mechanism) exists; it might be partially microbial.

**Table 1.2.** *Minor Components*

| | | |
|---|---|---|
| Water | $H_2O$ | 0–4% |
| Carbon dioxide | $CO_2$ | 325 ppm |
| Neon | Ne | 18 ppm |
| Helium | He | 5 ppm |
| Methane | $CH_4$ | 2 ppm |
| Krypton | Kr | 1 ppm |
| Hydrogen | $H_2$ | 0.5 ppm |
| Nitrous oxide | $N_2O$ | 0.3 ppm |
| Carbon monoxide | CO | 0.05–0.2 ppm |
| Ozone | $O_3$ | 0.02 – 10 ppm |
| Xenon | Xe | 0.08 ppm |

*Source:* https://chem.libretexts.org/Bookshelves/Environmental_Chemistry/Geochemistry_(Lower)/03%3A_The_Atmosphere/3.01%3A_Structure_and_Composition_of_the_Atmosphere.

**Table 1.3.** Trace Components

| | | |
|---|---|---|
| Ammonia | $NH_3$ | 4 ppb |
| Nitrogen oxide | NO | 1 ppb |
| Sulfur dioxide | $SO_2$ | 1 ppb |
| Hydrogen sulfide | $H_2S$ | 0.05 ppb |

*Source:* https://chem.libretexts.org/Bookshelves/Environmental_Chemistry/Geochemistry_(Lower)/03%3A_The_Atmosphere/3.01%3A_Structure_and_Composition_of_the_Atmosphere.

### 1.3.3. The Earth-Atmosphere Energy Balance

The balance between the energy Earth receives from the Sun and the energy it loses back into space is known as Earth's energy budget, or Earth's energy balance. Although they are taken into account, smaller energy sources—like the heat produced internally by the Earth—contribute very little in comparison to solar energy. The flow of energy within the climate system is also taken into consideration by the energy budget. The equatorial tropics receive more heat from the Sun than the polar regions. As a result, there is an uneven distribution of the solar irradiance received by different regions. The energy drives interactions in the Earth's climate system as it seeks equilibrium throughout the planet, i.e., the rocky crust, water, ice, atmosphere, and all living organisms on Earth. The climate of Earth is the outcome.

Clouds, land use patterns, surface albedo, greenhouse gases, and atmospheric aerosols are just a few of the variables that affect Earth's energy budget. Earth is in radiative equilibrium and the climate system is relatively stable when the incoming and outgoing energy fluxes are in balance. When the earth receives more energy than it returns to space, it warms up globally, and when it releases more energy, it cools down globally. A warming imbalance has been observed since at least 1970, according to a variety of measurements and observations (Trenberth Kevin, Fasullo John, von Shuckmann Karina, & Cheng LiJing, 2016).

There has never been a heating event with the current human-caused rate. The primary cause of variations in Earth's energy is due to alterations in the atmosphere brought about by human activity. The Earth's energy imbalance (EEI) averaged 460 TW, or $0.90 \pm 0.15$ W per $m^2$, globally between 2005 and 2019. Any substantial changes in the global surface temperature require time to materialize when adjustments are made to the energy budget. The cryosphere, land, and oceans' thermal inertia is to blame for this. This inertia, energy flows, and storage volumes are all accurately calculated by the majority of climate models.

- **Definition:** The primary energy flows that are significant to the climate system are included in Earth's energy budget. These include variations in the global energy inventory, the surface energy budget, the top-of-atmosphere energy budget, and internal energy flows within the climate system.

### *1.3.3.1. Earth's Energy Flows*

Earth maintains a relatively constant temperature despite massive energy transfers into and out of the planet because overall, there is little net gain or loss: roughly the same amount of energy is sent into space by Earth via solar insolation (all forms of electromagnetic radiation) and atmospheric and terrestrial radiation (shifted to longer electromagnetic wavelengths). Changes in the atmosphere caused by humans account for the majority of Earth's energy variations, which total roughly 460 TW, or 0.90 ± 0.15 W per $m^2$.

#### *1.3.3.1.1. Incoming Solar Energy (Shortwave Radiation)*

The solar constant multiplied by the cross-sectional area of the Earth that corresponds to the radiation yields the total amount of energy received per second at the top of Earth's atmosphere (TOA), which is expressed in watts. Given that a sphere's surface area is four times its cross-sectional area (i.e., the area of a circle), the average TOA flux over time, both globally and annually, is equal to one quarter of the solar constant, or about 340 $W/m^2$. Figure 1.2 given are multi-year averages derived from numerous satellite measurements because the absorption varies with location in addition to diurnal, seasonal, and annual variations.

Out of the approximately 340 $W/m^2$ of solar radiation that the Earth receives, clouds and the atmosphere reflect an average of 77 $W/m^2$, and surface albedo reflects about 23 $W/m^2$. This leaves approximately 240 $W/m^2$ of solar energy input to the Earth's energy budget. The term "absorbed solar radiation" (ASR) refers to this quantity. It suggests that Earth's mean net albedo, or Bond albedo (A), is approximately 0.3.

$$ASR = (1 - A) \times 340 \text{ W m}^{-2} \simeq 240 \text{ W m}^{-2}.$$

#### *1.3.3.1.2. Outgoing Longwave Radiation*

Longwave radiation (OLR) is the form of thermal energy that leaves the planet. The electromagnetic thermal radiation that the Earth's surface and atmosphere emit is known as longwave radiation. Infrared radiation is found in the longwave band. However, since infrared radiation can be either longwave or shortwave, the terms are not interchangeable. Significant amounts of shortwave infrared radiation are present in sunlight. Sometimes, to differentiate between longwave and shortwave radiation, a threshold wavelength of 4 microns is utilized.

Solar energy that is absorbed is typically transformed into various types of thermal energy. A portion of the solar radiation that the surface absorbs is transformed into thermal radiation at wavelengths in the atmospheric window. This radiation can directly escape to space and pass through the atmosphere without any problems, which adds to OLR. Through a number of heat transfer mechanisms, the remaining solar energy is transferred upward through the atmosphere and eventually released as thermal energy that can escape into space, which further contributes to OLR. For instance, radiative heat transport, evapotranspiration, latent heat fluxes, and conduction/convection processes are some of the ways that heat is transferred into the atmosphere. In the end, all energy that is released as longwave radiation into space.

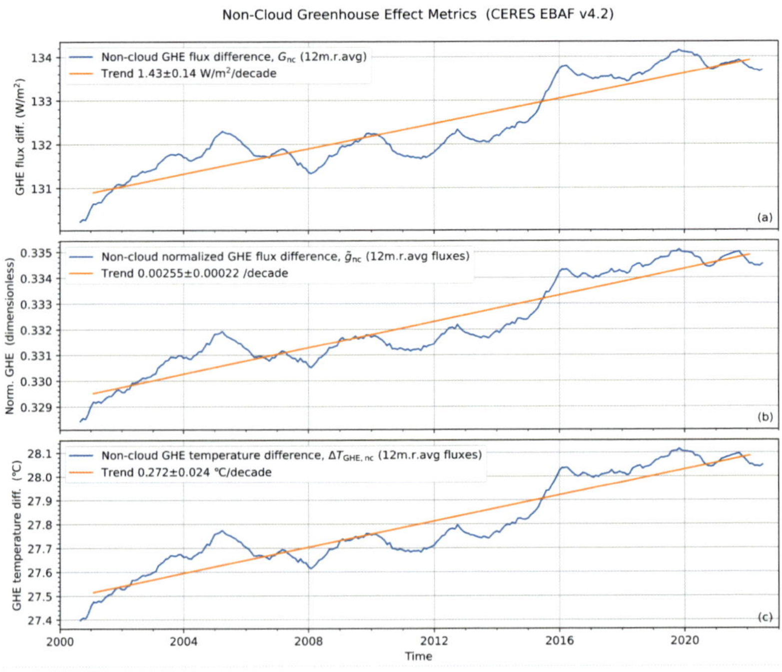


**Figure 1.2.** *Increase in the Earth's non-cloud greenhouse effect (2000–2022) based on satellite data.*

*Source: By Rhwentworth – Own work, CC BY-SA 4.0,* https://commons.wikimedia.org/w/index.php?curid=139002909.

Longwave radiation from the Earth's surface is transported through its multilayered atmosphere according to Kirchoff's law of thermal radiation and is governed by radiative transfer equations like Schwarzschild's equation for radiative transfer (or more complex equations if scattering is present).

Temperatures at the surface (Ts=288 Kelvin) and in the middle of the troposphere (Ta=242 Kelvin) are roughly in line with observed average values when described by a one-layer model. The Stefan-Boltzmann constant ($\sigma$) in this expression denotes the atmosphere's emissivity, which is less than 1 because the atmosphere does not emit within the atmospheric window. An effective value of approximately $\varepsilon$=0.78 is influenced by aerosols, clouds, water vapor, and trace greenhouse gases. Through slight variations in the planet's absolute temperature, the strong (fourth-power) temperature sensitivity keeps the outgoing energy flow close to the incoming flow.

Greenhouse gases affect the planet's atmospheric emissivity ($\varepsilon$), as observed from space. Thus, variations in the composition of the atmosphere can affect the overall balance of radiation. For example, an increase in heat trapping by a growing concentration of greenhouse gases (i.e., an *enhanced greenhouse effect*) forces a decrease in OLR and a warming (restorative) energy imbalance. In the end, as greenhouse gas concentrations rise or fall, so do in-situ surface temperatures, until incoming solar radiation (ASR) equals outgoing longwave radiation (OLR).

#### *1.3.3.1.3. Earth's Internal Heat Sources and Other Minor Effects*

There are an estimated 47 terawatts (TW) of geothermal heat flow from the Earth's interior (Davies, & Davies, 2010) and distributed roughly equally between heat produced by radioactivity and heat remaining after the Earth's creation. The 173,000 TW of incoming solar radiation dwarfs this, amounting to an average flux of 0.087 W/m$^2$, or only 0.027 percent of Earth's total energy budget at the surface (Archer David, 2012).

Even less energy is produced by humans; for the entire year 2019, it was estimated to be 18 TW, or 160,000 TW-hr. But as a result of rising consumption and the release of greenhouse gases into the atmosphere during the fossil fuel energy production process, there is an imbalance in the incoming and outgoing flows of solar radiation that is more than 20 times greater. Another important factor is photosynthesis, which is thought to absorb 140 TW, or 0.08%, of incident energy and use it to power plants' biomass production. When plants are used for fuel or food, they release a similar amount of thermal energy over the course of a year.

Other minor energy sources, such as solar wind and interplanetary dust accretion, light from stars other than the Sun, and space thermal radiation, are typically disregarded in the computations. In a paper that is frequently cited as the first on the greenhouse effect, Joseph Fourier had earlier asserted the significance of deep space radiation (Fleming James, 1999).

### *1.3.3.2. Budget Analysis*

Simply put, when the incoming flow and the outgoing flow are equal, Earth's energy budget is balanced. The balance can also be expressed as follows: absorbed incoming solar (shortwave) radiation equal to outgoing longwave radiation, since some incoming energy is directly reflected (Figure 1.3).

ASR = OLR

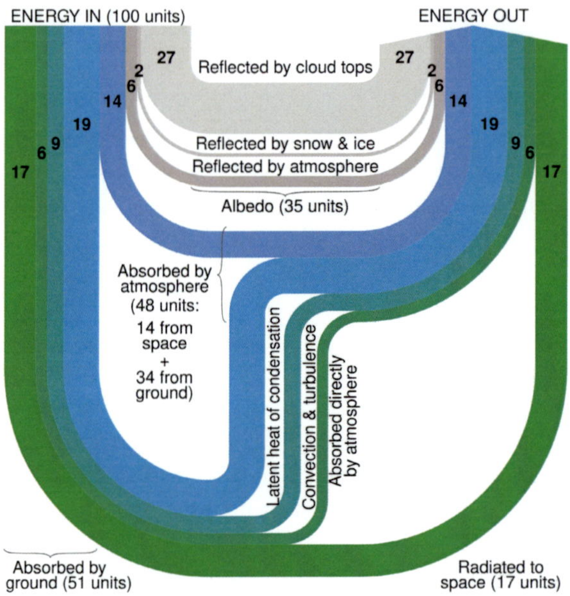


**Figure 1.3.** *A Sankey diagram illustrating a balanced example of Earth's energy budget. Line thickness is linearly proportional to relative amount of energy.*

*Source: By Cmglee – Own work, CC BY-SA 3.0,* https://commons.wikimedia.org/w/index.php?curid=45491898.

#### *1.3.3.2.1. Internal Flow Analysis*

As indicated by the accompanying Sankey diagram, let the insolation received at the top of the atmosphere be 100 units (= 340 W/m$^2$) to explain some of the internal flows within the budget. In this example, approximately 35 units—

known as Earth's albedo—are directly reflected back to space: two from snow- and ice-covered regions, six from other parts of the atmosphere, and 27 from the tops of clouds. Of the remaining 65 units (ASR=220 $W/m^2$), 51 are absorbed by the Earth's surface and 14 are absorbed by the atmosphere.

The 51 units that reach and are absorbed by the surface are released back into space using a variety of terrestrial energy sources: 34 are absorbed by the atmosphere, 9 are released into space by convection and turbulence, and 6 are absorbed infrared light by greenhouse gases. Of these, 17 are directly radiated into space. Finally, the 48 units—34 from terrestrial energy and 14 from insolation—absorbed by the atmosphere are radiated back into space. This oversimplified illustration omits some important details about the mechanisms that store, circulate, and cause additional heat buildup close to the surface. In the end, OLR is released from the 65 units (17 from the ground and 48 from the atmosphere). To ensure that Earth experiences a net-zero energy gain, they roughly balance the 65 units (ASR) absorbed from the sun (Sharma, (2008).

### *1.3.3.2.2. Heat Storage Reservoirs*

Together with the atmosphere, land, ice, and oceans are active material components of Earth's climate system. They have significantly more thermal inertia due to their much larger mass and heat capacity. Thermal energy will move as sensible heat via conduction and convection heat transfer processes into or out of the bulk mass of these components when radiation is directly absorbed or when the surface temperature changes. In addition, the movement of water between its solid, liquid, and vapor phases serves as a source or sink of latent heat, another type of potential energy. The surface conditions are protected by these processes from some of the abrupt radiative changes occurring in the atmosphere. As a result, there is not much of a surface temperature difference between day and night. Similar to this, changes in the atmospheric radiation balance cause the Earth's climate system as a whole to respond slowly.

More thermal energy is contained in the upper few meters of Earth's oceans than in the atmosphere. Vast amounts of this energy are transported across the surface of the planet by fluidic ocean waters, just like by atmospheric gases. In addition, sensible heat enters and exits deep waters in accordance with upwelling or downwelling circumstances. The oceans have stored more than 90% of the excess energy that has accumulated on Earth as a result of continuous global warming since 1970. A third or so have spread to depths of less than 700 meters. In recent decades, the overall growth rate has also increased; by 2020, it is expected to approach 500 TW (1 $W/m^2$). This resulted in a heat gain of roughly 14 zettajoules (ZJ) for the year, which is more than 20 times the 570 exajoules (= 160,000 TW-hr) of primary energy that humans consumed overall.

### *1.3.3.2.3. Heating/Cooling Rate Analysis*

Changes in the Earth's energy flux balance can generally be attributed to three factors: internal system variability, system feedbacks, and external forcings (both radiative and non-radiative, natural and human). Among other potential factors, these changes are mainly represented as detectable variations in temperature (T), clouds (C), water vapor (W), aerosols (A), trace greenhouse gases (G), land/ocean/ice surface reflectance (S), and as small variations in Insolation (I). Afterwards, the net change in energy (ΔE) related to the following characteristics can be used to analyze the rate of warming or cooling of Earth over chosen time periods (Δt):

$$\Delta E/\Delta t=(\Delta ET+\Delta EC+\Delta EW+\Delta EA+\Delta EG+\Delta ES+\Delta EL+\ldots)/\Delta t=ASR-OLR.$$

Here the term $\Delta E_T$, corresponding to the Planck response, is negative-valued when temperature rises due to its strong direct influence on OLR.

A positive ΔEG forcing term results from the enhanced greenhouse effect caused by the recent increase in trace greenhouse gases. By contrast, a large volcanic eruption (e.g., Mount Pinatubo, 1991; El Chichón, 1982) can inject sulfur-containing compounds into the upper atmosphere. Elevated levels of stratospheric sulfur aerosols have the potential to last for several years, resulting in a negative forcing effect on ΔEA. Other anthropogenic aerosol emissions contribute to ΔEA in both positive and negative ways. The $\Delta E_I$ produced by solar cycles is less than the ΔEG trends resulting from recent human activity.

Climate forcings are complicated because they can result in both direct and indirect feedbacks that strengthen the initial forcing (positive feedback) or lessen it (negative feedback). Frequently, these come after the temperature response. Due to evaporation shifts and the Clausius-Clapeyron relation, water vapor trends as a positive feedback in response to temperature variations. A rise in water vapor causes the greenhouse effect to intensify even more, which leads to a positive ΔEW. The ice-albedo feedback is a slower positive feedback. For instance, the melting of Arctic ice brought on by rising temperatures reduces the region's reflectiveness, increasing energy absorption and accelerating the rate of ice melt (Lindsey Rebecca, 2009).

This has a positive impact on $\Delta E_S$. When feedbacks come together, they often intensify global warming or cooling. About half of Earth's albedo is attributed to clouds, which are also strong indicators of the climate system's internal variability. They could also function as responses to forces or, in the case of cloud seeding activity, as forces in and of themselves. Contributions to $\Delta E_C$ differ by region and kind of cloud. Satellite measurements are collected in tandem with model simulations to enhance comprehension and minimize ambiguity.

### *1.3.3.3. Earth's Energy Imbalance (EEI)*

The continuous positive (downward) net top of atmosphere energy flux linked to greenhouse gas forcing of the climate system is known as the Earth's energy imbalance (EEI).

According to the law of energy conservation, the planet will either gain (warm) or lose (cool) net heat energy depending on whether Earth's incoming energy flux (ASR) is greater or smaller than its outgoing energy flux (OLR).

EEI≡ASR-OLR

Thus, the overall rate of planetary heating is defined by the positive EEI, which is commonly expressed in watts per square meter ($W/m^2$). The Earth's energy imbalance from 2005 to 2019 was on average 460 TW, or $0.90 \pm 0.15$ W per $m^2$.

Instruments atop satellites can measure shifts in Earth's energy imbalance (EEI) when they occur in sufficiently big increments. Long-term temperature changes in the atmospheric, oceanic, land, and ice components of the climate system will also be caused by imbalances that do not eventually reverse. Measures of EEI are thus also provided by temperature, sea level, ice mass, and related shifts. The largest variations in the Earth-Environment Interaction (EEI) result from human activity altering the composition of the atmosphere, which disrupts the natural flow of energy within the climate system. The primary causes of the changes are the rises in greenhouse gases, such as carbon dioxide, which cause pollution and heating (positive EEI). The latter is used to describe different types of atmospheric aerosols, some of which reflect energy and cause cooling (or a lower EEI) and others of which absorb energy. The absolute magnitude of the EEI cannot (yet) be measured directly at the top of the atmosphere, although changes over time as recorded by satellite-based instruments are believed to be reliable. Making a list of all the energy changes in the climate system is the only feasible method to determine the precise amount of EEI. The ocean is the largest of these energy reservoirs.

#### *1.3.3.3.1. Energy Inventory Assessments*

It is possible to calculate the planetary heat content contained in the climate system by considering the heat capacity, density, and temperature distributions of its constituent parts. Today, most regions—the deep ocean being the most notable exception—are sampled and monitored fairly well. As with the previous multi-decadal time intervals, estimates of the absolute magnitude of EEI have been computed from the measured temperature changes. The EEI increased significantly from the mean of $+0.48\pm0.1$ $W/m^2$ for the 1971–2020 period to approximately $+0.76\pm0.2$ $W/m^2$ for the 2006–2020 period.

Because temperatures have risen nearly everywhere for more than 50 years, the EEI has been positive. The average of air temperatures recorded over land and sea surface temperatures is used to compute the global surface temperature, or GMT. Since roughly 1970, the GST has increased steadily at a rate of 0. 18°C every ten years, according to trustworthy data dating back at least to 1880. Ocean waters have a far higher total heat capacity than the atmosphere, making them particularly good solar energy absorbers. Since before 1960, research vessels and stations have taken global and deepwater sea temperature samples. Furthermore, since the year 2000, the temperature anomaly, or more accurately the change in the ocean's heat content (ΔOHC), has been measured by an expanding network of almost 4,000 Argo robotic floats. OHC has grown at a constant or faster rate since at least 1990. Since the oceans have so far absorbed more than 90% of the net excess energy entering the system over time (Δt), ΔOHC represents the largest portion of EEI.

$EEI \geq \Delta OHC/\Delta t$.

Relatively little of the excess energy has been absorbed by the thickly ice-covered regions and the outer crust of Earth. This is due to the fact that surplus heat at their surfaces only enters through thermal conduction, penetrating only a few tens of centimeters on a daily cycle and a few tens of meters on an annual cycle. A significant portion of heat absorption is used to either evaporate more water from soils or melt ice and permafrost (Murphy, Solomon, Portmann, Rosenlof, et al., 2009).

### *1.3.3.3.2. Measurements at Top of Atmosphere (TOA)*

Numerous satellites monitor the energy that Earth receives and emits, thereby determining the energy imbalance. These are situated at the top of the atmosphere (TOA) and offer global data. Three of these satellites were part of the NASA Earth Radiation Budget Experiment (ERBE) project: the Earth Radiation Budget Satellite (ERBS), which was launched in October 1984; NOAA-9, which was launched in December 1984; and NOAA-10, which was launched in September 1986 (Figure 1.4). Since March 2000, NASA's Earth Observing System (EOS) has included the Clouds and the Earth's Radiant Energy System (CERES) instruments. The purpose of CERES is to measure radiation that is emitted from Earth (long wavelength) and reflected from the sun (short wavelength). The EEI increased from $+0.42 \pm 0.48$ W/m$^2$ in 2005 to $+1.12 \pm 0.48$ W/m$^2$ in 2019, according to the CERES data. Rising temperatures partially offset contributing factors such as declining ice, less clouds, rising greenhouse gases, and increased water vapor. A more thorough analysis of the behavior with the GFDL CM4/AM4 climate model revealed that the likelihood of the trend being caused solely by internal climate variability was less than 1%.

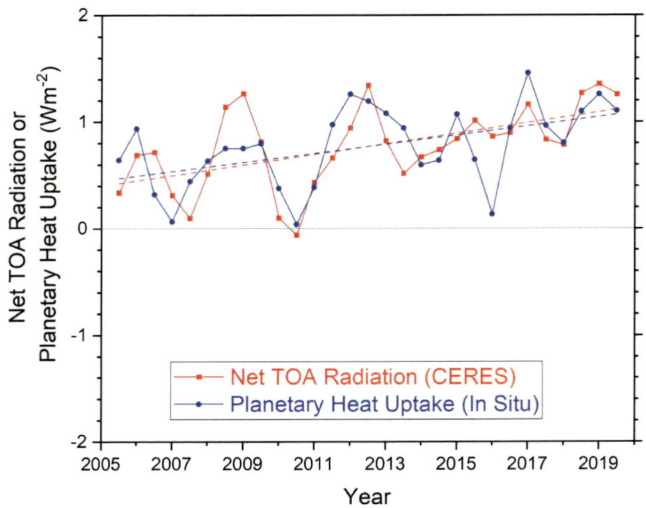


**Figure 1.4.** *The growth in Earth's energy imbalance from satellite and in situ measurements (2005–2019). A rate of +1.0 W/m² summed over the planet's surface equates to a continuous heat uptake of about 500 terawatts (~0.3% of the incident solar radiation).*

*Source: By Loeb, Norman G.; Johnson, Gregory C.; Thorsen, Tyler J.; Lyman, John M.; Rose, Fred G.; and Kato, Seiji – https://agupubs.onlinelibrary.wiley.com/doi/10.1029/2021GL093047, CC BY 4.0,* https://commons.wikimedia.org/w/index.php?curid=114504900.

Other researchers have searched for trends of radiative forcing embedded in the EEI data using data from CERES, AIRS, CloudSat, and other EOS instruments. According to their analysis, between 2003 and 2018, there was a forcing rise of $+0.53 \pm 0.11$ W/m$^2$. The increasing concentration of greenhouse gases, which lowered the amount of longwave radiation that was released into space, was responsible for about 80% of the increase. Additional precipitation has been observed by further satellite measurements, such as TRMM and CALIPSO data. This precipitation is maintained by increased energy evaporating from the surface (the latent heat flux), which partially offsets the increase in the longwave greenhouse flux to the surface. Notably, the otherwise stable and accurate current generation of satellite-based instruments is limited in its capabilities by radiometric calibration uncertainties. Because of this, relative changes in the EEI can be measured accurately, something that is not also possible for a single measurement of the absolute imbalance.

### *1.3.3.3.3. Geodetic and Hydrographic Surveys*

Since 1994, observations have revealed an accelerating retreat of ice from every region of the planet. Together with the general increase in ocean temperatures, the ice melt has also resulted in an increase in the mean global sea level. The planet's geometric shape and gravity have undergone quantifiable changes as a result of these shifts.

Gravimetric observations from the GRACE satellite instruments have been used to infer changes to the mass distribution of water within the hydrosphere and cryosphere. Using computer models that take into consideration salinity variations, thermal expansion, and other factors, these data have been compared against ocean surface topography and additional hydrographic observations. The resulting estimates for ΔOHC and EEI have, within uncertainties, agreed with the other (mostly) independent assessments.

### *1.3.3.3.4. Importance as a Climate Change Metric*

The monitoring of Earth's energy imbalance has been identified by climate scientists Kevin Trenberth, James Hansen, and colleagues as a crucial metric to assist policymakers in setting the pace for mitigation and adaptation measures. Longer-term trends in the Earth's energy imbalance, or EEI, are able to predict upcoming changes due to the inertia of the climate system.

Researchers discovered that the EEI is the most significant climate change metric. It is the overall outcome of all the interactions and processes that make up the climate system. Understanding the increasing frequency of weather extremes requires an understanding of how much additional energy affects rainfall and weather systems. NASA scientists stated in 2012 that, if all other climate forcings were eliminated, atmospheric $CO_2$ concentration would need to be lowered to 350 ppm or less in order to halt global warming. Due to ongoing increases in human emissions, atmospheric $CO_2$ reached 415 parts per million in 2020, and the concentration of all long-lived greenhouse gases exceeded 500 parts per million $CO_2$-equivalent.

## REFERENCES


1. Archer, D. (2012). *Global warming: Understanding the forecast* (2nd ed.). John Wiley & Sons. ISBN 978-0-470-94341-0.
2. Balmes, J., Fine, J., & Sheppard, D. (1987). Symptomatic bronchoconstriction after short-term inhalation of sulfur dioxide. *American Review of Respiratory Disease, 136*(5), 1117–1121. https://doi.org/10.1164/ajrccm/136.5.1117. PMID 3674573.
3. Davies, J. H., & Davies, D. R. (2010). Earth's surface heat flux. *Solid Earth, 1*(1), 5–24. https://doi.org/10.5194/se-1-5-2010. ISSN 1869-9529.
4. Fleming, J. R. (1999). Joseph Fourier, the 'greenhouse effect,' and the quest for a universal theory of terrestrial temperatures. *Endeavour, 23*(2), 72–75. https://doi.org/10.1016/S0160-9327(99)01210-7.
5. Hidy, G. (2012). *Aerosols: An Industrial and Environmental Science*. Elsevier. ISBN 978-0-323-14251-9.
6. Kim, K. H., Jahan, S. A., Kabir, E., & Brown, R. J. (2013). A review of airborne polycyclic aromatic hydrocarbons (PAHs) and their human health effects. *Environment International, 60,* 71–80. https://doi.org/10.1016/j.envint.2013.07.019. PMID 24013021.
7. Lindsey, R. (2009, January 14). Climate and Earth's energy budget (Part 7-Climate forcings and global warming). *Earth Observatory*. NASA Goddard Space Flight Center. Retrieved August 5, 2019, from https://earthobservatory.nasa.gov/features/EnergyBudget/page7.php.
8. Munsif, R., Zubair, M., Aziz, A., Nadeem Zafar, M., & Viskup, R. (2021). Industrial air emission pollution: Potential sources and sustainable mitigation. In *Environmental Emissions*. IntechOpen. https://doi.org/10.5772/intechopen.93104. ISBN 978-1-83968-510-1. S2CID 234150821. Retrieved June 7, 2022.
9. Murphy, D. M., Solomon, S., Portmann, R. W., & Rosenlof, K. H. (2009). An observationally based energy balance for the Earth since 1950. *Journal of Geophysical Research: Atmospheres, 114*(D17). https://doi.org/10.1029/2009JD012105.
10. Sharma, P. D. (2008). *Environmental biology & toxicology* (2nd ed.). Rastogi Publications. ISBN 9788171337422.
11. Singh, R., Kumar, S., Karmakar, S., Siddiqui, A. J., Mathur, A., Adnan, M., & Kumar, N., & Shukla, V. (Eds.). (2021). Causes, consequences, and control of persistent organic pollutants. In *Persistent Organic Pollutants in the Environment: Origin and Role* (pp. 31–54). CRC Press. ISBN 978-1-003-05317-0. Retrieved June 11, 2022.

12. Trenberth, K. E., Fasullo, J. T., von Shuckmann, K., & Cheng, L. (2016). Insights into Earth's energy imbalance from multiple sources. *Journal of Climate, 29*(20), 7495–7505. https://doi.org/10.1175/JCLI-D-16-0339.1. OSTI 1537015. S2CID 51994089.

Chapter 2

# Health and Environmental Effects of Air Pollution

## Contents

# INTRODUCTION

Air pollution is one of the biggest problems of our time because of its effects on both public and individual health due to rising morbidity and mortality rates, as well as its influence on climate change. Numerous pollutants play a significant role in human disease. Among these are Particulate Matter (PM), which are particles with varying but extremely small diameters that enter the respiratory system through inhalation and can lead to cancer, reproductive, cardiovascular, and central nervous system disorders, as well as respiratory and other diseases. Ozone protects against UV radiation in the stratosphere, but when it is concentrated too much below ground, it can be hazardous and negatively impact the cardiovascular and respiratory systems. Moreover, dioxins, sulfur dioxide, nitrogen oxide, volatile organic compounds (VOCs), and polycyclic aromatic hydrocarbons (PAHs) are all regarded as hazardous air pollutants for people. High levels of carbon monoxide can even cause direct poisoning through inhalation. Depending on the extent of exposure, heavy metals like lead can cause either acute poisoning or long-term intoxication when absorbed into the body. The diseases brought on by the substances mentioned above are mostly related to the respiratory system and include asthma, bronchiolitis, lung cancer, cardiovascular events, central nervous system dysfunctions, and cutaneous diseases. Last but not least, both natural disasters and climate change brought on by pollution in the environment have an impact on the geographic distribution of numerous infectious diseases. Public awareness and a multidisciplinary approach from scientific experts are the only ways to address this issue; national and international organizations must address the emergence of this threat and offer long-term solutions.

## 2.1. SOURCES OF EXPOSURE

It is well known that the majority of pollutants released into the environment come from large-scale human activities like driving cars, using combustion engines, power plants, and industrial machinery. Due to their massive scale, these activities are by far the main causes of air pollution; cars are thought to be responsible for about 80% of the pollution that exists today. In addition to a number of natural sources like soil and volcanic eruptions and forest fires, some other human activities like field cultivation methods, gas stations, fuel tanks, heaters, and cleaning practices also have a minor impact on our environment (Möller, Schuetzle, & Autrup, 1994).

The primary basis for classifying air pollutants is the source of the pollution. Therefore, it is worth mentioning the four main sources, following the classification system: Major sources, Area sources, Mobile sources, and

Natural sources (Jacobson & Jacobson, 2002). The chemical and fertilizer industries, metallurgical and other industrial plants, power plants, refineries, petrochemicals, and municipal incineration are some of the main sources of pollution emissions. Domestic cleaning operations, dry cleaners, printing businesses, and gas stations are examples of indoor area sources. Vehicles of all kinds, including cars, trains, and airplanes, are examples of mobile sources. Lastly, natural disasters include things like dust storms, volcanic erosion, forest fires, and agricultural burning (Stover, 2019). Nonetheless, a variety of classification schemes have been put forth. Another kind of classification is the following grouping based on the person who is receiving the pollution:

- Air pollution is defined as the large-scale, persistent presence of pollutants in the air. Particles in the air can include hydrocarbons, CO, $CO_2$, NO, $NO_2$, $SO_3$, and scattered particles.
- Water pollution is the high levels of biological, inorganic, and organic charge that have an impact on the water quality.
- Soil pollution happens as a result of waste disposal or chemical releases, including those containing pesticides, hydrocarbons, and heavy metals.

By contaminating precipitation and settling into soil and water environments, air pollution can have an adverse effect on the quality of soil and water bodies. Notably, acid precipitation can modify the chemistry of the soil by impacting water quality, plant life, and cultures. Furthermore, soil acidity promotes the movement of heavy metals, which leads to the metals' entry into wet environments. It is well known that fish and other wildlife are poisoned by heavy metals like aluminum. Given that acid rain increases the risk to soils with low calcium carbonate levels, soil quality appears to be important. Particulate matter, snow, and rain fall into bodies of water.

Lastly, pollution is classified following type of origin:

- Radioactive and nuclear pollution, radioactive fallout from nuclear weapons, handling or disposal of radioactive sewage, and nuclear explosions and accidents that release radioactive and nuclear pollutants into the air, water, and soil. Because they are harmful to the environment, plants, animals, and people, radioactive materials have the potential to contaminate surface water bodies. Numerous radioactive elements, including uranium and radium, are known to concentrate in the bones and can lead to cancer.
- Noise pollution is created by hearing-damaging machinery, cars, traffic noises, and musical installations (Stansfeld, 2015).

The term "DALYs" was first used by the World Health Organization. A disease or health condition's DALYs are calculated by adding the Years of Life Lost (YLL) resulting from population-wide premature mortality and the Years Lost due to Disability (YLD) for those who are affected by the illness or its aftereffects. Disability-adjusted life years lost (DALYs) are primarily caused by air pollution in Europe, with noise pollution coming in second. Research has been done on the possible links between air pollution and noise pollution and health. The effects of ambient noise on cardiovascular disease were found to be independent of air pollution, meaning that DALYs associated with noise were more significant than those associated with air pollution. One should consider ambient noise to be a separate public health risk.

Environmental pollution arises when variations in the environment's physical, chemical, or biological components (air masses, temperature, climate, etc.) are generated.

Either by introducing hazardous substances or by raising levels above normal, pollutants damage our ecosystem. The aforementioned sources directly produce primary pollutants, which are then released as secondary pollutants as a result of the primary ones. As previously mentioned, pollutants can be either naturally occurring or man-made, and they can also be biodegradable or non-biodegradable. Furthermore, their source may be distributed or a single point (point-source). The variations in the ability of pollutants to cause harmful effects can be attributed to differences in their physical and chemical characteristics. For instance, we state that because aerosol compounds are smaller (solid or liquid) in the atmosphere and have a higher penetration capacity, they are more toxic than gaseous compounds. Our respiratory system has an easier time getting rid of gaseous substances. Millions of people die young each year as a result of these particles' ability to harm the lungs and even enter the bloodstream. Additionally, it appears that the aerosol acidity ([$H^+$]) significantly increases the generation of secondary organic aerosols (SOA); however, other scientific groups do not concur with this last finding.

## 2.2. CLIMATE AND POLLUTION

Climate change and air pollution are closely associated. The other side of the same coin that degrades Earth's quality is climate. The amount of sunlight that enters the atmosphere is influenced by pollutants like aerosols, methane, black carbon, and tropospheric ozone. Ice, icebergs, and glaciers are melting as a result of the Earth's temperature rising (D'Amato, Pawankar, Vitale, & Maurizia, 2016). Similarly, the incidence and prevalence of imported and residual infections in Europe will be impacted by climate change. Climate and weather have a significant impact on the length, timing, and severity of outbreaks as well as altering the global distribution of infectious diseases. The transmission of parasitic or viral diseases by mosquitoes is highly susceptible to climate change, as it can shorten the incubation period of the pathogen and change the vector's geographic distribution. Similar to this, a high frequency of waterborne infections results from water warming brought on by climatic changes. Recently, it appears that eradicated diseases like malaria, poliomyelitis, cholera, and tick-borne encephalitis are resurfacing in Europe as a result of population migration. Storms and other natural climatic disasters are linked to the spread of epidemics, and they appear to be happening more frequently these days. The new infections impacting public health are also linked to malnourishment and immune system dysregulation.

Chikungunya outbreaks were reported in Italy, and autochthonous cases were reported in France, indicating that the virus "took the airplane" from the Indian Ocean to Europe. Following flooding, there appears to have been a rise in cryptosporidiosis in the Czech Republic and the United Kingdom. Aerosol compounds, as previously mentioned, are minuscule in size and have a significant impact on climate. By reflecting 25% of the sun's rays back into space, they can disperse sunlight (the albedo phenomenon), which has lowered global temperatures during the past 30 years (Schneider, 1989).

## 2.3. HAZARDOUS EFFECTS OF AIR POLLUTANTS ON HUMAN BODY SYSTEMS

Human health is directly impacted by air pollution. We have known throughout history that acute episodes of air pollution cause a significant number of fatalities from respiratory and cardiac conditions, bronchitis, and pneumonia. In the most dramatic of these incidents, a lethal concentration of air pollutants was created by a combination of local topography, particular meteorological conditions, and air pollution. The best known of these episodes were those that occurred in the Muese River Valley region of Belgium (1930), Donora, Pennsylvania (1948), London, England (1952, 1962), New York City (1953), and Bhopal, Indian (1984). Thousands of people lost their lives as a result of these episodes of air pollution.

### 2.3.1. Pollutant Movement Through the Body

The skin, eyes, nose, and respiratory system are examples of human receptors, whereas plants, animals, and natural resources are examples of environmental receptors. An organ like any other found in the body is a receptor, like the skin on humans. It acts as a barrier to protect the body from external pollutants and is incredibly durable against the majority of environmental harm. On the other hand, the eye functions as a receptor just like the skin, but it is far more vulnerable to harm and irritation. The skin can be permanently disfigured or scarred by many compounds, but if the same compounds were injected into the eye would become completely useless.

The main way that air pollutants enter the body is through breathing. They may also be absorbed through the skin or swallowed. Children who play or eat in contaminated soil, for instance, may develop chronic health issues that last a lifetime. Once a pollutant enters the body, it can stay in the lungs (like asbestos), be exhaled, or travel from the digestive system or skin to the lungs (like oxygen we breathe) and enter the blood. Pollutants can travel throughout the body if they get into the bloodstream. A pollutant can change chemically as it travels through the body, particularly as it passes through the liver, changing its level of toxicity. In addition to being exhaled, the pollutant can also leave the body through sweat, bowel movements, urine, breast milk, and storage in bone, fat, or hair.

Hazardous air pollutants have the potential to cause numerous health issues just by interfering with bodily functions. The most frequent way they tamper with things is by changing the chemical reactions that occur within individual cells, which are the fundamental units of all living things. These alterations have the power to destroy, damage, or reroute cell activity. The consequences can be extremely harmful to organs, lead to birth defects due to damage to an unborn child's cells, or cause cancer to spread uncontrollably within cells. The elderly, young children, expectant mothers, people with long-term chronic lung or heart disease, and pregnant women are the groups most immediately impacted by ambient air pollution, though the effects on human health can vary greatly from person to person. In addition, because of the volume of air exchange that occurs in deep lung tissue, people who exercise vigorously outside are also at risk.

Human exposure to air pollution typically results in either a short-term, acute health effect or a long-term, chronic health effect. Acute health effects are typically transient and disappear when the pollutant is no longer exposed to. Headaches, nausea, and eye irritation are common acute health effects. A chronic health effect usually manifests after prolonged exposure and has long-lasting, irreversible effects. Diminished lung function, heart disease, cancer, and even death are a few signs of long-term health effects. The chronic public health issues linked to prolonged exposure to air pollution are rarely explained by acute episodes of pollution. Chronic health issues are typically the result of ongoing exposure. Although many bodily systems are frequently impacted by pollutants, the respiratory system is arguably the most susceptible since it serves as both a filter and a means of transportation. The respiratory system comes into direct and instantaneous contact with raw pollutants because it is the primary pathway through which air pollution enters the body (Figure 2.1).

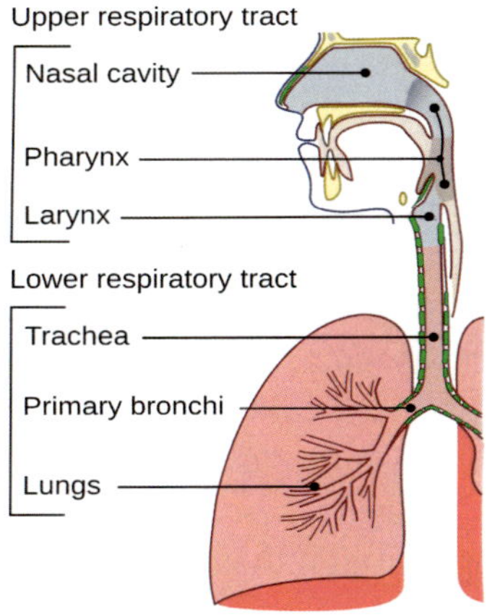


**Figure 2.1.** *Conducting passages of the human respiratory system.*

*Source: By Lord Akryl, Jmarchn – Vectorized version of File: Illu conducting passages.jpg, which was in the public domain as a work of the US Government, Public Domain, https://commons.wikimedia.org/w/index.php?curid=112619332.*

## 2.3.2. Upper Respiratory System

Pollutants usually enter the body through the mouth or nose and enter the upper respiratory system. The majority of larger particulate matter (>15µm diameter) is eliminated from the body by mucous membranes lining the respiratory tract from the nasal cavity through the nose and throat, or by tiny nasal hairs, before it reaches the lower respiratory system (Talukder & Sharma, 1991). Pollutant particles smaller than 10µm in diameter that evade these defense mechanisms end up stuck in the cilia. Via the bronchioles, the cilia—fine, hair-like structures—line the walls of the respiratory system. Mucus and trapped particulate matter are carried toward the upper respiratory tract by their wave-like motion, where they are expelled. The body has multiple defense mechanisms to combat and expel particulate matter, as evidenced by this.

The body's final line of defense in the bronchioles before it reaches the alveoli is the cilia. Particles with a diameter of approximately 1 µm to 0 µm that are not picked up by the body's defense mechanisms eventually become lodged in the alveoli, where they can remain for weeks, months, or even years before being expelled. In order to combat the foreign matter, alveolar tissue produces phagocyte cells, which eventually envelop the particulate and hold it in place indefinitely. But the soluble toxins are eliminated and carried to other areas of the body by the bloodstream as soon as the particulate matter is encircled. Even though the body naturally uses the alveolus to trap particulate matter, it is still dangerous for PM to penetrate so deeply into lung tissue. There are numerous ways that gaseous air pollutants impair lung function, but the primary one is by delaying the action of the tiny cilia. Breathing contaminated air on a regular basis can significantly reduce the body's capacity to carry out this necessary but normal cleansing task. When this type of normal bodily function breaks down, pollutant particles often manage to avoid being captured by the cilia and end up deeply embedded in lung tissue. Particulate matter has the potential to cause lung cancer, pulmonary emphysema, bronchitis, asthma, and other respiratory infections as long-term (chronic) health effects. The upper respiratory system can be significantly impacted by both gaseous and particulate pollutants because of the delicate nature of the body's structures. Pollutants are then transported to the lower respiratory system through the upper respiratory system.

### 2.3.3. Lower Respiratory System

The bronchi, bronchioles, and alveoli make up the lower respiratory system, which is the part of the lungs where most pollutant exchange takes place. Air exits the upper respiratory system and enters the left or right bronchi after passing through the trachea. The bronchioles are the smaller tubes that split off from each bronchus. The millions of tiny air sacs known as alveoli mark the end of the bronchioles. A large portion of man-made pollution never leaves the lower respiratory system, even with the natural cleansing process.

Chest pain, coughing, pneumoconiosis, lung cancer, bronchitis, and pulmonary emphysema are common outcomes of direct respiratory effects on the lungs. The indirect effects of lung damage are equally important, despite being less obvious. Reduced respiratory function, reduced pulmonary circulation, heart and blood vessel enlargement and weakening, skin and eye irritation, inflammation, and allergic reaction are all possible outcomes of indirect damage. Because air pollution prevents the lung from absorbing oxygen from the air and eliminating harmful carbon dioxide from the bloodstream, it poses a serious risk to both the upper and lower respiratory systems. When lung damage is permanent, heart and circulation issues become more common and respiratory diseases return more frequently. Although the respiratory system has numerous structural mechanisms to effectively remove and clear particulate matter from the airways, they are powerless to shield the body from long-term exposure to toxins and pollutants that are frequently found in urban and industrial areas. Criteria pollutants can swiftly outwit the body's defense mechanisms, depending on the aerodynamic size of the inhaled particles and the respiratory air movement velocity. Pollutants can travel to a distant organ or tissue and cause a range of negative effects once they are absorbed into the bloodstream as a chemical or gas.

## 2.3.4. Criteria Pollutants

Criteria air pollutants are those for which an ambient air quality standard has been established and for which acceptable exposure levels can be calculated. Ozone, carbon monoxide, nitrogen dioxide, sulfur dioxide, PM10, and PM2.5 are a few examples.

### *2.3.4.1. Particulate Matter (PM)*

- **Health Effects**: A mixture of solid particles and liquid droplets found in the air is referred to as particulate matter. Certain particles can be mistaken for soot or smoke due to their size or darkness. Others are so small they can hardly be detected with an electron microscope. The term PM2.5 refers to "fine" particles with a diameter of less than or equal to 2.5 micrometers. Particles larger than 2.5, but smaller than or equal to 10 micrometers in diameter, are referred to as "coarse" particles. All particles with a diameter of less than or equal to 10 micrometers are referred to as PM10. Ten micrometers is around one-seventh the diameter of a human hair, to put things in perspective.

  Both fine and coarse particles are included in particulate matter. These particles are closely linked to several health effects and can build up in the respiratory system when inhaled. Inhaling fine particulates is mainly linked to worsening respiratory ailments like asthma. Fine particles are also closely associated with such effects as premature mortality of elderly persons, increased hospital admissions and emergency room visits for heart and lung patients, increased respiratory symptoms and disease among the chronically ill, and decreased lung function in almost all population groups. However, it appears that children, the elderly, and those with cardiopulmonary diseases like asthma are most vulnerable to these effects.
- **Sources of the Pollutant**: Particulate matter comes from a wide range of natural sources as well as from stationary and mobile sources. The combustion of fuel in automobiles, power plants, and industrial facilities, as well as in wood stoves and home fireplaces, results in fine particles. The following are some of the sources of coarse particles: wind-blown dust, materials handling, vehicles operating on unpaved roads, and crushing and grinding operations.

Certain particles, like those from cars and smokestacks, are released straight from their sources. In other instances, fine particles are formed when gases like $SO_2$, NOx, and VOCs combine with other substances in the atmosphere. Their physical and chemical makeup varies with weather, season, and location.

- **Trends in PM Levels**: While PM10 emissions decreased by 16% between 1990 and 1999, average PM10 concentrations decreased by 18%. Although this research is ongoing, the EPA is currently unable to characterize the long-term trends for PM2.5 in urban areas. To start determining the extent of the PM2.5 issue, the EPA established a new monitoring network at the beginning of 1999. The network had over 1,100 sites and about 1,700 monitors by December 2000. These data are now being examined by EPA for upcoming reports.

### *2.3.4.2. Lead (Pb)*

- **Health Effects**: Lead (Pb) combines with other elements to form stable compounds that accumulate and linger in the environment and in living things. Lead enters the body through the mouth and nose, where it is subsequently absorbed into the bloodstream and distributed throughout all bodily tissues. Studies in toxicology, epidemiology, and clinical medicine have shown that lead exposure has a negative impact on people's health. It has been discovered that low-level lead exposure affects some enzyme systems and blood production. Lead exposure has also been linked to damage to kidney and nervous system cells. Research on animals has indicated that low fertility and congenital abnormalities may be associated with lead exposure. Youngsters are particularly vulnerable to many of the negative effects of lead. According to recent research, lead may also play a significant role in hypertension and the ensuing heart disease.

  Lead exposure may also have behavioral side effects that are quite serious. There is ample evidence of brain damage in children who have experienced severe lead poisoning. There have been reports of tremors, headaches, restlessness, and general signs of mental retardation. Although the brain appears to be especially vulnerable to lead poisoning, it is unknown if brain dysfunction will arise from low exposure levels. Studies have not yielded a clear conclusion, despite evidence suggesting that children with blood lead levels above normal are more likely to perform poorly academically.

- **Trends in Lead Levels**: Traditionally, burning leaded gasoline was the main source of atmospheric lead. Nonetheless, since 1975, the use of unleaded gas has resulted in a 90% reduction in lead emissions from mobile sources. Currently, regulatory action is concentrated on stationary sources, such as lead smelters, battery manufacturers, iron and steel producers, and others. Lead concentrations in the air are 94% lower in 1999 than they were in 1980. Lead emissions dropped by 95% during the course of the same 20-year period. The lead NAAQS are only broken these days in close proximity to sizable industrial sources, like lead smelters.

### *2.3.4.3. Sulfur Dioxide ($SO_2$)*

- **Health Effects**: The health effects of $SO_2$ include tissue irritation and inflammation where it comes into direct contact with tissue. When $SO_2$ is inhaled, it constricts the bronchi, which increases airflow resistance, lowers air volume, and increases heart rate and respiration rate. Children and adults with asthma who are active outdoors may experience a brief respiratory impairment due to high $SO_2$ concentrations. Breathing difficulties, including wheezing, tightness in the chest, and shortness of breath, may arise in asthmatic individuals after brief exposure to elevated $SO_2$ levels during moderate exertion. Longer-term exposure to high concentrations of $SO_2$, combined with high levels of PM, has also been linked to other effects such as respiratory illnesses, changes in the lungs' defenses, and exacerbation of pre-existing cardiovascular disease. The elderly and those with chronic lung diseases, as well as those with cardiovascular disease, are the population groups most vulnerable to these conditions.

  Pre-existing respiratory conditions (emphysema, bronchitis, and asthma) may get worse when exposed to $SO_2$. Conditions that would encourage the conversion of sulfur dioxide to sulfuric acid have been shown to enhance (synergistically) the toxic response to sulfur dioxide by particulate matter. The amount of improvement is correlated with the particulate matter concentration. When particulate matter that can oxidize sulfur dioxide to sulfuric acid is present, there is a two- to three-fold increase in the irritant response to sulfur dioxide. When sulfuric acid ($H_2SO_4$) is inhaled, the respiratory system secretes more mucus, which lessens the system's capacity to clear particles through mucociliary clearance. This may lead to a rise in the prevalence of respiratory infections.

- **Sources of $SO_2$:** Sulfur dioxide is a pollutant released into the atmosphere as a result of burning fossil fuels that contain sulfur compounds, refining petroleum, producing sulfuric acid, and smelting sulfur-containing ores. The majority of $SO_2$ emissions come from fuel combustion, mostly in coal-fired power plants. Generally speaking, low sulfur fuels or chemical sulfur removal systems can be used to reduce sulfur dioxide pollution levels. Sulfur dioxide can be converted to sulfur trioxide ($SO_3$) through oxidation, either photochemically or with the aid of a catalyst, once it has contaminated the atmosphere. When water vapor is present, $SO_3$ easily transforms into sulfuric acid mist. Sulfate aerosols are created when $SO_3$ combines with other basic oxides. Between 5–20% of the suspended particulate matter in urban air is believed to be made up of sulfates, including droplets of sulfuric acid. These substances have the ability to travel great distances and return to Earth as a significant component of acid rain. Numerous health issues that follow and are linked to $SO_2$ are actually the outcome of $SO_2$ oxidizing into other substances.

- **Trends in $SO_2$ Levels:** Average ambient $SO_2$ concentrations have dropped nationwide by 50% between 1980 and 1999, and by 36% during the more recent ten-year period between 1990 and 1999. Between 1980 and 1999, $SO_2$ emissions dropped by 28%, and between 1990 and 1999, they decreased by 21%. Controls put in place starting in 1995 as part of the EPA's Acid Rain Program are largely to blame for the decreases in $SO_2$ concentrations and emissions since 1994.

### *2.3.4.4. Carbon Monoxide (CO)*

- **Health Effects**: High CO concentrations are known to have harmful effects on the body. After being taken up by the lungs, carbon monoxide combines with hemoglobin, the blood's oxygen-carrying molecule, to create carboxyhemoglobin (COHb). Because hemoglobin has an affinity for carbon dioxide (CO) that is over 200 times greater than that of oxygen, this reaction lowers the blood's ability to carry oxygen. The severity of the health effect increases with the proportion of hemoglobin bound up as carboxyhemoglobin. The blood's level of COHb and the amount of CO inhaled air are directly correlated. After enough time, the blood's COHb level will reach an equilibrium concentration for any given ambient air CO concentration. As long as the level of CO in the ambient air does not change, the blood will maintain this COHb equilibrium level. However, as a new equilibrium of CO in the blood is established, the COHb level will gradually change in the same direction as the CO concentration (Emberson, Pleijel, Ainsworth, den Berg, Ren, Osborne, et al., 2018).

Cardiovascular disease frequently gets worse at the lowest CO concentrations that have been demonstrated to have harmful health effects. Research suggests that these concentrations have caused people with heart or circulatory diseases to exercise less before experiencing pain in their arms and chest. Exercise performance, time interval discrimination skills, and vigilance have all been found to decline with slightly higher CO levels. Additionally, there is evidence linking CO to the development of cardiovascular disease, heart attacks, and impaired fetal development.

- Sources of the Pollutant: Carbon monoxide (CO) is a colorless, odorless gas that becomes poisonous at high concentrations when fuel's carbon is not burned all the way through. It is a part of the exhaust from motor vehicles, which accounts for around 60% of all CO emissions in the country. The remaining CO emissions from the transportation sources category are attributed to non-road vehicles. High CO concentrations are typically found in places with significant traffic congestion. Up to 95% of CO emissions in cities may originate from vehicle exhaust. The burning of fuels used for purposes other than transportation, industrial operations, and natural sources like wildfires are additional sources of CO emissions. Peak CO concentrations usually happen in the winter months when nighttime inversion conditions—a situation in which air pollutants are trapped close to the ground beneath a layer of warm air—are more common and CO automobile emissions are higher.
- Trends in CO Levels: The ambient average CO concentration in 1999 was the lowest in the previous 20 years and was 57% lower nationally than it was in 1980. During the same time period, CO emissions fell by 22%. While the estimated number of exceedances of the national standard decreased by 93% between 1990 and 1999, ambient CO concentrations decreased by 36%. In the same time frame, CO emissions decreased by 7%. Despite a 30% rise in vehicle miles driven in the US during that time, this improvement happened.

### *2.3.4.5. Nitrogen Dioxide ($NO_2$)*

- **Health Effects**: Nitric oxide (NO) oxidizes to produce nitrogen dioxide ($NO_2$), a reddish-brown, extremely reactive gas that is released into the atmosphere. The combination of NO, $NO_2$, and other oxides of nitrogen is referred to as nitrogen oxides (NOx), and it is a major factor in the production of acid rain, particulate matter, and ozone. Short-term exposures (i.e., less than 3 hours) to low levels of nitrogen dioxide ($NO_2$) may lead to changes in airway responsiveness and lung function in individuals with pre-existing respiratory difficulties and increases in respiratory illness in children (5–12 years old). Long-term $NO_2$ exposure has the potential to permanently alter the lung and increase susceptibility to respiratory infections. Nitrogen oxides react with airborne particles to produce fine particle pollution and ground level ozone, both of which are harmful to human health.

- **Sources of the Pollutant**: High-temperature combustion processes, such as those found in cars and power plants, are the main causes of man-made NOx emissions. Significant amounts of $NO_2$ are also produced indoors by gas stoves and home heaters.
- **Trends in $NO_2$ Levels**: The measured $NO_2$ levels have dropped by 25% in the last 20 years. Every region in the nation that previously fell short of the national $NO_2$ air quality standard now satisfies it. Levels near urban monitors have decreased, but over the past 20 years, national emissions of nitrogen oxides—which include NO, $NO_2$, and other oxides of nitrogen—have actually increased by 1%. Numerous factors contribute to this increase, the main one being the rise in NOx emissions from diesel off-highway vehicles. This rise in NOx emissions is concerning because it leads to other environmental issues like acid rain and nitrogen loading of water bodies, as well as the creation of ground-level ozone, or smog.

#### *2.3.4.6. Ozone ($O_3$)*

- **Formation of Ozone**: Sunlight-induced atmospheric reactions lead to the formation of photochemical oxidants, which are intricate compounds. When reactive (non-methane) hydrocarbons and NOx accumulate in the atmosphere and are exposed to the ultraviolet component of sunlight, the formation of new compounds, including ozone and peroxyacetyl nitrate, takes place. UV light energy is absorbed by $NO_2$, which then splits into an oxygen atom and nitric oxide. Most of the time, the oxygen atoms combine with atmospheric molecular oxygen ($O_2$) to create ozone ($O_3$). Nitric oxide will typically react with $O_3$ to form $NO_2$ again, completing the cycle. When nitrogen oxide interacts with non-methane hydrocarbons, an accumulation of $O_3$ occurs above the equilibrium concentration indicated by the reaction cycle mentioned above. Without depleting ozone, nitric oxide is oxidized to $NO_2$ by oxygen atoms from the hydrocarbon radical. $O_3$ concentrations can therefore rapidly increase and are not depleted (Villányi, Turk, Franc, & Csintalan, 2010).

In addition, solar radiation and electrical discharge in the atmosphere and stratosphere can naturally produce ozone. Although the former process cannot produce appreciable concentrations of this pollutant in urban areas, some people think that, in certain meteorological circumstances, ozone intrusion from the stratosphere can greatly contribute to elevated ground level concentrations of $O_3$.

- **Health Effects:** Exposure to ambient $O_3$ for both short-term (1–3 hours) and long-term (6–8 hours) periods has been associated with several health effects that warrant concern. For instance, elevated exposure to ambient $O_3$ has been linked to a rise in hospital admissions and ER visits for respiratory-related reasons. Exposure to $O_3$ can exacerbate pre-existing respiratory conditions like asthma, cause lung inflammation, and increase a person's susceptibility to respiratory infections. Significant declines in lung function and an increase in respiratory symptoms like cough and chest pain are other health effects linked to $O_3$ exposures.

  These effects usually happen when people are exerting themselves. Children are particularly vulnerable to these effects because they are more likely to be outside during the summer, when ozone levels are at their highest. The adult population that engages in outdoor activity (i.e., workers who are outside) as well as people who already have respiratory conditions like asthma and chronic lung disease. Furthermore, prolonged exposure to moderate ozone levels raises the risk of permanent alterations to the structure of the lungs, which could accelerate the aging process of the lungs and exacerbate long-term respiratory conditions.

- **Environmental Effects**: Ozone also has an impact on vegetation and ecosystems, which results in decreased yields from commercial and agricultural forests, slower growth and survival of tree seedlings, and heightened vulnerability of plants to pests, disease, and other environmental stressors (i.e., severe weather). Long-term effects on forest ecosystems could result from these effects, which in long-lived species might not become noticeable for years or even decades. The natural beauty of our national parks and recreation areas, as well as the esthetic value of ornamental species, can be diminished by ground-level ozone damage to tree and other plant foliage.

- **Sources of the Pollutant**: Instead of being released into the atmosphere directly, ozone is created when VOCs and NOx react in the presence of heat and sunlight. The atmosphere easily produces ground-level ozone, usually in the summertime when temperatures are high. Vehicles, chemical plants, refineries, factories, consumer and commercial goods, and other industrial sources are some of the places where volatile organic compounds (VOCs) are released into the atmosphere. Power plants, automobiles, and other combustion sources all release nitrogen oxides into the atmosphere. Every year, there are significant regional variations in ozone concentrations due to shifting weather patterns. Pollution sources located hundreds of miles upwind can also carry ozone and the precursor pollutants that produce ozone into an area.
- **Trends in Ozone Levels**: According to 1-hour data, ambient ozone levels have dropped by 20% over the last 20 years, while 8-hour data shows a 12% decrease. The amount of VOC emissions dropped by 33% between 1980 and 1999. The same time frame saw a 1% increase in NOx emissions. Variations in weather patterns contribute to annual variations in ozone concentrations because heat and sunlight are key factors in the formation of ozone. The EPA can adjust analytically for this annual variability in meteorology to better reflect the effects that emissions have on measured air quality concentrations. The adjusted trend for 1-hour ozone levels for 53 metropolitan areas indicates a continuous improvement from 1980 to the mid-1990s. Through 1994, the adjusted ozone levels fell by 1% annually on average. But from 1994 onward, the progress seems to stall. Every major geographic region in the nation exhibits the same downward trend in 1-hour ozone levels from 1980 to 1999 as the national average. The reduction of ozone concentrations has been most rapid in the Northeast and West, but it has not been as rapid in the South.

### 2.3.5. Toxic Air Pollutants

Substances that cause or are suspected of causing cancer, birth defects, or other serious harms are referred to as toxic or hazardous air pollutants. They can be compounds and metals like asbestos, cadmium, mercury, and chromium, or they can be gases like hydrogen chloride, benzene, and toluene.

### 2.3.5.1. Nature and Sources

Pollutants that cause or may cause cancer or other major health effects, such as reproductive disorders, birth defects, or negative effects on the environment and ecology, are classified as toxic or hazardous air pollutants. Toxic air pollutants include things like gasoline-containing benzene, perchloroethylene released by some dry cleaners, and methylene chloride, which is a solvent used by many different industries. The majority of air pollutants are produced by humans, including mobile sources (i.e., vehicles, trucks, and heavy machinery) as well as stationary sources (i.e., refineries, power plants, and factories), in addition to interior sources (i.e., constructing supplies and tasks like cleaning). Additionally, some air toxics are released by natural occurrences like forest fires and volcanic eruptions.

### 2.3.5.2. Health and Environmental Effects

People exposed to toxic air pollutants at sufficient concentrations may experience various health effects including damage to the immune system, as well as neurological, reproductive (i.e., reduced fertility), developmental, respiratory and other health problems. Numerous of these toxins raise the possibility of cancer or other major health consequences. Apart from inhaling airborne pollutants, there are also hazards linked to the accumulation of toxic pollutants in soils or surface waters. These pollutants are absorbed by plants, consumed by animals, and ultimately intensified via the food chain. Animals exposed to air pollutants may have health issues similar to those experienced by humans.

### 2.3.5.3. Trends in Toxic Air Pollutants

In contrast to many other pollutants discussed, air toxics are not subject to a national monitoring network maintained by the EPA or state governments. To estimate and monitor national emissions trends for the 188 air toxics regulated by the Clean Air Act, the EPA created the National Toxics Inventory (NTI). The EPA classifies emissions in the NTI into four categories: (1) major (big industrial) sources; (2) area and other sources, which include natural and small-scale industrial sources like wildfires and small-scale gas stations and dry cleaners; (3) on-road mobile sources, such as highway vehicles; and (4) nonroad mobile sources, such as locomotives, aircraft, and construction equipment.

The sources of toxic air pollutants are distributed equally among the four categories of sources, according to estimates from 1996, the most recent year for which data are available. But the distribution differs in each city. The substances monitored and the monitors' geographic coverage vary from state to state, even though the EPA and the states gather monitoring data for a variety of hazardous air pollutants. The available monitoring data, in conjunction with the NTI's emissions data, aid in tracking trends in toxic air pollution at different points across the nation. The national monitoring network for a number of hazardous air pollutants is currently being expanded by the EPA. Estimates of the nation's air toxics emissions between 1990 and 1996 show a roughly 23% decrease, based on data from the NTI. Significant reductions in total air toxic emissions have been attained through EPA and state regulations, industry voluntary reductions, and changes in the way EPA compiled the national inventory over time. These factors may account for some variations in the data.

### 2.3.6. Assessing the Risks to Health from Air Pollution

Evaluations of the health risk posed by air pollution quantify and convey the effects of pollution exposure on the general public's well-being. The impact of modifications in air quality brought about by air quality improvement initiatives may also be included in these evaluations. The annual health risk assessment by the European Environment Agency (EEA) is included in the Air Quality in Europe report and attempts to:

- Provide an unbiased and comparable assessment of the effects of the current state of air pollution in Europe, broken down by pollutants but excluding any one particular source.
- Identify any changes over time.
- Contribute to the creation and execution of initiatives aimed at enhancing air quality.
- Increase understanding of the negative impacts of air pollution.

A health risk assessment involves three steps:

1. Assess the exposure of the population to the pollutants in question;
2. Estimate the health risk associated with that exposure;
3. Estimate the uncertainty of the calculation.

The three pollutants—ground-level ozone ($O_3$), nitrogen dioxide ($NO_2$), and particulate matter with a diameter of less than 2 μm ($PM_{2.5}$)—that the World Health Organization considers to be the most harmful and for which the strongest evidence of health effects is made the subject of the health risk assessment for long-term exposure. Since air pollution is a complicated mixture of various pollutants, some of the effects of one pollutant may also be caused by others. Experts in health believe that the pollutants listed above serve as stand-ins for the entire mixture. Specifically, it has been demonstrated that $PM_{2.5}$ is a reliable indicator of the risk connected to particulate matter exposure from a variety of sources and in a range of environments. Air pollution exposure has a wide range of effects, from subclinical ones like inflammation to fatal ones. As the most severe impact of air pollution and the one for which the evidence is strongest, mortality is the health outcome examined for the EEA's health risk assessment. "Premature deaths" and "years of life lost" are two metrics used to estimate mortality related to air pollution exposure.

### *2.3.6.1. Determining Population Exposure to Air Pollution*

Data on the spatial distribution of both air pollutant concentrations and the population are required in order to compute the exposure of the European population to varying levels of air pollution. The official reports of EEA member and cooperating countries are used to estimate concentrations. As a starting point, background stations are used in all areas—rural, suburban, and urban—and traffic stations are used for $NO_2$ and PM. Concentration maps are created from the data collected at these stations, encompassing the entirety of Europe. The maps are also created using supplementary data, which includes meteorological information, altitude, and the outcomes of air quality modeling provided by the European Monitoring and Evaluation Program. Concentrations are gridded at a scale resolution of 1x1 $km^2$ in the final map.

A picture of population exposure is created by superimposing population density maps at the same resolution over concentration maps. These maps can be used to estimate, for each country, the percentage of the population exposed to the whole range of concentrations, in increments of 1 $\mu g/m^3$. The population's age and sex distribution data are also used in addition to total population figures to calculate the attributable mortality.

### *2.3.6.2. Estimating the Risk to Health*

The following details are also required in order to calculate the health risks:

1. Concentration-response functions;
2. Baseline health statistics

The relationship between the concentration of an air pollutant to which a population is exposed and the risk of a health outcome is represented by concentration-response functions, which have been established by epidemiological studies. Therefore, concentration-response functions measure the health effect of each air pollutant concentration unit. The World Health Organization's concentration-response functions, which are based on relative risks, are used by the EEA. The increase in mortality that can be linked to a specific rise in the concentration of air pollutants is captured by relative risks. As statistical averages, relative risks are defined at the population level and cannot be attributed to any one person. As a result, it is impossible to pinpoint specific instances of mortality that are brought on by air pollution.

In general, concentration-response functions are linear; however, at very low or very high concentrations, this may not be the case. Furthermore, compared to "intermediate" concentrations, the evidence regarding the effects of very low or very high concentrations on human health may not be as strong. Because of this, an initial concentration, sometimes known as a "counterfactual concentration," is used when it is not possible to analyze the impact for the entire range of concentrations with the same degree of confidence. Therefore, the concentration above which the impacts are estimated is the counterfactual concentration. The EEA used the WHO's recommended counterfactual concentrations for Europe in their estimations.

- 0 $\mu g/m^3$ for $PM_{2.5}$, which means the full range of concentrations is considered;
- 20 $\mu g/m^3$ for $NO_2$, since the evidence of the concentration-response function for lower concentrations is not sufficiently robust;
- For $O_3$, the SOMO35 statistic is used, which implies a counterfactual concentration of 70 $\mu g/m^3$ as SOMO35 only considers concentrations above that level.

Baseline health statistics comprise total mortality data for each country, broken down by age and sex, as well as life expectancies specific to each country. These figures are indicative of the overall population and cannot be used on an individual basis. Consequently, the approximated quantity of preventable deaths obtained from these data serves as a gauge for the overall effect of air pollution on a specific population.

The percentage of the baseline incidence that can be linked to exposure to a pollutant can be ascertained thanks to the relative risks. The baseline incidence for mortality is the total number of deaths that occur in a nation each year. It is possible to estimate the expected burden of disease in terms of preventable deaths thanks to the relative risks. These outcomes are acquired at the grid level and totaled for every grid within a region (such as a nation or Europe as a whole). The age at which the premature deaths occurred is also taken into account when estimating years of life lost. The years of life lost as a result of a particular death are calculated by deducting the age at which the death occurred from the life expectancy at that age. The total years of life lost for the population is calculated by adding the years of life lost from all premature deaths.

### *2.3.6.3. Sensitivity Analysis of Health Risks*

Due to changing data, as evidenced by the Air quality in Europe – 2017 report and beyond, two more counterfactual concentrations are taken into consideration for additional estimations.

- 2.5 µg/m$^3$ for $PM_{2.5}$, This could be regarded as the minimum background concentration, or the natural concentration that would arise if all anthropogenic contributions were eliminated. It is the lowest annual mean concentration measured in Europe.
- 10 µg/m$^3$ for $NO_2$, as the lowest concentration at which effects were discovered in an epidemiological study published after the above-mentioned WHO recommendations.

### *2.3.6.4. Uncertainties of the Estimation*

The concentration-response functions that are employed in the health risk assessment are the primary source of uncertainty. This results from the presumptions made in the epidemiological research that served as the foundation for the concentration-response function. The majority of epidemiological research account for additional confounding variables, such as diet, lifestyle choices, and smoking, which may also affect mortality. A confidence interval surrounding a mean or central estimate is used to calculate the uncertainty related to the concentration-response function. A 95% confidence interval is used by the EEA. This suggests that there is a 95% chance that the true value falls within the interval's defined range. For instance, for every 10 µg/m$^3$ increase in concentrations of $PM_{2.5}$, there is a 95% probability that the risk will increase overall mortality by 4–8.3% (with a mean or central estimate of 6.2%).

The mean estimate and the confidence interval bounds are the values that the EEA includes in its reports.

Other recognized sources of uncertainty that need to be considered in a qualitative manner are:

1. The previously mentioned fact that air pollution is a complex mixture of multiple air pollutants. Because some pollutants have correlated health effects, it is not possible to simply add up the premature deaths linked to each pollutant. Specifically, it has been calculated that approximately 30% of premature deaths could be double counted if $PM_{2.5}$-attributed deaths are added to $NO_2$-attributed deaths.
2. the methods used to combine data from modeling and air quality measurement, as well as the uncertainties involved, to create maps of air quality.

3. The information on exposure. Everyone in the 1x1 $km^2$ grid is thought to be exposed to the same concentration. A further assumption is that people are exposed to the ambient concentrations of the areas in which they reside, but this ignores the fact that many commute for either work or play. For this reason, estimates are generated for a specific population rather than for each individual risk. A finer spatial resolution of exposure data is required for individual exposures.
4. The initial health data and any potential ambiguities in its gathering.
5. The concentration of the counterfactual when examining the total amount of premature deaths.

For population exposure estimation, concentration-response functions, or counterfactual concentrations, a variety of techniques are employed. These provide an explanation for the discrepancies in estimates between the EEA and those from other studies, including national and local studies or the Global Burden of Disease. For instance, the WHO used chemical transport models and satellite remote sensing in addition to ground air quality measurements from monitoring sites across the globe in its 2016 publication Ambient air pollution: A global assessment of exposure and burden of disease. A uniform distribution with lower and upper limits of 5.9 and 8.7 μg/$m^3$, respectively, was chosen as the counterfactual concentration. An estimated 190 000 deaths in Europe in 2012 were attributed to $PM_{2.5}$ air pollution.

The Lancet Commission used satellite-based estimates, chemical transport models, and ground-level measurements to estimate population-weighted mean concentrations of $PM_{2.5}$ at an approximate resolution of 11 × 11 $km^2$ in the 2015 Global Burden of Disease study. Instead of all cause mortality, the relative risk of dying from lower respiratory infections, lung cancer, chronic obstructive pulmonary disease, ischemic heart disease, and cerebrovascular disease was calculated. We assigned a uniform distribution of 2.4–5.9 μg/$m^3$ to the counterfactual concentration.

## 2.4. ENVIRONMENTAL EFFECTS OF AIR POLLUTION

Both terrestrial and aquatic ecosystems are negatively impacted by air pollution, which degrades ecosystems and lowers biodiversity. The exposure of vegetation to major air pollutants is examined in this section. It uses data from rural monitoring stations as well as estimated total exposure from models. Because ground-level ozone ($O_3$) reduces growth rates, yields, and has an adverse effect on biodiversity and ecosystem services, it harms plants, trees, and agricultural crops. An instance of the impact of ozone on the 2019 wheat harvest is provided.

Certain air pollutants cause ecosystems that receive them to deteriorate by depositing themselves on Earth's surface. Excess nitrogen is introduced to land and water bodies as a result of airborne nitrogen oxides (NOX) and ammonia ($NH_3$) deposition. This leads to eutrophication in water bodies, which is caused by an abundance of nutrients that promote algal blooms and decrease oxygen availability. The exceeding of critical loads for nitrogen deposition in sensitive terrestrial ecosystems, like grasslands, can lead to changes in the structure and function of the ecosystem, the extinction of sensitive species, and the increased growth of species that benefit from high nitrogen levels (Singh, Tiwari, & Agrawal, 2009).

Acidification—a process that alters the chemical composition of soils, lakes, rivers, and marine waters—is brought on by the deposition of sulfur dioxide ($SO_2$), nitrogen oxides (NOX), and $NH_3$. This process disturbs ecosystems and reduces biodiversity. Since $SO_2$ emissions have decreased dramatically in recent decades, $NH_3$ and NOX have contributed more in relation to surface water and soil acidification. Toxic pollutants known as heavy metals are transported over great distances in the atmosphere and settle in ecosystems, where they accumulate in soils and subsequently undergo bioaccumulation and biomagnification in the food chain.

**Table 2.1.** Air Quality Standards for Protecting Vegetation

| **Pollut-ant** | **Averaging Period** | **Standard Type and Concentration** | **Comments** |
|---|---|---|---|
| **$O_3$** | AOT40([a]) accumulated over May to July | EU target value: 18,000μg/m$^3$ per hour | Averaged over 5 years([b]) |
| | | EU long-term objective: 6,000μg/m$^3$ per hour | |
| | AOT40([a]) accumulated over April to September | Critical level for the protection of forests: 10,000μg/m$^3$ per hour | Defined under the UNECE Air Convention |
| **$NO_x$** | Calendar year | EU critical level: 30μg/m$^3$ | |
| **$SO_2$** | Winter | EU critical level: 20μg/m$^3$ | 1 October to 31 March |
| | Calendar year | EU critical level: 20μg/m$^3$ | |

*Source: https://www.eea.europa.eu/publications/air-quality-in-europe-2022/impacts-of-air-pollution-on-ecosystems.*

Notes: ([a]) AOT40 is an indication of accumulated $O_3$ exposure, expressed in µg/m$^3$ per hour, over a threshold of 40 parts per billion (ppb). It is the sum of the difference between hourly concentrations of >80µg/m$^3$ (40ppb) and 80µg/m$^3$ accumulated over all hourly values measured between 08:00 and 20:00 (Central European Time).

([b]) In the context of this chapter, only yearly AOT40 values, and not 5-year averages, are considered.

- **Ecosystem:** Pollution effects were observed in the environment long before they became apparent in human health. Rain, snow, or fog is left behind after sulfur dioxide ($SO_2$) and nitrogen oxides (NOx) in the atmosphere are converted into acids. This precipitation ends up on vegetation, in lakes, and in the soil. Since the leaf provides the mechanics for growth, damage to vegetation usually occurs in the leaf's structural mechanisms. Generally speaking, a leaf consists of three sections: the mesophyll, or spongy center section; the epidermis, or outer protective layer; and the veins, which are a dense network of tubes that carry nutrients throughout the plant. Plants that are subjected to acid rain, $SO_2$, $O_3$, and $NO_2$ irreversibly damage all three structures.
- **Property Damage:** Damage to property is yet another important consequence of air pollution. According to research, buildings and outdoor artwork that are priceless cultural assets can be destroyed by acid rain. Furthermore, rubber goods, nylon, polymer plastics, paint dyes, and other textiles are split and cracked by acid rain. Acidification erodes the metal and stone in our structures and monuments, in addition to contaminating clothes, cars, and buildings.
- **Quality of Life**: Particulate matter and ozone formation also have a negative impact on quality of life. Air pollution carried hundreds of miles across can dramatically change once-pristine vistas. For instance, pollution has a significant impact on leisure activities and outdoor sports. According to a survey conducted on 61 native trout streams in southwest Pennsylvania, acidic conditions prevented 26% of the streams from supporting healthy brook trout populations. Similar to how maintaining the esthetic appeal of our homes and vehicles while avoiding the negative effects of air pollution requires a substantial time and financial investment.

### 2.4.1. Acid Rain

When emissions of sulfur dioxide ($SO_2$) and nitrogen oxides (NOx) in the atmosphere combine with water, oxygen, and oxidants to form acidic compounds, the result is acidic deposition, also known as "acid rain." These substances either descend to Earth in a wet state (rain, snow, and fog) or a dry one (gas and particles). Some cross state and national borders and are carried hundreds of miles by the wind. The burning of fossil fuels in electric utility plants generates approximately 26% and 64% of the nation's annual emissions of NOx and $SO_2$ respectively. Prior to hitting the Earth, gases such as $SO_2$ and NOx, along with associated particulate matter like sulfates and nitrates, affect public health and cause poor visibility. The effects of these substances on breathing and the respiratory system, lung tissue damage, and early mortality are the main health risks to humans. Acid rain harms some trees at high altitudes and increases the acidity of soils and water bodies, rendering the water unfit for certain fish and other wildlife. Additionally, it hastens the deterioration of structures, statues, and artwork that are a part of our country's history. Public health is significantly improving as a result of the start of a major reduction in these harmful environmental effects caused by reductions in $SO_2$ and NOx (Zuhara & Isaifan, 2018).

### 2.4.2. Visibility

One of the most evident consequences of air pollution is the reduction of visibility. Air pollution, which includes particles and gases, scatters and absorbs light, causing impairment of visibility. Air pollution not only reduces our ability to see farther, but it can also deteriorate the color, contrast, and clarity of scenes by scattering and absorbing light. The same particles that have been connected to detrimental health effects can also have a major impact on our vision. Impaired vision is caused by both primary emissions and secondary particle formation. "Primary" particles are released into the atmosphere straight from the source, like road dust or soot (elemental carbon) from burning wood. Primary gaseous emissions in the atmosphere give rise to "secondary" particles. Examples are nitrates, which are created from nitrogen oxide (NOx) emissions from power plants, cars, and other combustion sources, and sulfate, which is created from sulfur dioxide ($SO_2$) emissions from power plants and other industrial facilities. Secondarily formed sulfates are the primary cause of reduced visibility in the eastern United States. Primary emissions from sources such as wood smoke contribute a higher percentage of the total particulate loading in the West than in the East, even though these secondary particles still make up a sizable portion of the particle load. Additionally, humidity can greatly worsen the way pollution affects visibility. Certain particles, like sulfates, gain size and water content, which increases their ability to scatter light and impair vision. The East experiences 70–80% relative humidity on an annual average, while the West experiences 50–60%. High humidity and sulfate concentrations are the main causes of poor summer visibility in the eastern United States.

### 2.4.3. The Greenhouse Effect

Another effect of air pollution on the environment is the greenhouse effect, also known as global climate change. Normally, solar radiation makes its way into the earth's atmosphere, where it is converted to heat radiation. Within the earth's atmosphere, some of this radiation is absorbed, reflected, or reradiated. But as "greenhouse gases" in the atmosphere rise, long-wave infrared heat radiation is reradiated and trapped in the earth's atmosphere, eventually warming the planet's surface. According to recent scientific findings, the Earth's temperature is rising due to the continuous release of "greenhouse gases" into the atmosphere, which is intensifying the greenhouse effect. We refer to this as "global warming." Approximately 81% of the greenhouse gases emitted in the US are carbon dioxide ($CO_2$). The burning of fossil fuels for the production of electricity, as well as in the industrial and transportation sectors, is the main source of carbon dioxide emissions. The next biggest source of greenhouse gas emissions in the US and around the world is methane ($CH_4$) emissions, which come from landfills, farming, and other sources.

Other industrial activities that release greenhouse gases include the production of foam, refrigeration, dry cleaning, chemicals, and semiconductors. These processes also release hydro fluorocarbons (HFCs). Perfluorinated compounds are a different type of greenhouse gas that are produced when aluminum is melted (PFCs). While ground-level ozone, or smog, is also a greenhouse gas, it is formed in part by NOx and VOC emissions from industrial processes and vehicle exhaust. To evaluate the scientific and economic data on climate change that was then available, the Intergovernmental Panel on Climate Change (IPCC) was established in 1988. A report compiled by more than 2,000 of the top scientists in the world was released by the IPCC in 1995. According to the IPCC, "climate change is likely to have wide-ranging and mostly adverse impacts on human health, with significant loss of life." This means that humans are causing the climate on Earth to change."

The IPCC estimates that if atmospheric levels are not lowered, greenhouse gas emissions could result in a 2° to 6° Fahrenheit rise in temperature over the course of the next century. Even though this change might not seem like much, it could cause more extreme weather events, like floods and droughts, which could endanger wetlands and coastal resources by increasing sea levels. Because these effects create new breeding grounds for pathogens and pests, they may also raise the risk of contracting specific diseases. Forests and agricultural areas are vulnerable to climatic shifts that may lead to an increase in insect populations and plant diseases. Degradation of natural ecosystems may eventually result in a decline in biological diversity.

### *2.4.3.1. International Developments*

The Framework Convention on Climate Change (FCCC), which was ratified by 150 nations in 1992, aims to stabilize greenhouse gas concentrations in the atmosphere at levels that will not jeopardize the climate system. Industrialized nations committed to bringing down greenhouse gas emissions to 1990 levels by the year 2000 as part of the FCCC. The majority of industrialized nations, including the US, did not, however, reach this goal. At a 1997 meeting in Kyoto, Japan, parties to the FCCC developed the "Kyoto Protocol" in response to the IPCC's 1995 scientific findings and the ongoing increase in greenhouse gas emissions. Industrialized nations have set targets for their greenhouse gas emissions between 2008 and 2012 as part of the Kyoto Protocol. For this time frame, the average reduction goal for all industrialized nations is to reduce emissions by 5% from 1990 levels. The United States has set a reduction target of 7% below 1990 levels, although other countries have different targets to reflect their unique circumstances. Additionally, the Kyoto Protocol offers market-driven initiatives, like global emissions trading, to assist nations in fulfilling their obligations at the lowest feasible cost. In 2001, the U.S. and other industrial countries rejected ratification of the Kyoto Protocol.

## 2.4.4. Stratospheric Ozone Depletion

Living things are shielded from the sun's damaging ultraviolet radiation (UV-b) by a layer of ozone gas found in the stratosphere, which is situated between 6 and 30 miles above Earth. Nevertheless, this barrier of defense has been compromised during the previous 20 years. Every year, ozone levels drop to 70% below average due to the formation of a "ozone hole" over the Antarctic. Ozone levels are roughly 5% below average in the summer and 10% below average in the winter, even over the United States. Based on current data, the average total ozone over the mid-latitudes of the northern hemisphere has been declining by 3.4% per decade since 1979.

An increasing amount of UV-b radiation reaches Earth as the ozone layer thins. For the first time, researchers showed in 1996 that UV-b levels have risen over the majority of populated areas. Scientists discovered a connection between ozone depletion and a number of human-caused substances in the 1970s, including methyl chloroform, carbon tetrachloride, halons, and chlorofluorocarbons (CFCs). Commercial air conditioners, refrigerators, insulating foam, and other industrial processes release these chemicals into the air. They are carried into the stratosphere by strong winds that pass through the troposphere, the lowermost layer of the atmosphere. There, intense solar radiation causes the release of chlorine and bromine atoms, which damage ozone-producing molecules. Scientists calculate that 100,000 ozone molecules can be destroyed by one chlorine atom.

Even at normal ozone levels, some UV-b radiation reaches the Earth's surface. However, ozone depletion is predicted to increase the negative effects of UV-b radiation because the ozone layer normally absorbs the majority of UV-b radiation from the sun. UV-b radiation has been related to skin cancer in humans, including melanoma, which is the most deadly type of the disease. UV-b also weakens the immune system and causes cataracts. It is unclear how UV-b radiation affects aquatic and plant ecosystems. However, too much UV-b radiation can hinder the growth of some food plants. Furthermore, some researchers contend that UV-b radiation is already stressing marine phytoplankton, which forms the foundation of the ocean food chain. The ocean's food supply for humans may suffer as a result of this stress. Global warming could be exacerbated by serious harm to phytoplankton populations because they absorb $CO_2$ from the atmosphere.

### *2.4.4.1. Programs to Restore the Stratospheric Ozone Layer*

The Montreal Protocol, which committed the world to restricting the production of ozone-depleting substances and acknowledged the international nature of ozone depletion, was signed by 27 countries in 1987. The Protocol, which has been strengthened five times and now asks for the removal of chemicals that deplete stratospheric ozone, has been signed by over 170 countries. The 1990 Clean Air Act Amendments established a U.S. regulatory program to protect the stratospheric ozone layer. In January 1996, U.S. production of many ozone-depleting substances virtually ended, including CFCs, carbon tetrachloride, and methyl chloroform. The last halons were produced in January 1994. These days, a lot of novel products that are less harmful to the ozone layer or have no effect at all are becoming more and more popular. For instance, ozone-safe solvents are used by computer manufacturers to clean circuit boards, and new car air conditioners use ozone-safe refrigerant HFC-134a. The move away from ozone-depleting substances has already been finished in some sectors. The UV Index, a daily forecast of the amount of UV radiation people may be exposed to outside, is one of the new initiatives the EPA is highlighting to inform the public about the health risks associated with UV radiation overexposure and the precautions they can take to lower those risks.

### *2.4.4.2. Trends in Stratospheric Ozone Depletion*

The strategy used in accordance with the Montreal Protocol has proven successful thus far, according to scientific data. Methyl chloroform concentrations had begun to decline in 1996, according to measurements, suggesting that emissions had significantly decreased. Other ozone-depleting compounds, such as CFCs, are starting to be found in lower concentrations in the upper atmosphere. These materials release bromine and chlorine into the stratosphere over the course of several years. Because of this, it is anticipated that stratospheric chlorine levels will peak in early 2000 and then gradually decrease. Chlorine will be released into the stratosphere for many years due to the stability of most ozone-depleting substances, and the ozone layer won't fully recover until well into the century. If the ozone layer is to fully recover, all the signatory countries to the Protocol need to finish implementing their ozone protection policies.

## 2.5. Risk Assessment

To put it simply, health risks are an indicator of your likelihood of developing health issues. Your risk of illness may increase if you are exposed to harmful air pollutants. For instance, your risk of developing cancer may rise if you reside close to a factory that releases chemicals known to cause cancer and breathe in contaminated air. Inhaling air pollutants may also raise your chance of developing non-cancerous side effects like emphysema or reproductive problems. The method the government employs to control this health risk is called risk management.

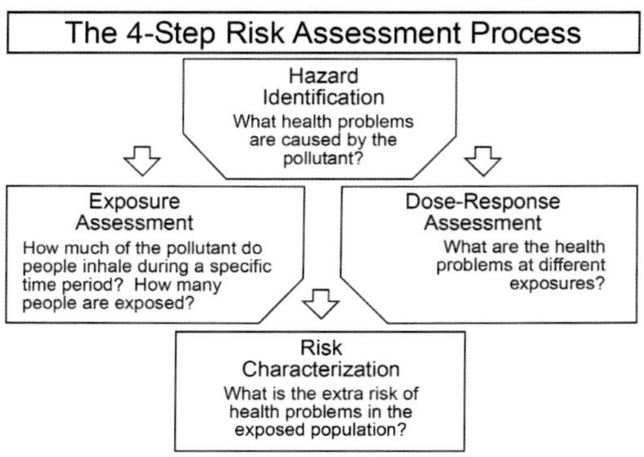


**Figure 2.2.** *The risk assessment process.*

*Source: Dr. William Franek, Mr. Lou DeRose (2023), Principles and Practices of Air Pollution Control. https://www.4cleanair.org/wp-content/uploads/APTI/452combined.pdf.*

Risk management makes decisions about the control of toxic air pollutants by combining this health risk information with other non-risk factors, like control options. One tool in risk management is risk assessment. It is a method used by scientists and public health officials to calculate the higher risk of health issues in individuals exposed to varying concentrations of harmful substances. The results of studies estimating the level of human exposure at various distances from the source of the pollutant are combined with findings on the health effects of various animal and human exposures to the pollutant to create a risk assessment for that toxic air pollutant.

Even though these risk assessments' estimates are far from ideal, they do assist scientists in assessing the dangers connected to the release of harmful air pollutants. The government can establish regulatory standards to lower people's exposure to harmful air pollutants and lower their chance of developing health issues by using risk estimates and other considerations (Figure 2.2).

## 2.5.1. Hazard Identification

Toxic air pollutants of greatest concern are those that cause serious health problems or have a widespread effect. Hazard identification provides an answer to the question, "What health problems are caused by the toxic air pollutant?" Birth defects, respiratory irritation, cancer, and nervous system issues are examples of health issues. Certain health issues manifest immediately following inhalation of a toxic air pollutant. Watery eyes are one of these minor but immediate effects.

Alternatively, they might be grave, like lung damage that could be fatal. It could take several months or years for additional health issues to manifest following an individual's initial exposure to the harmful air pollution. One instance of a health issue that develops later is cancer.

In order to determine the probability that a chemical will have a particular effect on humans, scientists assess all available data regarding the effects of toxic air pollutants. The stronger the evidence, the more confident scientists can be that a particular health issue is caused by a hazardous air pollutant. Evidence's quantity, nature, and caliber are all crucial factors. Studies involving humans provide the best kind of evidence. Case reports, such as medical professionals' accounts of an unusually high number of cases of a particular ailment, may be one type of this evidence.

Additional, more rigorous research can be conducted to compare the incidence of a specific disease across populations with varying exposure levels (e.g., leukemia cases among rubber industry workers). Scientists frequently use laboratory animals, like rats, in their studies because human knowledge regarding the majority of toxic air pollutants is extremely limited. Studies on animals are carried out in carefully monitored lab settings. By subjecting animals to different concentrations of pollutants for different lengths of time, scientists can investigate a wide range of health effects. Scientists must be certain that the health effects on humans are likely to be the same as those in the tested animals when relying solely on animal studies. Scientists attempt to use animal species with human-like bodily functions in order to maximize the validity of the study.

## 2.5.2. Exposure Assessment

Examining the effects of exposure to different pollutants on the environment and human health is another step in the risk assessment process. The amount or degree to which a receptor may come into contact with a chemical, physical, or biological agent is known as pollutant exposure. There are numerous ways to be exposed, including by eating, breathing in, absorbing, or injecting. The amount or concentration of a pollutant and the frequency or number of times it is exposed greatly increase the hazardous effect that exposure has on a receptor, regardless of how it is ultimately exposed. The length of the exposure, or duration, also has a significant impact on how severe the outcome is. As a result, exposure assessment provides an answer to the questions of how many people are exposed and how many pollutants people inhale over a given period of time. In the end, a complex interaction between receptor, dose, toxicity, and exposure determines the harmful effects of air pollution on both human health and the environment. This principle must be understood when attempting to quantify or qualify the behavior of pollutants on the environment or public health. Toxic air pollution come from a variety of sources. A factory smokestack or the thousands of cars that pass a busy intersection every day are two examples of potential sources of pollution. Determining the sources of the pollutant of concern is therefore the first stage in an exposure assessment. Determining the amounts of the toxic air pollutant released in a given time period and how it moves away from the sources is the next step after the sources' identities and locations are established. To estimate the amount of each pollutant released from the source and the amount of these pollutants at various distances from the source, engineers use either computer models or monitors. The amount of the pollutant in the air is measured using monitors that sample the air. The processes that take place when a facility releases a pollutant as well as the movement of pollutants through the air are represented by mathematical equations in computer models. These estimates are influenced by variables like the height of the smokestack, wind direction and speed, and the distance from the source to those who are exposed. Computer models that utilize data from maps and the census can be used to estimate the number of people exposed at various distances from the release site. Certain models have the ability to calculate exposures for the various locations that people visit on a daily basis, such as indoor, outdoor, car, and workplace settings. Estimating the quantity of pollution that each individual breathes in is the last stage in an exposure assessment process. In order to achieve this, scientists combine estimations of the average person's breathing rates, lifespan, and air pollution levels.

### 2.5.3. Dose-Response Assessment

The dose-response relationship for a particular pollutant describes the relationship between exposure and the observed response (health effect), and provides an answer to the question, What are the health problems at different exposures? Stated differently, it calculates the variations in the probability and intensity of health impacts resulting from varying degrees of exposure to a pollutant. To determine dose-response relationships, scientists use data from studies on humans and animals, just as they do in the identification of hazards. The EPA makes the assumption that there are no exposures with zero risk in the absence of convincing evidence to the contrary. Even very low exposure to a pollutant that causes cancer can raise the risk of cancer, albeit slightly. The Environmental Protection Agency (EPA) also makes the assumption that the relationship between dose and response is linear, meaning that for every unit increase in exposure (dose), the cancer response will also increase. There might be a dose below the minimal level of health effect for which side effects don't materialize. However, the EPA usually assumes that there are no negative effects at low doses because the body's natural defenses repair any harm the pollutant may have caused. On the other hand, low dosages of certain substances may have non-cancer effects. The response that happens with increasing dose is known as the dose-response relationship, and it depends on the type of health effect, individual sensitivity, and pollutant.

### 2.5.4. Risk Characterization

Risk information is presented in various ways to illustrate how individuals or populations may be affected. Risk characterization answers the question, What is the extra risk to health? Here is a description of a few of the most popular risk measures. Maximum Individual Lifetime Cancer Risks: An estimate of the increased lifetime risk of cancer for a person exposed to the maximum predicted long-term concentration is obtained by combining the findings of the exposure assessment and the dose-response assessment.

Maximum Lifetime Exposure × Dose-Response Relationship = Maximum Lifetime Exposure

- **Distribution of Individual Risks:** It's possible that many will be exposed to less than the upper limit. A person's risk of developing cancer will differ based on the level of exposure. The number of individuals thought to be at different risk levels is typically used to express the distribution of individual risk.
- **Population Cancer Risks:** To compute population risk, individual risk distributions are used. The expected increased incidence of

cancer, or the number of new cases annually, for every person exposed to the pollutant is typically used to express the population cancer risk. For instance, the number of new cancer cases anticipated annually among people living within 30 miles of a specific large source may be used to estimate the population's cancer risk.

- **Non-Cancer Risks:** Exposure levels that do not pose a significant risk of non-cancer health effects are referred to as health reference levels. It is assumed that long-term exposure to levels below these thresholds will not have any negative effects.

One index that government agencies use to describe health risks other than cancer is the health reference level. These thresholds are typically established from exposure levels that in experimental animals do not result in negative outcomes. The animal-human differences (breathing rate, for example) and underlying uncertainties (sensitivity differences between healthy adults and more sensitive groups, such as children and the elderly) are taken into consideration when adjusting these exposure levels. The number of individuals exposed to concentrations above the health reference level is then ascertained by risk analysts comparing the exposure estimates with the health reference levels. Some of these individuals may suffer negative consequences.

### 2.5.5. Uncertainty of Risk Estimates

While it is possible for scientists to estimate the risks associated with toxic air pollutants in humans exposed to unusual levels of exposure or in animals used in experiments, extrapolating these estimates to human populations under a wide range of conditions is challenging and sometimes misleading. It is impossible for risk estimates to be totally accurate by nature. The primary issue is a lack of knowledge among scientists regarding actual exposure and the damage that harmful air pollutants do to human cells. When it is difficult to measure the amount of pollution that is being transferred from the source to people, computer models are frequently used in the exposure assessment. In order to translate results of animal experiments at high doses to human exposures at low doses, dose-response relationships frequently rely on assumptions about the effects of pollutants on cells. Risk analysts typically make assumptions that tend to prevent them from underestimating the potential risk when information is lacking or unclear; in other words, these assumptions offer a margin of safety when it comes to safeguarding public health.

## REFERENCES


1. D'Amato, G., Pawankar, R., Vitale, C., & Maurizia, L. (2016). Climate change and air pollution: effects on respiratory allergy. *Allergy Asthma Immunol Res, 8*, 391–395. https://doi.org/10.4168/aair.2016.8.5.391.
2. Emberson, L. D., Pleijel, H., Ainsworth, E. A., den Berg, M., Ren, W., Osborne, S., et al. (2018). Ozone effects on crops and consideration in crop models. *European Journal of Agronomy, 100*, 19–34. https://doi.org/10.1016/j.eja.2018.06.002.
3. Jacobson, M. Z., & Jacobson, P. M. Z. (2002). *Atmospheric Pollution: History, Science, and Regulation*. Cambridge University Press. https://doi.org/10.1256/wea.243.02.
4. Möller, L., Schuetzle, D., & Autrup, H. (1994). Future research needs associated with the assessment of potential human health risks from exposure to toxic ambient air pollutants. *Environmental Health Perspectives, 102*(Suppl. 4), 193–210. https://doi.org/10.1289/ehp.94102s4193.
5. Schneider, S. H. (1989). The greenhouse effect: Science and policy. *Science, 243*, 771–781. https://doi.org/10.1126/science.243.4892.771.
6. Singh, E., Tiwari, S., & Agrawal, M. (2009). Effects of elevated ozone on photosynthesis and stomatal conductance of two soybean varieties: A case study to assess impacts of one component of predicted global climate change. *Plant Biology, 11*(Suppl. 1), 101–108. https://doi.org/10.1111/j.1438-8677.2009.00263.x.
7. Stansfeld, S. A. (2015). Noise effects on health in the context of air pollution exposure. *International Journal of Environmental Research and Public Health, 12*, 12735–12760. https://doi.org/10.3390/ijerph121012735.
8. Stover, R. H. (1979). Flooding of soil for disease control. In D. Mulder (Ed.), *Developments in Agricultural and Managed Forest Ecology* (pp. 19–28). Elsevier. https://doi.org/10.1016/B978-0-444-41692-6.50009-4. Retrieved July 1, 2019.
9. Villányi, V., Turk, B., Franc, B., & Csintalan, Z. (2010). Ozone pollution and its bioindication. In V. Villányi (Ed.), *Air Pollution*. Intech Open. https://doi.org/10.5772/10047.
10. Zuhara, S., & Isaifan, R. (2018). The impact of criteria air pollutants on soil and water: A review. *Journal of Environmental Science and Pollution Research, 133*, 278–284. https://doi.org/10.30799/jespr.133.18040205.

*Chapter* **3**

# Transport and Dispersion of Air Pollution

## Contents

# INTRODUCTION

Air pollutants can spread through the air, water, soil, living things, and food once they are in the environment. The pathways of dispersion differ significantly based on the pollutant in question as well as the source of the emissions. Environmental factors also have a significant impact on dispersion rates and patterns. Numerous factors influence the dispersion of pollution in the air:

- Meteorological conditions, particularly air stability, wind direction, and speed.
- The height of emission (e.g., sources at different levels (e.g., tall chimneys or ground level sources like traffic on the roads).
- Local and regional geographical features,
- The source (e.g., fixed point, such as a chimney, or a diffuse number of sources such as cars and solvents).

Pollutants experience a variety of changes and transfers during dispersion. Dilution happens as a result of air mixing. Based on the physical properties of the pollutants, they separate or accumulate. When chemical reactions take place, the original pollutant is broken down or transformed into new compounds. Certain pollutants can also be eliminated from the transporting medium by deposition; these include settling out due to gravity, rain-wash, and scavenging by plants and other obstacles. Because of this, many pollutants exhibit incredibly complex dispersion patterns, particularly in settings like cities and towns where there are numerous emission sources and significant variations in the surrounding environment. Because of this complexity, modeling or measuring pollutant patterns and trends—and consequently estimating levels of human exposure—can be exceedingly challenging. Variations in pollution levels over time are significant. Long-term trends are frequently present, indicating underlying variations in the rates of emission (e.g.as a result of policy intervention, technological advancements, or economic shifts). There might be yearly variations on top of these, which would represent variations in the climate or source activity from year to year. Due to activity cycles, short-term climatic effects, and other factors, many pollutants also exhibit distinct seasonal, weekly, and daily patterns. Sudden, unintentional releases can also result in significant, transient pollution episodes. As a result, exposure measurements will differ depending on the time, location, and duration of air monitoring.

## 3.1. VARIOUS METHODS OF AIR POLLUTION TRANSPORT AND DISPERSION

The main purpose of air pollution models is to simulate the movement and dispersion of different types of pollutants, such as carbon monoxide (CO), sulfur dioxide ($SO_2$), nitrogen oxides (NOX), and suspended particulate matter (SPM). being emitted into the atmosphere as a result of vehicle movement, home fuel burning during cooking, and industrial combustion processes. The models are widely used in urban planning and regulation for air quality surveillance program design, pollution episode forecasting, impact assessment of new or existing sources, and strategy evaluation.

Four geographic divisions are used to characterize the time and space scales used in air pollution dispersion: site-specific (local), regional, national, and global. Together, these create a sensible classification scheme for the air quality models' horizontal spatial and temporal scales. Site-specific or local situations, at the lowest subdivision level, include factors like emissions, source characteristics, initial plume rise, initial mixing phase, local terrain, and initial transport. An unseen column of airborne pollutants is referred to as a plume. These are columns of smoke, gas, or other materials released into the atmosphere that can be seen or invisible. The site-specific category at higher spatial resolutions is focused on interacting plumes from sources that are 10–20 km apart. The regional scales span from a large industrial complex or urban area to a region where the air quality models represent urban areas as point sources.

Modeling the dispersion of air pollutants in the atmosphere is one of the most significant and challenging scientific problems. Passive or chemically active substances are released into the atmosphere by a number of man-made and natural events. These chemical species may have detrimental effects on both human health and the environment. Modeling the dispersion of air pollutants allows for the prediction of this effect. Remaining pollutants from the burning of materials may be released into the atmosphere. An essential component of estimating potential health risks to humans associated with these releases is knowledge of the concentrations and rates of deposition of air pollutants in the neighborhoods surrounding the combustion facility. Air dispersion models are typically used to estimate air concentrations and deposition rates. Mathematical constructs known as air dispersion models are used to try and explain how physical processes in the atmosphere affect the rates at which emissions from a source, like a combustor stack, disperse. Both vertical and horizontal directions are involved in the dispersion of air pollution.

### 3.1.1. Horizontal Dispersion

The extent to which pollution disperses in the atmosphere at a particular level is known as horizontal dispersion (here at NCDAQ, we concentrate on pollution at the ground level). Although topography can have an impact, wind speed and direction are the main factors that drive it. Where and how quickly the wind blows will determine where air pollution ends up, whether it's smoke from a big wildfire or an ozone plume produced locally. Take a small controlled burn, for instance, that releases smoke particles for a few hours. The image below illustrates how wind direction and speed can impact dispersion. It was produced using the HYSPLIT dispersion model, which simulates the dispersion of a hypothetical controlled burn using actual meteorological data. When there are mild easterly winds (i.e., Smoke concentrations are highest near and downwind (the direction the wind is blowing) of the fire in the west, with concentrations falling farther out. These winds are coming from the east. The smoke plume will spread not only westward but also north and south as it moves westward, drifting outward from its centerline—an imaginary line drawn down the middle of the plume as viewed from above. The greatest concentrations will still be in the middle of the plume on a day with moderate northeasterly winds, but faster wind speeds will prevent it from spreading as far from the centerline, resulting in a more elongated plume that reaches farther away. Keep in mind that topography can have an impact on horizontal dispersion. For instance, pollution can occasionally be blown into a valley's entrance and become trapped there if wind speeds are insufficient to push it over the ridges that define the valley. In a city like Los Angeles, which has mountains on three sides, this occurs on a larger scale (Figure 3.1).

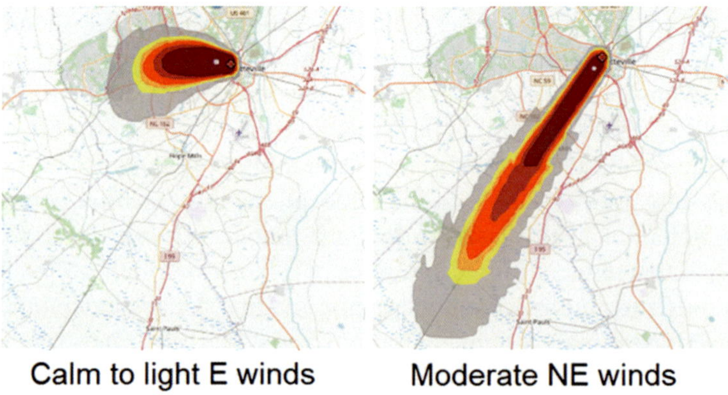


**Figure 3.1.** *The difference in dispersion with different wind speed and direction.*

*Source:* https://airquality.climate.ncsu.edu/2021/06/06/atmospheric-dispersion-and-pollution-transport/.

### 3.1.2. Vertical Dispersion

It all comes down to thermodynamics when it comes to dispersion in the vertical. The adage "warm air rises" is something to bear in mind when considering vertical dispersion. An "atmospheric sounding," or vertical temperature and wind profile of the atmosphere, is one of the factors that air quality forecasters examine when creating their forecasts. Either forecast models or measurements made in the field with weather balloons are used to gather this data. Think about this: picture yourself picking up a "parcel of air," which is simply an imaginary box filled with air at the surface. The air parcel will rise as the morning sun heats the surface.

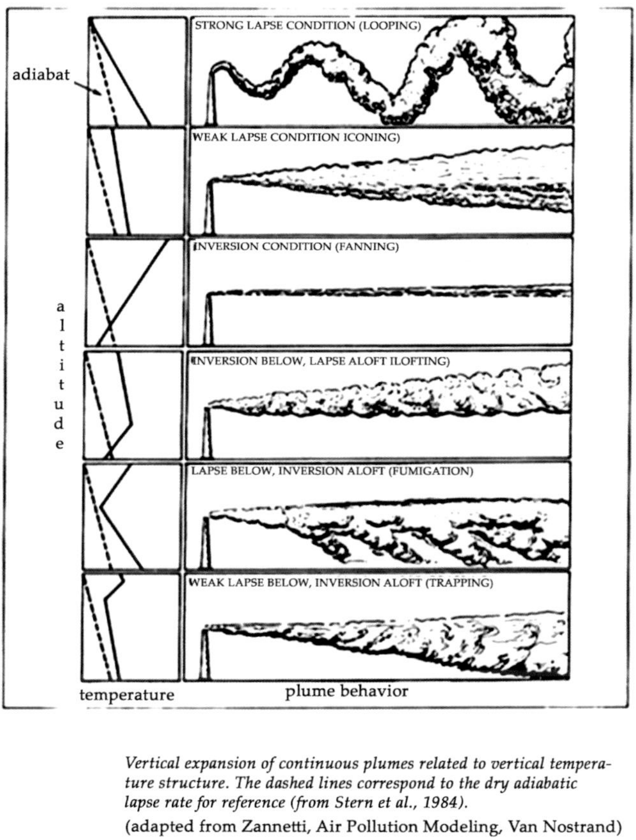


**Figure 3.2.** *Different atmospheric temperature profiles result in different plume dispersion patterns.*

*Source:* https://airquality.climate.ncsu.edu/2021/06/06/atmospheric-dispersion-and-pollution-transport/.

The atmosphere normally cools with height, so as the air parcel rises, it will continue to rise because it is warmer than the air around it. The term "surface temperature inversion" refers to the situation where air near the surface, usually at night, cools below the temperature of the air above it, causing temperatures to rise with height. During a temperature inversion, if a piece of air is lifted from the surface and falls back to the surface, it is because the air surrounding it is colder. Consider the air pollution parcel that is essentially trapped close to the surface during a surface temperature inversion and unable to easily disperse vertically. In essence, the "mixing height" is the point below which air will easily mix, and it is indicated by the top of the inversion. Thanks to the sun's continued warming of the lower atmosphere, an overnight temperature inversion will erode and mixing heights will rise, improving air mixing. If you have ever watched New Year's Eve or July 4th fireworks outside in the early morning, you may have noticed this phenomenon: occasionally, the air appears hazy or smokey. This is typically caused in part by a low-level temperature inversion that traps the celebratory smoke. The following graphic illustrates how pollution can spread vertically depending on the temperature profile of the atmosphere (temperature on the X-axis and height on the Y-axis). It shows six distinct atmospheric temperature profiles (Figure 3.2).

### 3.1.3. Atmospheric Profiles

Two distinct atmospheric soundings from forecasts are displayed in the image below. Each sounding's right side displays how the wind's direction and speed vary with altitude. The longer the barb on the end, the faster the wind; the wind barbs indicate the direction of the wind. Which, in your opinion, will provide superior dispersion? (Figure 3.3).

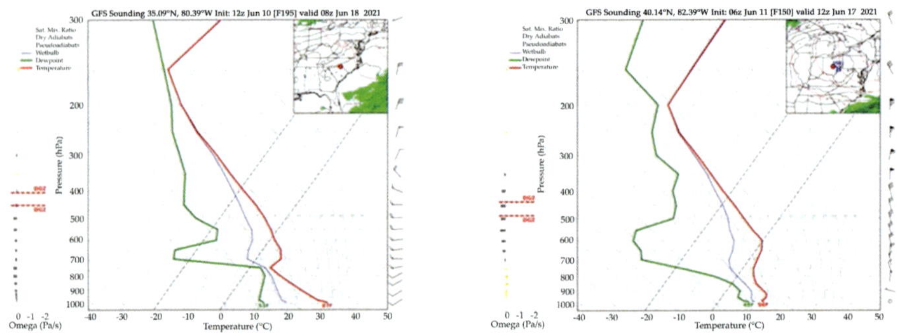


**Figure 3.3.** *Two different modeled atmospheric soundings.*

*Source:* https://airquality.climate.ncsu.edu/2021/06/06/atmospheric-dispersion-and-pollution-transport/.

The sounding on the left depicts moderate winds from the surface to the upper atmosphere, and an atmosphere that readily cools with height (albeit with an inversion around 750mb). A tiny low-level inversion can be seen in the sounding on the right, along with calm to light winds up to a height of 700 mb. You guessed it right—the faster wind speeds will increase horizontal dispersion at all levels through the atmosphere, aiding in air mixing and lowering pollution concentrations.

### 3.1.4. Long-Range Transport

Upper-level winds can help carry air pollution over great distances, sometimes even across the continent, when it originates from a large source or rises high above the surface. The western states' wildfire season can have a significant impact on local communities, but when fires burn for days or weeks, they produce a significant amount of heat and smoke. Since "warm air rises," as was previously mentioned, the intense heat helps to raise the smoke high and away from the surface. This explains also why there can occasionally be a "fire tornado" or "fire whirl" in large fires; the fire whirl is a good way to visualize the phenomenon of warm air rising as upward air currents lift the burning flames. When the massive amounts of smoke reach the mid-upper levels of the atmosphere, they can travel great distances, across countries or even continents, thanks to upper-level winds and the jet stream!

**Figure 3.4.** *A dust storm visible on satellite in the TX Panhandle.*

*Source:* https://airquality.climate.ncsu.edu/2021/06/06/atmospheric-dispersion-and-pollution-transport/.

Dust has the same long-range properties as smoke. The Permian Basin and portions of the Chihuahuan Desert are located in West Texas, in the United States. Large amounts of blowing dust can be produced when the land is dry and the winds are strong enough (typically as a strong, dry cold front advances). The finer the dust, the longer it will take for it to settle out of the air mass. The city of El Paso is located east/northeast of Lake Palomas, a dried-up lakebed covered in fine dust, in remote West Texas. The fineness of the dust makes it susceptible to wind disturbance, and in this area, blowing dust events can occur several times a year. Blowing dust frequently accumulates along a frontal boundary as it moves forward, occasionally bringing high levels of fine particle pollution throughout the entire state! Other desert states like Arizona and New Mexico also experience similar events. Daily fine particle pollution AQIs frequently reach Code Red or even Code Purple levels in areas close to and directly downwind of the dust source region (Figure 3.4).

### 3.1.5. Super Long-Range Transport

That's right, dust from Africa's Sahara Desert can (and often does) travel the whole length of the Atlantic Ocean! The Sahara Desert region was once submerged under water, and the sand there is incredibly fine and rich in nutrients and minerals from long-gone, decomposing marine life. Huge amounts of blowing dust are produced by strong summer winds, far more than what is produced in these American desert regions.

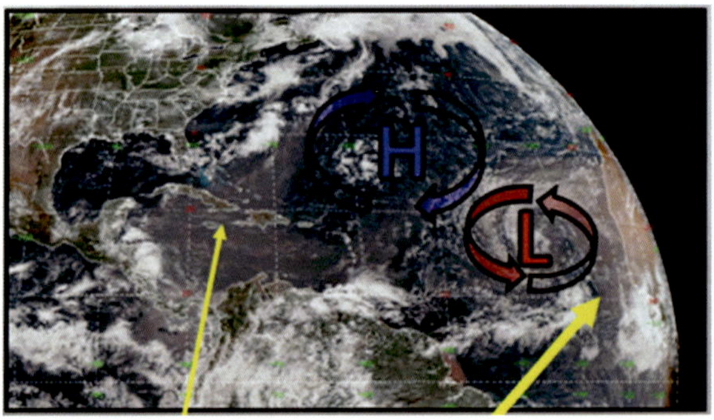


**Figure 3.5.** *Dust from the Saharan Desert swirling around high and low pressure systems over the Atlantic Ocean.*

*Source:* https://airquality.climate.ncsu.edu/2021/06/06/atmospheric-dispersion-and-pollution-transport/.

The dust is transported across the ocean by the dominant easterly winds, reaching both North and South America. Although the Amazon Rain Forest benefits from the dust in that it supplies nutrients, the southeastern United States may experience higher than average levels of fine particle pollution (Figure 3.5).

While Florida and the states bordering the Gulf Coast are more likely to experience high dust concentrations, occasionally the weather pattern will be such that the dust spreads into the Great Plains, or even into our own backyard! As an example, in June 2020, readings close to the Code Yellow/Code Orange threshold were recorded throughout the state of North Carolina.

## 3.2. DISPERSION MODELING USED IN AIR QUALITY MANAGEMENT (AQM)

The technique of air dispersion modeling is used to forecast the concentration of one or more air pollution sources at ground level. Predicted ground level ambient concentrations of pollutants are derived from data such as:

- Land use (urban/rural)
- Meteorological data (wind speed, temperature)
- Pollutant source information (mass emission rate/stack gas temperature/flow rate)
- Terrain characteristics

The modeling's output is used to assess whether an industrial source of air pollution complies with ambient air quality standards.

### 3.2.1. Types of Air Pollution Dispersion Modeling

The following are some variations on air pollution dispersion modeling.

#### *3.2.1.1. Box Models*

One of the earliest and most basic types of pollutant dispersion modeling is box modeling. Because box models' simplified spatial and temporal dispersion allowed for a greater focus on the chemical aspects, they were traditionally particularly useful in situations requiring the simulation of chemical interactions between pollutants. When using a box model, all emissions are released into a hypothetical simple box with defined dimensions that represents the airshed. It is assumed that the emissions will be dispersed equally throughout the box after they are released. As one might anticipate, this kind of model has very little accuracy. The box model's primary benefit is its simplicity, which means it uses very little processing power and enables very quick simulation runtimes. Furthermore, very little in the way of input data is needed.

Examples of Simple Box Models

- EKMA: The Empirical Kinematic Modeling Approach (EKMA) model was initially employed to evaluate the probability of photochemical smog formation in urban environments. The VOC and NOx concentrations in this model were taken to be constant from the readings obtained early in the morning. In actuality, EKMA is a kind of Lagrangian simulation—albeit one that is restricted to box model systems. EKMA is still occasionally used, despite its age, to examine the relationships between ozone, NOx, and VOCs in straightforward settings.

- Uses: Box models are not frequently used in modern regulatory settings due to their obvious simplicity, with the exception of initial assessment. They do, however, continue to have a role in pollutant dispersion modeling, especially in small, enclosed anthropogenic spaces. As an instance, Lin et al. investigated the photodegradation-induced removal of formaldehyde from indoor air spaces using a box model. The use of a box model is quite appropriate in these situations because of the small air spaces inside and the frequent lack of ventilation to the outside environment. However, more intricate models have also been used in the indoor setting. Modified versions of the box model, such as a two-box model, have been utilized in modeling photochemical pollutant levels in street canyons (i.e., a street enclosed by tall buildings on each side), amongst other uses. Many other models designed specifically for street canyons are based off the box model, albeit typically modeling each street as an individual box. Examples include the STREET and STREET-BOX models (Johnson, Ludwig, Dabberdt, & Allen, 1973).

### *3.2.1.2. Eulerian Models*

When modeling pollution, eulerian models employ a purely mathematical approach. The average pollutant concentration in each grid cell, created by dividing the study area into multiple horizontal and vertical grid cells, is determined at each time point. Reynolds and associates introduced the concept of Eulerian dispersion modeling. More recent versions may be used for longer periods of time, even though they were initially only intended to model a few days per simulation.

Eulerian models readily take into account the removal of the constituent particles through deposition or chemical reactions because they follow average grid concentrations instead of a full plume.

**Examples**

- **TAPM:** Unlike other dispersion models, the CSIRO-developed Air Pollution Model (TAPM) can compute dispersion using either a Lagrangian particle model or an Eulerian grid. In areas near the emission source, the latter is thought to be more accurate. Its capacity to extract meteorological conditions from synoptic charts—past, present, or forecast—is another outstanding feature. Measurements of the surface can also be included. TAPM works especially well in complex environments, like places with a sea breeze or uneven topography. Other dispersion models have also used meteorological input data from the integrated prognostic meteorological model. TAPM requires a lot of processing power, which is to be expected for a simulation that is based on mathematics. Applications of TAPM in recent times have included modeling heavy metal deposition surrounding a copper smelter, assessing health risks associated with VOC emissions from municipal waste, and using it in a complex, mountainous terrain.

- **Variable K-Theory Model:** When simulating $NO_2$ and $SO_2$ concentrations across 17 sites, it has been found that an Eulerian Variable K-Theory model provides the highest accuracy when compared to box, Gaussian plume, and Lagrangian models (Gronwald & Chang, 2018).

### 3.2.1.3. Gaussian Models

Gaussian models assume the pollutant concentrations follow a normal (Gaussian) distribution in both the horizontal and vertical aspects, as determined through experimental measurements of plume spread. This assumption stems from the idea that plume spread is caused by the diffusion of the constituent pollutants. In the USA, these models have been used for regulatory purposes for nearly 60 years. Gaussian plume models represent the pollutants as a single, continuous plume and assume that they are released at a constant rate (Figure 3.6). Gaussian plumes grow over time in both y and z dimensions. The Gaussian plume models necessitate the following presumptions: wind speeds must consistently equal or surpass 1 $ms^{-1}$, no chemical transformations can occur, and emission and meteorological conditions must stay constant.

The rate at which pollutants are released, the height at which they are released, wind speed at the reference height—which is frequently the height at which emissions are released—the mixing/inversion height, and the horizontal and vertical dispersion factors are examples of inputs. Furthermore, the plume's rise or fall can be simulated. It is believed that the plume will quantitatively reflect from the ground and the upper boundary layer of the atmosphere when it reaches these surfaces. This could eventually give the impression that pollutants are building up at ground level, which the model can account for.

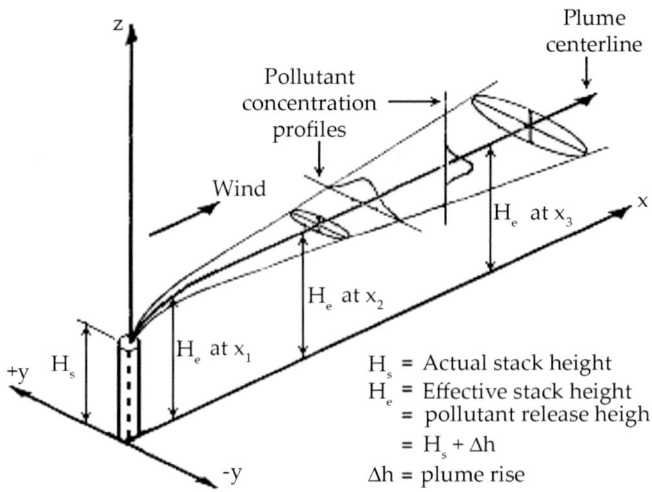


**Figure 3.6.** *A representation of a Gaussian plume model.*

*Source: By Milton Beychok; reproduced under Creative Commons 3.0 license.* https://www.mdpi.com/2673–4931/19/1/18.

### *3.2.1.3.1. Gaussian Plume Models*

- **AEOLIUSF:** Modeling dispersion in urban street canyons is the purpose of AEOLIUSF (Assessing the Environmental Of Locations In Urban Streets Full version). In modern environments, AEOLIUS/AEOLIUSF is not as frequently utilized as other street models. Its accuracy is mediocre, higher than that of STREET models but lower than that of SEUS models.
- **AERMOD:** The US Environmental Protection Agency's (EPA) preferred Gaussian plume model, AERMOD, replaced the ISC model. A single wind field is calculated and used in the simulation based on the meteorological conditions, terrain, and upper atmospheric conditions entered. Meanwhile, factors like turbulence, stability class, and Monin-Obukhov length (a continuous measure of near-surface atmospheric stability) are calculated based on the terrain, elevation, surface roughness, and land use. AERMOD works well in both simple and complex terrain, with ground or elevated sources. When there is no wind, AERMOD is not the right modeling tool. These days, AERMOD is used for a variety of tasks such as estimating near-road pollution levels, modeling emissions from gas-fired power plants and cement factories, and evaluating complex

industrial emissions. AERMOD is particularly useful for modeling pollutant sources that are surrounded by complex terrain.

- **AUSPLUME:** In New Zealand and Australia, this model is widely utilized. Although AUSPLUME's level of detail is thought to be a little lower than AERMOD's, it can still be used to model scenarios in which there is no wind. This model has been used recently to model the dispersion of radon released from a Romanian uranium mine and the dispersion of smells from Malaysian rubber factories.
- **CALINE3:** A line, as opposed to a point, serves as the emission source in the modified Gaussian plume model known as CALINE. Its primary application is in the modeling of pollutant dispersion from roads, where the geometry of the road is not limited to a straight line. The revised CALINE4 model is also utilized in some modern applications, but CALINE3 is still the Gaussian line model that the EPA recommends. Based on CALINE3, which is intended for comparatively simple terrain, are models like CAL3QHC and CAL3QHCR.
- **CAL3QHC and CAL3QHCR:** Although both models are based on CALINE3, they are especially made to identify the accumulation of CO hotspots brought on by traffic snarls, especially at intersections. Since CAL3QHCR is the "improved" form of CAL3QHC, it needs more data, especially localized meteorological data. The epidemiological investigation of the birth impacts of vehicle pollution and the evaluation of the dispersal of CO, NOx, and VOCs from Indian intersections have both made use of CAL3QHC in recent years.
- **CTDMPLUS:** The Complex Terrain Dispersion Model Plus Algorithms for Unstable Situations (CTDMPLUS) is intended for use in complex terrain scenarios, as its name would imply. All stability conditions, though, including neutral and stable ones, can employ it. A survey of the literature suggests that it is not widely applied in modern contexts.
- **ISC:** The model known as the Industrial Source Complex (ISC) was examined and reported by Bowers et al. was previously the US Environmental Protection Agency's (EPA) authorized Gaussian plume model. The ISC model has been used recently in modeling volatile organic compounds (VOCs) downstream of a petrochemical manufacturing facility and in monitoring various pollutants (CO, VOC, NOx, and PM10) in Italian agricultural land. To increase the accuracy of the results, some authors also combine the AEROMOD

and ISC model (Esbrí, López-Berdonces, Fernández-Calderón, Higueras, & Díez, 2015).

- **OCD:** A straight-line Gaussian model, the Offshore and Coastal Dispersion (OCD) model is intended to forecast the dispersion of pollutants over marine or coastal regions. We incorporate changes that occur as the plume moves along the coast. The Gulf of Mexico is one of the many environments in which this model has been applied.

### *3.2.1.4. Lagrangian Models*

Pollutant "puffs" are simulated by Lagrangian models as they are released from the source, typically on a regular basis. The most popular is the "Gaussian puff" model, in which the expansion and downwind motion of each puff is thought to follow a Gaussian distribution. The puffs are three-dimensional objects that grow over time in all three dimensions (x, y, and z) while also traveling downwind from the source of the emission. Hundreds to hundreds of thousands of these hypothetical puffs can make up a model. Since each puff is handled separately, it is possible for them to move in different directions and have different rates of dispersion, which enables the simulation to model local conditions more accurately. The Lagrangian random walk model is a related model of this kind in which the plume is discretized as multiple independent tracer particles. A stochastic "random walk" algorithm is used to account for local turbulence while the particles are moved by the mean wind field. Lagrangian models, in instance, exhibit increased accuracy when modeled with intricate topography or flow patterns (e.g., recirculation of the pollutants) and variations in meteorology or emissions over time. Longer time periods and distances (up to several years) are commonly modeled using Lagrangian models. On the other hand, predictions from Gaussian plume models are usually limited to a distance of 50 km from the point source. Since Rodhe and Rodhe first introduced the field of Lagrangian modeling, it has quickly gained traction. One can use time steps ranging from 1 to 180 seconds. Similar to Gaussian plume modeling, if one does not account for the assumption of complete pollutant reflection from the ground and upper atmospheric boundary layer, one may arrive at misleading conclusions.

Examples

- **AFTOX:** Kunkel developed the chemical dispersion model known as AFTOX (Air Force Toxic). The assumption is that the source emits four Gaussian puffs per minute. Compared to other model types, AFTOX frequently yields higher pollutant concentrations and lower accuracy because it does not take into account the pollutants' settling or decay. Although limited to gases that are neutrally buoyant, AFTOX is especially helpful in simulating liquid spills that eventually evaporate. Recently, modeling examining the connection between raindrop size and aerosol particle scavenging efficiency was conducted using AFTOX as the foundation.

- **CALPUFF:** The US EPA's authorized long-range (>80 km) atmospheric emissions model is called CALPUFF. However, CALPUFF is also utilized in short-range simulations with intricate surface topography; in Australia and New Zealand, it is a commonly used regulatory model. When it comes to short- and long-range simulations, CALPUFF is flexible. More modern software, like VISTAS version 6, allows for shorter run times for CALPUFF, often less than an hour. Applications include studying shipping emissions in a port in Western Australia, modeling the dispersion of odors, and modeling the dispersion of different pollutants produced by industrial facilities. A noteworthy application involved simulating the atmospheric mercury emissions from a coal-fired power plant located in Mexico. CALPUFF has also been used in many studies to model dispersion over complicated terrain.
- **Hybrid Eulerian–Lagrangian Dispersion Models (HDMs):** Andrén's description of hybrid models that combine the Eulerian and Lagrangian approaches is still quite common, especially when it comes to modeling dispersion near point-line sources. Using the Lagrangian method, the emissions are first modeled as puffs. Afterward, the puffs are assumed to approximate a volume emission after expanding to a specified level or traveling a specified distance. The long-range pollutant dispersal is then determined using the Eulerian approach. The main benefit of this approach is that it requires less processing power at all scales than utilizing a Lagrangian model.

Hybrid models have been used to predict PM10 levels and other pollutants in complex scenarios, particularly in complex urban and mixed industrial-residential environments. They have also been used in conjunction with CFD modeling.

### *3.2.1.5. Computational Fluid Dynamics Models*

Engineering fields have long used computational fluid dynamics (CFD) as a reliable modeling technique. The application of modeling atmospheric pollutants is a relatively recent development, though. The foundation of CFD is the Navier-Stokes equations, which are partial differential equations that are three-dimensional, unsteady, non-linear, and precisely model the flow of atmospheric gases. Different methods are used to model unknown turbulence terms in order to find a solution because there are more unknowns than discretized equations. To compute the movement and dispersion of pollutants through the atmosphere, a Langrangian or Eulerian framework is integrated into the model. Depending on the simulation's fidelity and resolution, CFD models can demand extremely high processing power. Large Eddy Simulation (LES), Direct Numerical Simulation (DNS), and Reynolds-Averaged Navier Stokes (RANS) are the three main categories of CFD models.

- **Examples and Uses:** Brown, et al. compared the Quick Urban and Industrial Complex (QUIC) dispersion model, which simulates wind flow in an urban setting and is based on empirical parameterizations of the flow around and between buildings. Although the model's results in this case were just as accurate as those of standard CFD modeling, the benefit of CFD models is that wind flow through novel building combinations and/or configurations can be simulated as opposed to depending solely on empirical data. Mocho, et al. investigated the movement of formaldehyde in an indoor environment using a CFD model. As anticipated, the results' accuracy exceeded that of a straightforward box model. The near-field dispersion of pollutants in the presence of plumes with different buoyancies has been modeled using CFD. Reactive chemical component movement and PM10 movement modeling are two more applications.

### *3.2.1.6. Street Network Models*

Right now, the street network model is the least used modern modeling technique. Typically, this method treats each street as a line source of emissions, with the quantity of emissions calculated from the traffic volume along that street. It is intended for the analysis of vehicle emissions in built-up urban environments.

Examples

- **SIRANE:** The SIRANE model, developed by Soulhac, et al., is intended especially for simulating the dispersion of pollutants from traffic in urban areas. As of right now, the literature reports on the main street network model. Every street is represented as a box, and pollution is transferred along the box (i.e., between boxes (at street intersections), between boxes and the street, and between boxes and the atmospheric boundary layer. While there could be hourly variations in the atmosphere, between these time intervals, it is assumed that nothing changed. The model has been verified using data from wind tunnels and one year's worth of $NO_2$ emissions. But according to research done by Wang and co. has proposed that there is a stronger correlation between the average $NO_2$ values and the SIRANE output and near-road $NO_2$ concentrations. However, SIRANE has been employed in a number of epidemiological investigations, mainly in France. Based on SIRANE, derivative models of street pollution have been developed. One such model is MUNICH (Model of Urban Network of Intersecting Canyons and Highways), which makes use of a grid modeling technique.

#### *3.2.1.7. Other Models*

For modeling the dispersion of pollutants over large distances, in intricate terrain, and for pollutants that are photochemically reactive, specialized models are available. The Operational Street Pollution Model (OSPM), which simulates the chemistry of photochemical smog formation, is one example. Other widely used photochemical models are CALGRID, UAM (Urban Airshed Model®), CAMx (Comprehensive Air quality Model with extensions), and CMAQ (Community Multiscale Air Quality). The most popular photochemical air quality model, UAM, and the open-source CAMx model are particularly noteworthy. Based on recent and ongoing air quality measurements, statistical models are available for short-term air quality forecasting. These models only attempt to connect patterns in emission trends to the air quality; they do not attempt to prove cause and effect. Analogously, $O_3$, $NO_2$, and $SO_2$ concentration predictions have also been tested using machine-learning algorithms. Owing in part to the short half-lives and distinct characteristics of odorous compounds, there are air quality models like ModOdor that are made expressly for projecting their dispersion.

### 3.2.2. Importance of Air Pollution Dispersion Modeling

It is necessary to comprehend how emissions affect human health and air quality. It supports risk assessment, urban planning, and regulatory decision-making. There are various reasons why air pollution dispersion modeling is crucial. A few of the most crucial points are listed below.

- **Health Impacts:** Understanding the effects of air pollution on health requires an understanding of air pollution dispersion modeling. It assists in determining the levels of pollution exposure for various regions and populations by forecasting how pollutants spread throughout the atmosphere. Millions of Indians suffer from respiratory illnesses as a result of air pollution, which has serious negative health effects. By using modeling to better understand the dispersion of pollutants, authorities can pinpoint high-risk areas and carry out focused interventions.
- **Industrial Regulation:** Industrial regulation design and enforcement

depend on air pollution dispersion modeling. Regulatory bodies can set emission limits and guidelines for industries by precisely forecasting the dispersion of pollutants in the atmosphere. One of the main causes of air pollution is industry. Regulatory agencies use modeling to evaluate the possible effects of new industrial facilities and create emission standards that guarantee their operations stay within legal bounds.

- **Climate Change Mitigation:** For a number of reasons, air pollution dispersion modeling is important when it comes to mitigating climate change. Through precise simulation of pollution dispersion, researchers and decision-makers can evaluate the relative contributions of different pollutants to global warming. Climate change is also exacerbated by some pollutants. By simulating their dispersion, we can better understand how they contribute to global warming and develop mitigation strategies for its effects.

### 3.2.3. Challenge Faced During Air Pollution Dispersion Modeling

Accurate emissions data collection, localized meteorological data analysis, and taking into account the distinct combination of pollution sources found in densely populated urban areas are among the challenges. India faces a number of difficulties when modeling the dispersion of air pollution.

- **Data Availability:** A big obstacle is the lack of reliable and sufficient data on the sources of emissions, the weather, and the concentrations of pollutants. It is hard to do accurate modeling without trustworthy data. Large volumes of data, such as those pertaining to atmospheric chemistry, topography, meteorological conditions, and emission sources, are needed for accurate modeling. Inaccurate projections resulting from inadequate or obsolete data can impair the efficacy of regulatory decisions and pollution control tactics.
- **Limited Monitoring Infrastructure:** There are few data points because India's current infrastructure for monitoring air quality is not very large. This restriction lowers the accuracy of dispersion models and makes it more difficult to validate them. Air pollution dispersion modeling is severely hampered by the inadequate monitoring infrastructure. It is difficult to precisely evaluate air quality and forecast pollution dispersion patterns when there are insufficient monitoring stations in some areas, which leads to sparse data points.

- **Resource Constraints:** Significant resources are needed for the development and upkeep of sophisticated air pollution dispersion models, including highly qualified workers, powerful computers, and constant data inputs. Resource limitations may prevent advanced modeling techniques from being widely used. Accurate model development and implementation demand a significant investment of funds, highly qualified workers, and cutting-edge technology. Insufficient resources can impede the advancement of complex modeling methodologies, the availability of superior data, and the ability to continuously validate and enhance models. The understanding, monitoring, and mitigation of the detrimental effects of air pollution on public health, the environment, and climate depend heavily on air pollution dispersion modeling. These models identify emission sources, encourage the dispersion of pollutants in the atmosphere, and offer important insights into pollution patterns and the efficacy of regulatory actions. The downwind ambient concentration of air pollutants or toxins released from sources like industrial plants, moving cars, or accidental chemical releases is determined using dispersion models. They can also be used to predict concentrations in the future under different conditions (e.g., shifts in the sources of emissions).

## 3.3. MAJOR METEOROLOGICAL FACTORS AFFECTING POLLUTION DISPERSION

The fate of air pollutants is influenced by air movements. Thus, a study of local weather patterns (meteorology) should be a part of any investigation into air pollution. Pollutant concentrations will rise if the atmosphere is still and they are unable to disperse. Conversely, when there are strong, erratic winds, the pollutants diffuse quickly and the concentrations of the pollutants are reduced.

Meteorological data helps:

- identify the source of pollutants
- predict air pollution events such as inversions and high-pollutant concentration days
- simulate and predict air quality using computer models.

The following variables should be measured when researching air quality because they can provide insight into the chemical reactions that take place in the atmosphere:

- wind speed and direction

- temperature
- humidity
- rainfall
- solar radiation

### 3.3.1. Wind Speed and Direction

Wind data records can be used to identify the general direction and area of emissions when high concentrations of pollutants are detected at a monitoring station. Planning can be done to lessen the effects on air quality once the sources have been identified. Wind speed is measured by a device known as an anemometer. A sonic anemometer is the kind of instrument we use at our monitoring stations. The idea behind a sonic anemometer is that the speed of the wind influences how long it takes sound to travel between two points. It will take less time for sound to travel with the wind than it will to travel against it. Sonic anemometers measure wind speed and direction simultaneously by measuring the speed of sound waves in two different directions.

### 3.3.2. Temperature

Temperature measurement aids in modeling, forecasting, and assessment of air quality. The chemical reactions that take place in the atmosphere to create photochemical smog from other pollutants are significantly influenced by temperature and sunlight, or solar radiation. Higher smog concentrations can result from favorable circumstances. Using a substance whose resistance varies with temperature, like platinum wire, is the most popular method of taking temperature readings. This change is detected by a sensor, which then provides a temperature reading.

### 3.3.3. Humidity

In addition to temperature and solar radiation, water vapor is a key player in a variety of atmospheric thermal and photochemical processes. Water molecules can bind strongly to a wide range of substances due to their small size and high polarity. They can greatly increase the amount of light scattered by airborne particles if they are attached to them (measuring visibility). Corrosive gases, like sulfur dioxide, will dissolve in water and form an acidic solution that can harm people's health and property if the water molecules attach to them. The percentage of water's saturation vapor pressure at a specific temperature that makes up the water vapor content of air is reported. The relative humidity is shown here. The amount of water vapor in the atmosphere varies greatly depending on factors such as air temperature, wind direction, proximity to water bodies, and geographic location. In general, summertime brings higher relative humidity along with higher temperatures and rainfall. The absorption characteristics of a polymer film are used to measure humidity. As the relative humidity of the surrounding air fluctuates, water vapor is either absorbed or lost by the film. These variations are detected by a sensor, which then provides a humidity reading.

### 3.3.4. Rainfall

Rain has the ability to dissolve gaseous pollutants and wash particulate matter out of the atmosphere, an effect known as "scavenging." Particle removal increases visibility. The quality of the air is usually better in areas with frequent heavy precipitation. Rain may dissolve gaseous pollutants, like sulfur dioxide, and create acid rain, which may cause harm to vegetation or materials. A tipping bucket rain gauge is a popular tool for measuring rainfall. Small amounts of rain are collected and counted by the gauge to record rainfall. Rainwater entering the funnel is directed into the tipping bucket, which is separated into two equal sections by a partition. When the tipping bucket is empty, its design causes one compartment to tilt downward and come to a stop, allowing the other compartment beneath the funnel to be prepared to collect rainwater. The bucket tilts in the opposite direction so that the rain-containing compartment rests against the stop on the opposite side after a predetermined amount of rain has flowed from the funnel into the upper compartment. The other compartment then begins to fill as the collected water empties out. Using the area of the funnel, the number and rate of bucket movements, and other data, the device determines the amount and intensity of rainfall.

### 3.3.5. Solar Radiation

Since the rate at which the chemical reactions that cause smog are accelerated by sunlight, solar radiation monitoring is crucial for modeling photochemical smog events. The amount of sunlight varies depending on the sky's cloud cover, the time of day, and the geographical location. Solar radiation is measured by a device known as a pyranometer using the output of a particular kind of silicon cell sensor.

## 3.4. ATMOSPHERIC STABILITY

Usually, when you hear the word "stable," you picture something balanced or that is unlikely to change. When something is "unstable," the opposite is true. An unstable object will probably eventually fall or shift in position. With clouds, it is also the case. You may notice that a fluffy cumulus cloud is changing from minute to minute in appearance. These clouds are a constant representation of the unstable state of the atmosphere because they are constantly changing. The concept of atmospheric instability is closely related to vertical motion, cumulus formation, and thunderstorms. Imagine a boulder at the bottom of a canyon surrounded by steep hills to help you visualize the idea of stability. This is what the blue circle in the Figure 3.7 represents.

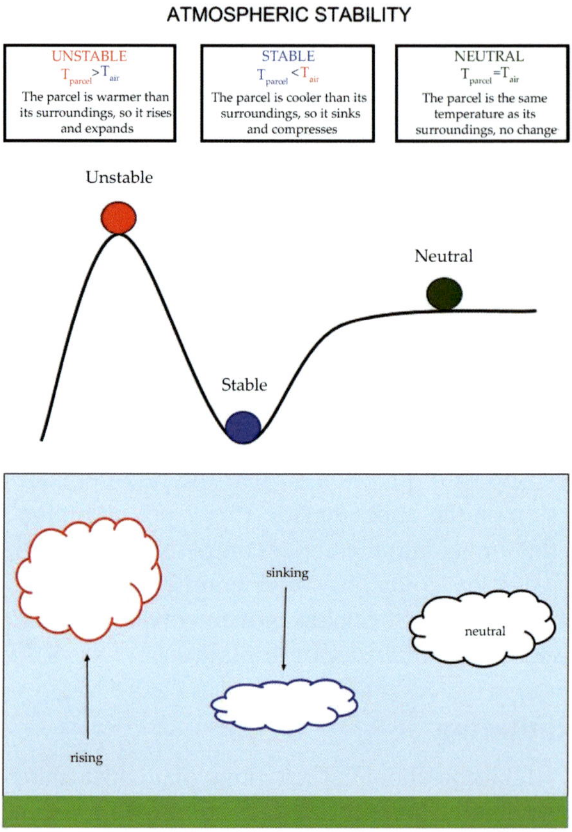


**Figure 3.7.** *Examples of stability and instability in relation to air and parcel temperatures.*

*Source: By Britt Seifert, https://pressbooks-dev.oer.hawaii.edu/atmo/chapter/Chapter 5-atmospheric-stability/.*

The boulder would roll back to the bottom if you released it after pushing it from its starting position halfway up one of the hills with sufficient strength. Even though you applied force to the boulder and caused an initial displacement, the boulder would eventually return to its starting position, leaving no net displacement. Imagine the same boulder at the top of a hill (red circle below) to get an idea of how instability is conceptualized. The boulder would roll downhill and accelerate away from its starting position if you could push it just a little bit in any direction. The same boulder would, however, move but stay in its new location if it were placed on level ground (see the green circle below) and pushed. An illustration of neutral stability is this. These ideas can all be used to explain how air parcels move through the atmosphere. Because cloud formation is intimately tied to atmospheric stability or instability, the topic of stability is significant to atmospheric science. This section will teach us how to use thermodynamic diagrams to visualize movement and relate these ideas to the buoyancy of air parcels.

## 3.4.1. Atmospheric Stability & Lapse Rates

### *3.4.1.1. Adiabatic Processes*

In atmospheric sciences, when we talk about stability, we usually picture air parcels, which are fictitious blobs of air that are able to expand and contract without interacting with their surroundings or disintegrating. The most important finding is that air parcel movement in the atmosphere can be roughly described as an adiabatic process. Adiabatic processes are reversible and do not involve heat exchange. Assume you have a piece of air at the surface of the Earth.

The temperature and pressure of the air parcel are identical to those of the ambient air, as we shall refer to it. Since we know that pressure drops with height, if you were to lift the air parcel, it would end up somewhere with lower environmental air pressure. The air molecules inside the parcel will effectively push outward on the walls of the parcel and expand adiabatically because the ambient air pressure outside the parcel is lower than the pressure inside the parcel. The temperature inside the air parcel drops as the internal energy drops because the air molecules inside the parcel have to expend some of their own energy to expand the walls of the air parcel. In summary, rising air parcels do not exchange heat with the surrounding environment as they expand and cool adiabatically.

Imagine now that you are returning the same air parcel to the surface of the Earth. The air parcel is traveling into an area where air pressure is higher. The parcel walls will compress as a result of the increased external pressure pushing inward, increasing the interior temperature. Once more, there is no heat exchange with the environment because the process is adiabatic. Even so, temperature variations within the air parcel are still possible; however, these variations are caused by variations in the air parcel's internal energy rather than by mixing.

### *3.4.1.2. Dry Adiabatic Lapse Rate*

An air parcel's temperature will change at a constant rate as long as it is unsaturated (relative humidity < 100%). While the temperature drops with altitude, it is defined as positive because it is a lapse rate, which is a decrease in temperature with height. $9.8\ K\cdot km^{-1} = 9.8°C\cdot km^{-1}$ is the value of the dry adiabatic lapse rate, or Γd. The decrease in internal energy and adiabatic expansion are the causes of this temperature drop (Figure 3.8).

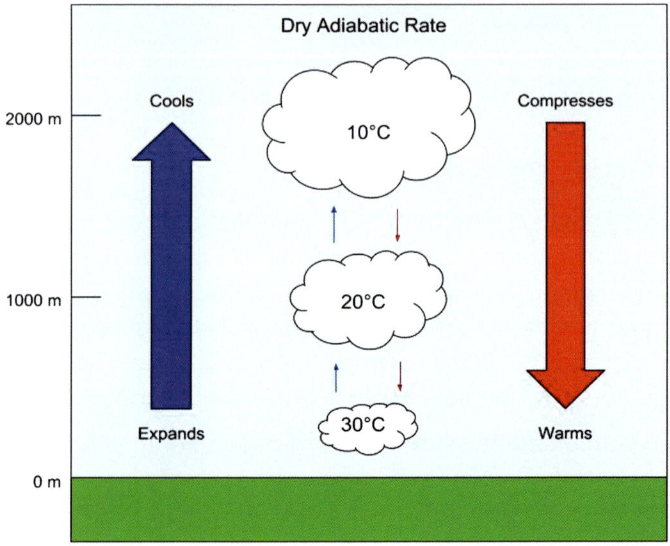


**Figure 3.8.** *Air rises, expands, and cools at the dry adiabatic lapse rate, approximated as a 10°C decrease per km.*

*Source: By Britt Seifert, https://pressbooks-dev.oer.hawaii.edu/atmo/chapter/Chapter 5-atmospheric-stability/.*

Back to the subject of atmospheric stability now. An equilibrium state is referred to as atmospheric stability. The boulder on a hill or valley served as an example for discussing how some initial movement could produce more (unstable), less (stable), or no change at all (neutral). If the air is in stable equilibrium, an air parcel will generally return to its initial position after being forced to rise or sink, given an initial change in elevation. An air parcel that has been pushed will accelerate away from its starting position in an unstable equilibrium. Although the motion could be downward or upward, unstable atmospheres typically encourage vertical motions. Last but not least, a small initial change in an air parcel's elevation will not cause any more movement in a neutral equilibrium.

### *3.4.1.3. Determining Stability*

After some initial displacement, how can one determine whether an air parcel will remain stable? One way to do this is to compare the temperature of a rising or sinking air parcel to the air temperature in the surrounding environment. Consider the following: an air parcel initially has the same pressure and temperature as its surroundings. The dry adiabatic lapse rate, or 9.8 $K \cdot km^{-1}$, is the temperature drop that occurs when the air parcel is raised a certain distance. The air parcel will have a higher density and gravitate toward its initial position if it is colder than the surrounding area in its new location. Since vertical motion is resisted in this instance, the air is stable. The rising air will keep rising until it reaches a new equilibrium where its temperature equals that of the surrounding air if it is warmer and less dense than the surrounding air. In this instance, the air parcel is unstable due to the amplified initial change. We need to know the temperature of the environment at various altitudes as well as the rising air in order to determine whether or not the air parcel is unstable. Using a weather balloon is one practical method for doing this. The environmental lapse rate can be vertically profiled by releasing a radiosonde that is affixed to a weather balloon. Plotting of the temperature, humidity, wind, and position data received from a radiosonde results in a thermodynamic diagram. A sounding is a vertical plot of temperature and other variables.

#### *3.4.1.3.1. Dry Stability*

Stability of an unsaturated air parcel is relatively simple if it is dry. An atmosphere is deemed neutrally stable when its environmental lapse rate equals that of the dry adiabatic lapse rate, which implies that the environment's temperature decreases by 9.8 $K \cdot km^{-1}$. There won't be another change in position after a certain initial vertical displacement because the air parcel's temperature will always be the same as the surrounding environment. The air parcel will initially be displaced vertically and become either colder (if lifted) or warmer (if pushed downward) than the surrounding environment if the environmental lapse rate is lower than the dry adiabatic lapse rate. This is due to the fact that if the air parcel were lifted, its temperature would decrease more than the ambient temperature. This is a typical atmospheric scenario and a stable state for a dry air parcel. At 6.5 $K \cdot km^{-1}$, the global average tropospheric lapse rate is stable during dry lifting. Lastly, if the environmental lapse rate exceeds the dry adiabatic lapse rate, the air parcel will initially be displaced vertically and become either warmer (if lifted) or colder (if pushed downward) than the surrounding environment. This is due to the fact that if the air parcel were lifted, its temperature would decrease more slowly than the surrounding air. A dry air parcel is in an unstable situation in this instance.

In general for a dry air parcel, the following is true.

$\Gamma_d = \Gamma_{env}$ *NEUTRAL*

$\Gamma_d < \Gamma_{env}$ *STABLE*

$\Gamma_d > \Gamma_{env}$ *UNSTABLE*

### *3.4.1.4. Moist Adiabatic Lapse Rate*

Add moisture, and the whole thing becomes more complex. Temperature and, of course, moisture content are the two main factors that determine whether or not an air parcel is saturated. The Clausius-Clapeyron relationship graph indicates that air is more likely to be saturated at a lower temperature when there is an equal amount of moisture present.

We are aware that the dry adiabatic lapse rate causes an air parcel to cool as it is raised. The short answer is that water vapor will condense to liquid water to form a cloud if the air parcel cools down to the point where the air is saturated with respect to water vapor. Water vapor transitions from a higher energy state to a lower energy state during condensation. Since energy can never be created or destroyed—especially during phase transitions—what happens to all that extra energy? It is released as latent heat. The latent heat of condensation is roughly equal to $2.5 \times 10^6$ J·kg$^{-1}$, meaning that $2.5 \times 10^6$ Joules of energy are released for each kg of water vapor that condenses to form liquid water.

This separates the moist adiabatic lapse rate from the dry adiabatic lapse rate and has significant implications for the lapse rate of an air parcel. Part of the adiabatic cooling caused by expansion is counteracted by the latent heat that is added during the condensation process. This means that the air parcel will cool at a slower rate, referred to as the moist adiabatic lapse rate, rather than at the dry adiabatic lapse rate. In summary, a parcel cools at the dry adiabatic rate up until saturation, at which point condensation slows down the rate of cooling. Although the moist adiabatic lapse rate varies slightly with temperature, for the sake of simplicity in this class, we will treat it as a constant: $\Gamma m = 4.5$ K·km$^{-1}$ = 4.5°C·km$^{-1}$.

#### *3.4.1.4.1. Moist Stability*

Because moisture alters the air parcel's lapse rate, it has an impact on stability. The ideas remain the same, though, and we continue to compare the temperature of the air parcel to the surrounding air. There's only one more thing for us to worry about: we need to determine whether the air parcel is moist or dry. Below are some definitions that account for adiabatic lapse rates that are both dry and moist (Figure 3.9).

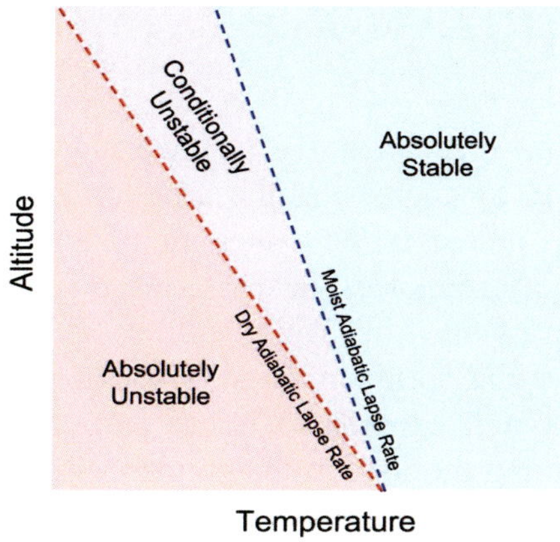


**Figure 3.9.** *A thermodynamic diagram showing the stability of the atmosphere based on the dry ($\Gamma_d = 9.8\ K\ km^{-1}$) and moist ($\Gamma_m = 4.5\ K\ km^{-1}$) adiabatic lapse rates (Created by Britt Seifert).*

*Source: By Britt Seifert, https://pressbooks-dev.oer.hawaii.edu/atmo/chapter/Chapter 5-atmospheric-stability.*

If the environmental lapse rate is smaller than the moist adiabatic lapse rate, the atmosphere is considered to be completely stable. This implies that, even after it reaches saturation, a rising air parcel will always cool more quickly than the surrounding air. Even after becoming saturated, when latent heating will offset some cooling, an air parcel will not be able to rise if it is colder throughout. If the environmental lapse rate exceeds the dry adiabatic lapse rate, the atmosphere is considered completely unstable. This implies that, even in the unsaturated state, a rising air parcel will always cool more slowly than its surroundings.

This implies that it will be permitted to rise because it will be warmer (and less dense) than the surroundings. If the environmental lapse rate falls between the moist and dry adiabatic lapse rates, the atmosphere is considered to be conditionally unstable.

This implies that an air parcel's buoyancy, or its capacity to rise, is dependent upon its saturation level. When an air parcel is unsaturated in a conditionally unstable atmosphere, it will cool more quickly than the surrounding air at the dry adiabatic lapse rate, preventing vertical motion. However, it will cool at the moist adiabatic lapse rate if it is forced to rise and is able to become saturated. In this instance, it will rise, warm more than the surroundings, and cool more slowly than the surroundings.

## 3.5. CHARACTERIZATION OF ATMOSPHERIC TURBULENCE

Impact of turbulence on dispersion: Turbulence lowers the concentration of pollutants in the plume by increasing the entrainment and mixing of clean air into the plume (i.e., increases the dispersion of the plume. Classifying the amount of atmospheric turbulence that exists at any given time is therefore crucial. The dispersion of this kind varies with scale. Such that there will be mixing for flows where the pollutant cloud is smaller than the biggest eddies present. Larger clouds will experience larger and stronger mixing motions because the size of mixing motions in the atmosphere is infinite. This kind of dispersion is therefore scale dependent.

### 3.5.1. The Pasquill Atmospheric Stability Classes

Pasquill atmospheric stability classes: Pasquill's 1961 method was the first and, for a very long time, the most widely used way to classify the amount of atmospheric turbulence present. He divided the atmospheric turbulence into six classes of stability, class A being the most unstable or turbulent class and class F the most stable or least turbulent. The classes are A, B, C, D, E, and F.

- Table 3.1 lists the six classes
- Table 3.2 gives the weather conditions that characterize each class. A few important concepts are illustrated in the stability classes. Because solar radiation warms the Earth's surface, warm air rises to the surface and mixes with cooler, denser air, causing atmospheric instability. As the ground cools more quickly on clear nights, more stable conditions and inversions are established. Wind accelerates vertical mixing, dismantling stratification in all directions and moving the stability class in the direction of neutral (D).

**Table 3.1.** *The Pasquill Stability Classes*

| Stability Class | Definition | Stability Class | Definition |
|---|---|---|---|
| A | very unstable | D | neutral |
| B | unstable | E | slightly stable |
| C | slightly unstable | F | stable |

*Source:* https://en.wikipedia.org/wiki/Outline_of_air_pollution_dispersion.

**Table 3.2.** Meteorological Conditions that Define the Pasquill Stability Classes

| **Surface Windspeed** | | **Daytime Incoming Solar Radiation** | | | **Nighttime Cloud Cover** | |
|---|---|---|---|---|---|---|
| m/s | mi/h | Strong | Moder-ate | Slight | > 50% | < 50% |
| < 2 | < 5 | A | A – B | B | E | F |
| 2 – 3 | 5 – 7 | A – B | B | C | E | F |
| 3 – 5 | 7 – 11 | B | B – C | C | D | E |
| 5 – 6 | 11 – 13 | C | C – D | D | D | D |
| > 6 | > 13 | C | D | D | D | D |
| Note: Class D applies to heavily overcast skies, at any windspeed day or night | | | | | | |

*Incoming solar radiation is based on the following: strong (> 700 W m$^{-2}$), moderate (350–700 W m$^{-2}$), slight (< 350 W m$^{-2}$).*

*Source:* https://en.wikipedia.org/wiki/Outline_of_air_pollution_dispersion.

### 3.5.2. Other Parameters that can Define the Stability Class

The stability class can be defined also by using the:

- Temperature gradient
- fluctuations in wind direction
- Richardson number
- Bulk Richardson number
- Monin–Obukhov length

### 3.5.3. Advanced Methods of Categorizing Atmospheric Turbulence

Simplified meteorological parameters, which are typically used to define the six Pasquill classes as indicated in Table 3.2 above, are not used by advanced air pollution dispersion models to classify atmospheric turbulence. Some variation of Monin-Obukhov similarity theory is used in the more sophisticated models. Here are a few instances:

- **AERMOD:** The most sophisticated model used by the US EPA no longer classifies atmospheric turbulence using the Pasquill stability classes. Rather, it employs the Monin-Obukhov length and the surface roughness length.
- **ADMS 4:** The most sophisticated model used in the UK classifies atmospheric turbulence using the Monin-Obukhov length, windspeed, and boundary layer height.

## 3.6. TOPOGRAPHY

Topography is another factor that influences the movement and dispersal of air pollution. The depiction of surface features like mountains, hills, rivers, and valleys is known as topography. For instance, placing a metropolis inside or close to a mountain range can seriously harm the populace's health due to pollution. Although both vertical and horizontal airflow can help disperse pollutants, a mountain range acts as a natural barrier.

### 3.6.1. Valleys

Furthermore, air naturally flows downward and collects on a valley floor in a similar way to water. Huge populations may be at high risk of excessive pollution exposure when cool, dense air descends into a valley. The impact on human health can be disastrous when an inversion of temperature occurs along with this naturally occurring cool air flow. Air pollution is trapped at ground level in some valleys due to several hundred-meter-thick inversion layers.

### 3.6.2. Lakes and Oceans

The movement and dispersal of air pollution are influenced by large bodies of water as well. In comparison to lakes and oceans, which absorb more heat but do so more slowly, air above land warms more quickly. The heated air cools as it ascends into the upper atmosphere and moves toward other cold air masses, the majority of which are located over sizable bodies of water. This cycle starts when a warm air mass falls on top of water, quickly cools, and weakens the circulation cell. When night falls, sea breezes recirculate air back toward the cooling landmasses, completing the cycle.

### 3.6.3. Mountains

The mere fact that cities like Los Angeles, Denver, and Mexico City are situated in naturally occurring basins surrounded by mountain ranges makes them all susceptible to high pollution concentrations. The topography of the surrounding region has a strong and persistent influence, even though there are many other factors that also contribute to the air pollution issues in these cities. Natural barriers like mountain ranges are an easy way to trap escaping pollutants, even in the best of atmospheric conditions. Though the effects of topographic features are unavoidable, the detrimental effects of air pollution on population centers can be greatly mitigated by intelligent planning that makes use of recorded data on seasonal patterns of wind speed and direction, atmospheric stability, and topography.

# REFERENCES

1. Andrén, A. (1990). A meso-scale plume dispersion model: Preliminary evaluation in a heterogeneous area. *Atmospheric Environment Part A: General Topics, 24*, 883–896. https://doi.org/10.1016/0960-1686(90)90290-4.
2. Arya, S. P. (1998). *Air pollution Meteorology and Dispersion* (1st ed.). Oxford University Press. ISBN 0-19-507398-3.
3. Barrat, R. (2001). *Atmospheric Dispersion Modelling* (1st ed.). Earthscan Publications. ISBN 1-85383-642-7.
4. Bartzis, J. G. (Ed.). (2000). Harmonization within atmospheric dispersion modelling for regulatory purposes (5th Workshop). *International Journal of Environment and Pollution, 14*(1–6). ISSN 0957-4352.
5. Batchvarova, E. (Ed.). (2003). Harmonization within atmospheric dispersion modelling for regulatory purposes (8th Workshop). *International Journal of Environment and Pollution, 20*(1–6). ISSN 0957-4352.
6. Beychok, M. R. (2005). *Fundamentals of Stack gas Dispersion* (4th ed.). Self-published. ISBN 0-9644588-0-2.
7. Borrego, C., & Incecik, S. (2004). *International Technical Meeting on Air Pollution Modeling and Its Application (26th)*. NATO Committee on the Challenges of Modern Society [by] Kluwer Academic/Plenum Press. ISBN 0-306-48464-1.
8. Brown, M. J., Gowardhan, A. A., Nelson, M. A., Williams, M. D., & Pardyjak, E. R. (2013). QUIC transport and dispersion modelling of two releases from the Joint Urban 2003 field experiment. *International Journal of Environment and Pollution, 52*, 263–287. https://doi.org/10.1504/IJEP.2013.058458.
9. Colls, J. (2002). *Air pollution* (1st ed.). Spon Press. ISBN 0-415-25565-1.
10. Cooper, J. R., Randle, K., & Sokh, R. G. (2003). *Radioactive releases in the environment* (1st ed.). John Wiley & Sons. ISBN 0-471-89924-0.
11. Coppalle, A. (Ed.). (2001). Harmonization within atmospheric dispersion modelling for regulatory purposes (6th Workshop). *International Journal of Environment and Pollution, 16*(1–6). ISSN 0957-4352.
12. De Wispelaere, C. (1984). *International Technical Meeting on Air Pollution Modeling and Its Application* (13th). NATO Committee on the Challenges of Modern Society [by] Plenum Press. ISBN 0-306-41491-0.
13. Esbrí, J. M., López-Berdonces, M. A., Fernández-Calderón, S., Higueras, P., & Díez, S. (2015). Atmospheric mercury pollution around a chlor-

alkali plant in Flix (NE Spain): An integrated analysis. *Environmental Science and Pollution Research, 22*, 4842–4850. https://doi.org/10.1007/s11356-014-3305-x.

14. Forago, I., Georgiev, K., & Havasi, A. (2005). *Advances in air Pollution Modeling for Environmental Security (NATO Workshop).* Springer. ISSN 0957-4352.
15. Godish, T., Davis, W. T., & Fu, J. S. (2014). *Air Quality*. CRC Press. https://doi.org/10.1201/9780429105036.
16. Gronwald, F., & Chang, S.-Y. (2018). Evaluation of the precision and accuracy of multiple air dispersion models. *Journal of Atmospheric Pollution, 6*, 1–11. https://doi.org/10.12691/jap-6-1-1.
17. Gryning, S., & Batchvarova, E. (2000). *International Technical Meeting on Air Pollution Modeling and Its Application* (23rd). NATO Committee on the Challenges of Modern Society [by] Kluwer Academic/Plenum Press. ISBN 0-306-46188-9.
18. Hanna, S. R., & Drivas, P. J. (1996). *Guidelines for Use of Vapor Cloud Dispersion Models* (2nd ed.). Wiley-American Institute of Chemical Engineers. ISBN 0-8169-0702-1.
19. Hanna, S. R., & Strimaitis, D. G. (1989). *Workbook of Test Cases for Vapor Cloud Source Dispersion Models* (1st ed.). Center for Chemical Process Safety, American Institute of Chemical Engineers. ISBN 0-8169-0455-3.
20. Johnson, W., Ludwig, F., Dabberdt, W., & Allen, R. (1973). An urban diffusion simulation model for carbon monoxide. *Journal of the Air Pollution Control Association, 23*, 490–498. https://doi.org/10.1080/00022470.1973.10469794.
21. Kretzschmar, J. G., & Cosemans, G. (Eds.). (1997). Harmonization within atmospheric dispersion modelling for regulatory purposes (4th Workshop). *International Journal of Environment and Pollution, 8*(3–6). ISSN 0957-4352.
22. Mocho, P., Desauziers, V., Plaisance, H., & Sauvat, N. (2017). Improvement of the performance of a simple box model using CFD modeling to predict indoor air formaldehyde concentration. *Building and Environment, 124*, 450–459. https://doi.org/10.1016/j.buildenv.2017.08.033.
23. Pasquill, F. (1961). The estimation of the dispersion of windborne material. *The Meteorological Magazine, 90*(1063), 33–49. https://doi.org/10.11298/taiki1995.37.141.
24. Perianez, R. (2005). *Modelling the dispersion of Radionuclides in the Marine Environment: An Introduction* (1st ed.). Springer. ISBN 3-540-24875-7.

25. Pielke, R. A. (2001). *Mesoscale modeling* (2nd ed.). Elsevier. ISBN 0-12-554766-8.

26. Reynolds, S. D., Roth, P. M., & Seinfeld, J. H. (1973). Mathematical modeling of photochemical air pollution—I: Formulation of the model. *Atmospheric Environment, 7*, 1033–1061. https://doi.org/10.1016/0004-6981(73)90214-X.

27. Rodhe, H. (1975). Some aspects of the use of air trajectories for the computation of large-scale dispersion and fallout patterns. In *Advances in Geophysics* (Vol. 18, pp. 95–109). Elsevier. https://doi.org/10.1016/S0065-2687(08)60573-7.

28. Schnelle, K. B., & Dey, P. R. (1999). *Atmospheric Dispersion Modeling Compliance Guide* (1st ed.). McGraw-Hill Professional. ISBN 0-07-058059-6.

29. Soulhac, L., Salizzoni, P., Cierco, F.-X., & Perkins, R. (2011). The model SIRANE for atmospheric urban pollutant dispersion; part I, presentation of the model. *Atmospheric Environment, 45*, 7379–7395. https://doi.org/10.1016/j.atmosenv.2011.07.008.

30. Suppan, P. (Ed.). (2005). Harmonization within atmospheric dispersion modelling for regulatory purposes (8th Workshop). *International Journal of Environment and Pollution, 24*(1–6) & 25(1–6). ISSN 0957-4352.

31. Turner, D. B. (1994). *Workbook of atmospheric dispersion estimates: An introduction to dispersion modeling* (2nd ed.). CRC Press. ISBN 1-56670-023-X.

32. Walton, J. (1973). Scale-dependent diffusion. *Journal of Applied Meteorology, 12*(3), 547–548. https://doi.org/10.1175/1520-0450(1973)012<0547:>2.0.co;2.

33. Wang, A., Fallah-Shorshani, M., Xu, J., & Hatzopoulou, M. (2016). Characterizing near-road air pollution using local-scale emission and dispersion models and validation against in-situ measurements. *Atmospheric Environment, 142*, 452–464. https://doi.org/10.1016/j.atmosenv.2016.08.020.

34. Zannetti, P. (1990). *Air Pollution Modeling: Theories, Computational Methods, and Available Software*. Van Nostrand Reinhold. ISBN 0-442-30805-1.

35. Zannetti, P. (Ed.). (1993). International Conference on Air Pollution (1st, Mexico City). Computational Mechanics. ISBN 1-56252-146-2.

*Chapter* **4**

# Indoor Air Pollution

## Contents

# INTRODUCTION

The release of dangerous pollutants inside causes indoor air pollution. These can include different types of toxins, carbon monoxide, and fine particulate matter. Because people frequently burn fuel indoors for heating and cooking, indoor air pollution is a major issue in developing nations. It is a worry for those who reside in energy-efficient houses as well. Since the air inside these typically airtight buildings can quickly become stagnant, the levels of pollutants can rise quickly. Although pollution can be harmful in any form, indoor air pollution is especially dangerous because people are frequently exposed to it for extended periods of time.

Indoor air pollution has several detrimental effects that can be fatal. It is dangerous for both young children and elderly people. The following are a few consequences of indoor air pollution.

- If asbestos is found in your home, it can cause serious health problems like cancer, mesothelioma, asbestosis, and lung cancer.
- Let's say that contaminants like dust mites, pet dander, or other microbes enter the house. Then, they may have disastrous effects like symptoms of asthma, sore throats, the flu, and other infectious diseases.
- Lead can be harmful to your health if it is discovered in the home. It can damage the nervous system and brain, as well as result in anemia, renal failure, and cardiovascular problems.
- Formaldehyde is the most common indoor air contaminant and can be dangerous to your health. You may experience nasal, throat, and eye irritation. It has been repeatedly connected to cancer and is recognized as a human carcinogen by the World Health Organization.
- Smokers suffer from emphysema, heart disease, lung cancer, bronchitis, severe pneumonia, and respiratory irritation.
- Chemicals found in paints and cleaning products, for instance, can harm the kidneys, liver, and brain. They can also cause loss of coordination. It is also accountable for various forms of cancer.
- Respiratory issues may arise from the toxic nitrogen dioxide and carbon monoxide that wood and gas stoves in homes produce.
- The second biggest cause of lung cancer in the US, according to the US National Cancer Institute, is radon in the home.
- It has been established that highly volatile and semi-volatile compounds found in artificial fragrances, perfumes, and deodorizers can lead to cancer, allergic reactions, diseases of the central nervous system, skin irritation, birth defects, and reproductive issues.

## 4.1. INDOOR AIR POLLUTION: SOURCES, EFFECTS, AND CONTROLS

Indoor air pollution is caused by the release of chemical and biological agents into a closed space. According to Brooks and Davis (1992), there are currently over 900 compounds on the list of possible indoor air pollution sources. These substances' detrimental effects on health are amplified in an enclosed indoor space. Pollutant effects on human health and the environment can be exacerbated by environmental factors like temperature, humidity, lighting, and noise, in addition to pollutant-specific characteristics like dose, rate, and concentration. Despite the fact that a wide range of dangerous pollutants can enter buildings through ventilation intakes from the outside, ventilation system managers also have to deal with a multitude of indoor air pollution sources.

### 4.1.1. Biological Contaminants

The following are examples of biological contaminants: house dust, mites, cockroaches, pollen, animal dander, mold, mildew, viruses, and saliva from cats. These pollutants come from a variety of sources. Plants are the source of pollen; people and animals can spread viruses; soil, plant debris, animals, and people can carry bacteria; and household pets can spread animal dander and saliva. Rat and mouse urine contains a protein that is a strong allergen. It can take flight once it dries. Mold, mildew, and other biological contaminants can grow in contaminated central air handling systems, which can then spread these contaminants throughout the house.

A home's relative humidity level can be adjusted to minimize the growth of some biological sources. For homes, a relative humidity of 30–50% is usually advised. Molds, mildews, bacteria, and insects can also grow and reproduce in standing water, water-damaged materials, and damp surfaces. One of the strongest biological allergens, house dust mites, thrive in warm, humid conditions.

- **Health Effects of Biological Contaminants:** Hypersensitivity pneumonitis, allergic rhinitis, and certain forms of asthma are among the allergic reactions brought on by certain biological pollutants. Airborne transmission occurs for infectious diseases like chicken pox, measles, and influenza. Toxins that cause disease are released by molds and mildews. The following are signs of biological pollution-related health issues: fever, diarrhea, coughing, dizziness, shortness of breath, sneezing, and watery eyes.
- **Ways to Reduce Exposure:** Installing and using fans that are vented outdoors in bathrooms and kitchens is one effective method to lessen

exposure to indoor pollutants. Additionally, vent clothes dryers outside. Make sure you follow the manufacturer's instructions when using clean mist and ultrasonic humidifiers, and that you regularly replenish them with clean water. Regularly empty the water trays in refrigerators, dehumidifiers, and air conditioners. Water-damaged carpets should be cleaned, dried, or removed. Finally, only use leak-proof, well-ventilated basements as living spaces. If necessary, use dehumidifiers to keep interior humidity levels between 30–50%.

### 4.1.2. Radon (Rn)

When uranium decays, a naturally occurring radioactive gas called radon is created. Radon is released during uranium decay and permeates the air and water supplies for human consumption. As they continue to degrade, radon and other radium decay products that are breathed into the lungs will continue to release lethal radiation. Elevated radon levels have the potential to cause lung cancer as they accumulate in the human body. The U.S. According to EPA estimates, radon causes between 14,000 and 30,000 fatalities annually.

The most frequent cause of indoor radon is uranium found in the rock or soil that houses are constructed upon. Radon gas is a colorless, odorless, radioactive material that is released when uranium decays naturally. Homes can be exposed to radon gas through floor drains, sump pumps, floor drains, and fissures in concrete walls and floors. Exposure to radon becomes problematic when radon gets trapped in buildings and concentrations increase inside. Any type of home, whether it is new or old, well sealed or drafty, or has a basement or not, may have a radon problem. Radon can sometimes find its way into homes through well water, and in certain cases, it can even come from the building materials.

#### *4.1.2.1. Health Effects of Radon*

Lung cancer is the main health consequence linked to exposure to high radon levels. Although the risks associated with ingesting high radon water levels are thought to be much lower than those associated with breathing radon-contaminated air, research suggests that they may still exist. Estimates that radon contributes to thousands of avoidable lung cancer deaths annually are supported by prominent health organizations such as the American Medical Association, the American Lung Association (ALA), and the Centers for Disease Control and Prevention. Despite the fact that the EPA estimates that radon causes roughly 14,000 deaths annually in the US, your risk of developing lung cancer is increased if you smoke and your home has high radon levels.

#### *4.1.2.2. Ways to Reduce Exposure*

It is estimated that the average indoor radon level is approximately 1.3 picocuries per liter (pCi/L), whereas the average outdoor radon level is typically about 0.4 pCi/L. The U.S. congress established a long-term objective of having indoor radon levels equal to those outside. While most homes today can be reduced to 2 pCi/L or less, this goal is not yet technologically achievable in every situation. If the radon level in your home is 4 pCi/L or higher, make repairs. Less than 4 pCi/L of radon still poses a risk, though it may be mitigated in many situations.

Radon in homes can be decreased using a number of techniques. A fundamental component of the majority of radon reduction strategies is sealing foundation cracks and other openings. In order to reduce radon entry, the EPA does not advise using sealants alone. Radon levels cannot be consistently or significantly reduced by sealing alone. Radon is typically reduced using a system that includes a vent pipe(s) and fan(s). Not much needs to be changed in your house for these sub-slab depressurization systems to function. Homes with crawl spaces can also install similar systems. These systems stop radon gas from penetrating the house from outside the foundation and below the concrete floor. Contracts for radon mitigation may employ additional techniques that could be effective in your house. The layout of your house and other elements determine which system is best. Lowering your radon level will lower your risk of developing lung cancer.

While businesses and homeowners have made significant efforts to prevent radon poisoning, government regulation has not kept up with these efforts. This is true for a variety of reasons, the main one being that laws and regulations have not been adequately defined because indoor air pollution and the effects of radon on human health have only recently come to the public's attention since the Clean Air Act Amendments of 1990. Furthermore, radon-related public education has only recently started to spread. Most importantly, though, is that there is still disagreement in science regarding the extent of the health issues this substance directly causes.

### 4.1.3. Environmental Tobacco Smoke (ETS)

The combination of smoke exhaled by the smoker and smoke emitted from the burning end of a cigarette, pipe, or cigar is known as environmental tobacco smoke (ETS). It is a complex mixture of more than 4,000 compounds, many of which are potent irritants and more than 40 of which are known to cause cancer in humans or animals. ETS exposure is frequently referred to as passive smoking, and ETS is frequently referred to as secondhand smoke. The EPA finished a thorough analysis of the respiratory health risks associated with

ETS in 1992. According to the report's conclusion, exposure to ETS damages hundreds of thousands of children's respiratory systems and causes about 3,000 lung cancer deaths annually among nonsmoking adults.

### 4.1.4. Stoves, Heaters, Fireplaces, and Chimneys

Apart from ambient tobacco smoke, unvented kerosene and gas space heaters, woodstoves, fireplaces, and gas stoves are additional sources of combustion products. Particles, nitrogen dioxide, and carbon monoxide are the main pollutants that are released. Acid aerosols may also be produced by unvented kerosene heaters. Inadequately installed or maintained chimneys and flues, as well as fractured furnace heat exchangers, are additional sources of combustion gases and particles. Particularly in weatherized homes, pollutants from woodstoves and fireplaces can be back-drafted into the living area through the chimney if there is no dedicated outdoor air supply.

- **Ways to Reduce Exposure:** When using unvented space heaters that burn fuel, exercise extra caution. Generally speaking, a persistent flame with a yellow tip indicates maladjustment and higher emissions of pollutants. Over gas cooking stoves and ranges, install and operate exhaust fans while maintaining proper burner adjustment. Get annual inspections and prompt repairs for all central air handling systems, including furnaces, flues, and chimneys.

### 4.1.5. Asbestos

The mineral fiber known as asbestos has long been utilized as a fire retardant and insulation in a wide range of building supplies. The CPSC and EPA have outlawed a number of asbestos products. Additionally, producers have freely restricted the applications of asbestos. These days, the most common places to find asbestos are older homes, millboard, textured paints and other coating materials, asbestos shingles in pipes and furnaces, and floor tiles. When asbestos-containing materials are disturbed by cutting, sanding, or other remodeling activities, elevated concentrations of asbestos dust may result. Inadequate efforts to eliminate these substances may cause asbestos fibers to be released into indoor air, raising asbestos concentrations and putting occupants in danger.

- **Health Effects of Asbestos:** The tiniest asbestos fibers are invisible and extremely dangerous. They have the potential to stay and build up in the lungs after being inhaled. Asbestos can lead to asbestosis, which is an irreversible and potentially fatal lung scarring, mesothelioma, a cancer of the linings of the chest and abdomen, and lung cancer. Regretfully, these diseases don't manifest their symptoms for many years after exposure.

- **Ways to Reduce Exposure:** Asbestos that is in good condition is usually best left alone. Asbestos fibers are typically not released by materials that are in good condition. If fibers are not released and breathed into the lungs, there is no risk. Items containing asbestos should not be cut, torn, or sanded.

## 4.1.6. Pesticides

According to a recent survey, 75% of U.S. households used at least one pesticide product indoors during the past year. Insecticides and disinfectants are the most frequently used products. According to another study, the majority of people are exposed to pesticides 80% of the time indoors, and the air inside homes has been found to contain detectable amounts of up to 12 different pesticides. Pesticide use within the last few years does not seem to account for the amount of pesticides detected in homes; other potential sources include contaminated dust or soil that comes in from the outside or floats, pesticide containers that are kept in storage, and surfaces that gather and subsequently release pesticides. Termiticides, insecticides, rodenticides, fungicides, and disinfectants—products that regulate microorganisms—are among the pesticides used in and around homes. As sprays, liquids, balls, crystals, sticks, powders, and foggers, they are available for purchase.

- **Health Effects of Pesticides:** Pesticides can contain both organic compounds as their active and inert ingredients, which increases the amount of organic compounds in the air inside of homes. High concentrations of cyclodiene pesticide exposure—which is frequently linked to improper application—have been linked to a number of symptoms, including headaches, nausea, tingling sensations in the muscles, dizziness, and twitching. Moreover, EPA is worried that cyclodienes may raise the risk of cancer and inflict long-term harm to the liver and central nervous system.
- **Ways to Reduce Exposure:** Examine the label and adhere to the instructions. (Using any pesticide in a way that deviates from the instructions on its label is prohibited. Ventilate the space thoroughly after using pesticides. When feasible, control pests without the use of chemicals; dispose of unwanted pesticides properly. Lastly, minimize your contact with moth repellents.

## 4.1.7. Organic Chemicals

Many household products contain ingredients that are organic. Organic solvents are found in many cleaning, disinfecting, cosmetic, degreasing, and hobby products as well as paints, varnishes, and wax. The components of fuels are organic compounds. When you use these products and, to some extent, when they are stored, all of them have the potential to release organic compounds. Regardless of whether the homes were in highly industrial or rural areas, the EPA's Total Exposure Assessment Methodology (TEAM) studies revealed that levels of approximately a dozen common organic pollutants were 2 to 5 times higher inside homes than outside. According to more TEAM research, people who use products containing organic chemicals run the risk of exposing others and themselves to extremely high pollutant levels, and even after the activity is over, there may still be elevated concentrations in the air.

- **Health Effects of Organic Chemicals:** Organic compounds range widely in their potential to have negative effects on health, from extremely toxic ones to ones that have no known negative effects. Like with other pollutants, a number of factors, such as exposure level and duration, will determine the extent and type of health effects. Some people have reported experiencing immediate symptoms such as headaches, dizziness, visual disorders, memory impairment, and irritation of the eyes and respiratory tract after being exposed to certain organic substances. Little is currently known about the potential health effects of the typical household organic content levels. Numerous organic compounds have been shown to cause cancer in animals, and some are known or suspected to cause cancer in people.
- **Ways to Reduce Exposure:** Pay close attention to the label's instructions. Products that may be dangerous frequently come with warnings meant to lessen the user's exposure. Safely dispose of partially filled containers containing outdated or unnecessary chemicals.

### 4.1.8. Formaldehyde

An essential chemical used extensively by industry to produce a wide range of household goods and building materials is formaldehyde. In addition, it is a byproduct of burning and a few other organic processes. As such, it could exist in significant amounts both indoors and outdoors. Building supplies, tobacco use, household goods, and the use of fuel-burning appliances without vents, such as gas stoves or kerosene space heaters, are among the home's sources of formaldehyde. Formaldehyde is used in manufactured goods for a variety of purposes, either alone or in conjunction with other chemicals. For instance, it is a constituent of glues and adhesives, adds permanent press properties to garments and draperies, and functions as a preservative in certain paints and coating products.

Pressed wood products manufactured with adhesives containing urea-formaldehyde (UF) resins are probably the biggest sources of formaldehyde in homes. Particleboard (used as subflooring, shelving, cabinetry, and furniture); hardwood plywood paneling (used as decorative wall covering, cabinetry, and furniture); and medium density fiberboard (used as drawer fronts, cabinets, and furniture tops) are examples of pressed wood products intended for indoor use. Generally speaking, medium density fiberboard is the highest formaldehyde-emitting UF pressed wood product because it has a higher resin-to-wood ratio than any other product. The rate of formaldehyde release from goods like textiles and pressed wood can fluctuate. As products age, formaldehyde emissions usually decrease. High indoor temperatures or high levels of humidity can accelerate the formaldehyde release from new products. As a way to save energy, urea formaldehyde foam insulation (UFFI) was installed in the wall cavities of many homes during the 1970s. However, shortly after the UFFI installation, it was discovered that formaldehyde concentrations inside of many of these homes were rather high. Nowadays, this product is being used to insulate very few homes. Research indicates that the emissions of formaldehyde from UFFI decrease over time, meaning that homes with UFFI installed many years ago are unlikely to have high formaldehyde levels today.

- **Health Effects of Formaldehyde:** Formaldehyde is a colorless, strong-smelling gas that, in certain people exposed to high concentrations (above 0.1 parts per million), can result in watery eyes, burning in the eyes and throat, nausea, and difficulty breathing. Individuals who have asthma may experience attacks due to high concentrations. Although formaldehyde has been demonstrated to cause cancer in animals and may cause cancer in humans, there is some evidence that exposure to it can cause sensitivity in some people.
- **Ways to Reduce Exposure:** Before making a purchase, find out how much formaldehyde is in pressed wood furniture, cabinetry, and building supplies. According to certain studies, applying a polyurethane coating to pressed wood products may temporarily lower formaldehyde emissions. Heat accelerates the release of formaldehyde, and the amount of humidity can also have an impact. As a result, using air conditioning and dehumidifiers to regulate humidity and keep a comfortable temperature can aid in the reduction of formaldehyde emissions. Reducing formaldehyde levels in your house can also be achieved by increasing ventilation rates.

## 4.1.9. Lead (Pb)

It has long been known that lead is a dangerous environmental contaminant. Lead was identified as the top environmental risk to children's health in the United States by the Secretary of the Department of Health and Human Services in late 1991. Humans can be exposed to lead in a variety of ways, including the air, food, water, soil, crumbling paint, and dust. When lead dust or particles have settled, they can be inhaled or swallowed, resulting in airborne lead exposure. Lead was utilized in paint, gasoline, water pipes, and many other products before it was realized how dangerous the metal could be. Old lead-based paint is the most significant source of lead exposure in the U.S. today. When lead-based paint is improperly removed from surfaces by sanding, dry scraping, or open-flame burning, harmful exposures to lead may result. Additionally, lead dust from outside sources, such as contaminated soil carried inside, and the use of lead in specific indoor activities, like soldering and stained glass creation, can contribute to high concentrations of airborne lead particles in homes.

### *4.1.9.1. Health Effects of Lead*

Lead has an impact on almost every system in the body. When taken in excess, it can result in coma, convulsions, and even death. Reduced lead exposure can have negative effects on the kidneys, brain, central nervous system, and blood cells. Lead exposure can have serious consequences for young children and fetuses. These consist of reduced IQs, shorter attention spans, delayed physical and mental development, and an increase in behavioral issues. Since lead is more easily absorbed into growing bodies and small children's tissues are more sensitive to the harmful effects of lead, fetuses, infants, and children are more susceptible to lead exposure than adults. Since children are more likely to get lead dust on their hands and then put their fingers or other lead-contaminated objects in their mouths, they may be exposed to higher levels of lead. As a result, kids pose a serious risk and ought to have their lead exposure evaluated.

### *4.1.9.2. Ways to Reduce Exposure*

Steer clear of paints containing lead as one method of lowering the risk of lead exposure. The majority of houses constructed before 1960 have a lot of lead paint. Significant amounts of lead may also be present in certain houses constructed as recently as 1978. This paint may be on walls, windows, the exteriors of buildings, or other surfaces. If lead-based paint is in good condition, don't touch it. Keep lead dust out of the house; either change into clean clothes while at work or wash any contaminated clothing right away. It is possible for you to unintentionally bring lead into your home on your hands or clothing if you work in construction, demolition, painting, working with batteries, radiator repair shops, lead factories, or if lead is part of your hobby. Lead may also be ingested by you from the surrounding soil. Lead paint on the exterior of buildings may contaminate the soil in close proximity to residential areas. The leaded gasoline used by cars and trucks for years has caused exhaust fumes that may have contaminated the soil near roads and highways.

Find out if drinking water contains lead. The majority of well and city water is lead-free, but lead-containing plumbing materials in homes can quickly contaminate water that is transported from them. Testing the water is the only reliable method to determine whether lead is present in it. Consume healthfully. A child will absorb less lead if they receive adequate calcium and iron. Red meats, beans, and eggs are foods high in iron. Dairy products are also high in calcium, but avoid storing food or liquids in old or imported pottery or lead crystal glasses. Issues with indoor air quality are not exclusive to residences. In actuality, a large number of office buildings are major sources of air pollution. A few of these structures might not have enough ventilation.

**Table 4.1.** *Common Indoor Pollutants and Their Effects on Human Health*

| Pollutants | Sources | Health Impacts | Refs. |
|---|---|---|---|
| PM | Outdoor environment, cooking, combustion activities (burning of candles, use of fireplaces, heaters, stoves, fireplaces and chimneys, cigarette smoking), cleaning activities | Premature death in people with heart or lung disease, nonfatal heart attacks, irregular heartbeat, aggravated asthma, decreased lung function, increased respiratory symptoms | [Brook R.D., et al., 2010; Hamanaka R.B., et al., 2018; Miller M.R., et al., 2012; USEPA Indoor Particulate Matter., 2020] |
| VOCs | Paints, stains, varnishes, solvents, pesticides, adhesives, wood preservatives, waxes, polishes, cleansers, lubricants, sealants, dyes, air fresheners, fuels, plastics, copy machines, printers, tobacco products, perfumes, dry-cleaned clothing, building materials and furnishings | • Eye, nose and throat irritation<br>• Headaches, loss of coordination and nausea<br>• Damage to liver, kidney and central nervous system<br>• Some organics can cause cancer | [Brickus L.R., et al., 1998; Tang X., et al., 2015; USEPA Pesticides' Impact on Indoor Air Quality, 2020; Weschler C.J., Nazaroff W.W., 2014; C.J., Nazaroff W.W., 2012] |

| | | | |
|---|---|---|---|
| $NO_2$ | Gas-fueled cooking and heating appliances | • Enhanced asthmatic reactions<br>• Respiratory damage leading to respiratory symptoms | [Bernstein J.A., et al., 2008] |
| $O_3$ | Outdoor sources, photocopying, air purifying, disinfecting devices | DNA damage, lung damage, asthma, decreased respiratory functions | [Huang Y., et al., 2019; Salonen H., et al., 2018] |
| $SO_2$ | Cooking stoves; fireplaces; outdoor air | • Impairment of respiratory function<br>• Asthma, chronic obstructive pulmonary disease (COPD), and cardiovascular diseases | [Seow W.J., et al.,2016] |
| $CO_x$ | Cooking stoves; tobacco smoking; fireplaces; generators and other gasoline powered equipment; outdoor air | Fatigue, chest pain, impaired vision, reduced brain function | [Raub J.A., et al., 2000; USEPA Carbon Monoxide's Impact on Indoor Air Quality, 2020] |
| Heavy metals | Pb, Cd, Zn, Cu, Cr, As, Ni, Hg, Mn, Fe Outdoor sources, fuel-consumption products, incense burning, smoking and building materials | • Cancers, brain damage<br>• Mutagenic and carcinogenic effects: respiratory illnesses, cardiovascular deaths | [Komarnicki G.J.K., 2005; Madany I.M., 1994; Rashed M.N., 2008] |
| Aerosols | Tobacco smoke, building materials, consumer products, incense burning, cleaning and cooking | Cardiovascular diseases, respiratory diseases, allergies, lung cancer, irritation and discomfort | [Koivisto A.J., et al., 2019; Kulmala M., et al., 1999; Oh H.-J., et al., 2019] |
| Radon (Rn) | Soil gas, building materials, and tap water<br>Outdoor air | Lung cancer | [Bruno R.C., 1983; USEPA, 2003; WHO, 2009] |

| | | | |
|---|---|---|---|
| Pesticides | • - Termiticides, insecticides, rodenticides, fungicides, disinfectants and herbicides<br>• - Building materials: carpet, textiles, and cushioned furniture<br>• - Outdoor environment | Irritation to eye, nose and throat; Damage to central nervous system and kidney; Increased risk of cancer | [Colt J.S., et al., 2004; Holt E., 2017; Hwang H.-M., et al., 2008; USEPA, 2020] |
| Biological allergens | House dust, pets, cockroaches, mold/dampness, pollens originating from animals, insects, mites, and plants | Asthma and allergies Respiratory infections, sensitization, respiratory allergic diseases and wheezing | [Baldacci S., et al., 2015; WHO, 1988] |
| Microorganism | Bacteria, viruses, and fungi are carried by people, animals, and soil and plants | Fever, digestive problems, infectious diseases, chronic respiratory illness | [Hospodsky D., et al., 2012; WHO, 1988] |

*Source: Tran VV, Park D, Lee YC. Indoor Air Pollution, Related Human Diseases, and Recent Trends in the Control and Improvement of Indoor Air Quality. Int J Environ Res Public Health. 2020 Apr 23;17(8):2927. doi: 10.3390/ijerph17082927. PMID: 32340311; PMCID: PMC7215772.*

For instance, it's possible that mechanical ventilation systems aren't built or run to supply enough outdoor air. Furthermore, compared to their homes, people typically have less control over the indoor environment in their offices. Consequently, the frequency of reported health issues has received more attention.

## 4.2. SICK BUILDING SYNDROME (SBS)

A condition known as sick building syndrome (SBS) occurs when a person contracts a chronic illness or develops illness symptoms as a result of exposure to the building where they live or work. SBS is also referred to as Idiopathic Environmental Intolerance (IEI), Building Related Symptoms (BRS), or Building Related Illness (BRI) in scientific literature. The primary distinguishing feature is a rise in the frequency of reports of symptoms like headache, fatigue, nausea, and irritation of the eyes, nose, and throat. That is how SBS is defined in the 1989 Oxford English Dictionary (Passarelli & Guiseppe Ryan, 2009). When SBS was solely linked to non-organic causes in 1984, the World Health Organization wrote a 484-page manual on indoor air quality and indicated that the book might serve as the foundation for future legislation or legal action.

The outbreaks might or might not be directly related to improper or insufficient cleaning techniques. Staff concerns in post-war buildings with improperly planned building aerodynamics, flaws in the building materials or assembly process, and/or insufficient maintenance have also been described using SBS. As people spend more time in the building, some symptoms tend to get worse; they frequently get better over time or even go away when people leave the building (Stolwijk, 1991). In order to focus the name of the condition on the symptoms of the patients rather than a sick building, SBS is also sometimes referred to as building-related symptoms.

Many causes have been identified in an attempt to link sick building syndrome, including volatile organic compounds (VOC), light industrial chemicals used within, improper exhaust ventilation of ozone (a byproduct of some office machinery), contaminants produced by outgassing certain types of building materials, and inadequate fresh-air intake/air filtration. Another factor linked to sick building syndrome is HVAC (heating, ventilation, and air conditioning) systems. Regarding the question of whether or not air conditioning systems cause SBS, there have been conflicting results (Shahzad Sally, Brennan John, Theodossopoulos Dimitris, Hughes Ben, & Calautit John Kaiser, 2016).

## 4.2.1. History

There was a notice in the late 1970s that tenants in newly built homes, offices, and nurseries were reporting vague symptoms. It was dubbed "office illness" in the media. The World Health Organization (WHO) first used the term "sick building syndrome" in 1986, estimating that 10–30% of newly constructed office buildings in the West had indoor air quality issues. Symptoms were reported in early Danish and British studies. Unhealthy interior spaces garnered attention. As anticipated, sick buildings were identified by the Swedish allergy study (SOU 1989:76) as a contributing factor to the allergy epidemic. Thus, a great deal of research on sick building was done in the 1990s. A wide range of chemical and physical aspects of the buildings were investigated.

The issue was being called a ticking time bomb and was receiving more and more attention in the media. Numerous investigations were carried out in separate structures. Sick buildings and healthy buildings were compared in the 1990s. The emphasis was on the chemical composition of building materials. Numerous producers of building materials were making concerted efforts to replace controversial additives and regain control over the chemical composition. Above all, the ventilation industry promoted more efficient ventilation. Some saw the use of natural materials, ecological building practices, and straightforward methods as a solution.

A growing mistrust of the sick building concept emerged at the end of the 1990s. The methodology of earlier research was questioned in a 1999 dissertation at the Karolinska Institute in Stockholm, and these shortcomings were experimentally demonstrated in a 2005 study conducted in Denmark. It was proposed that sick building syndrome was actually a grouping of up to twelve semi-related illnesses rather than a cohesive syndrome that needed to be diagnosed separately. The Swedish National Board of Health and Welfare advised against using sick building syndrome as a clinical diagnosis in the medical journal Läkartidningen in 2006. Subsequently, the usage of phrases like sick buildings and sick building syndrome in research has decreased significantly. Nonetheless, the idea is still prevalent in popular culture and is used to identify a group of symptoms associated with subpar engineering of the home or workplace. Thus, the phrase "sick building" is particularly used in reference to workplace health.

The media quickly moved from covering sick building syndrome to the courtroom, where licensed architects and engineers were named as defendants and were defended by their respective professional practice insurance. Expert witnesses, medical and technical specialists, building managers, contractors, and manufacturers of finishes and furnishings were always called upon to testify as to cause and effect in these proceedings. None of these actions were particularly dramatic; the majority led to sealed settlement agreements. The insurers were required to present a defense based on Standards of Professional Practice in response to a court ruling stating that the HVAC systems in a contemporary, nearly airtight building have to generate enough breathing air for human consumption. The American Society of Heating, Refrigeration and Air Conditioning Engineers, or ASHRAE, set out to codify its indoor air quality (IAQ) standard. ASHRAE currently has over 50,000 international members.

ASHRAE empirical research used carbon dioxide as an accurate indicator of occupant presence and activity and found that acceptability was a function of outdoor (fresh air) ventilation rate. This dilution technique would adequately control contaminants and building odors. In order to ensure compliance, ASHRAE established a carbon dioxide level of 1,000 parts per million and mandated the use of widely accessible sense-and-control equipment. The federally set goal of minimizing HVAC system energy consumption was achieved by aiming for a ventilation level of 5,000 ppm of carbon dioxide (the OSHA workplace limit). The reasons and methods behind this decision were published in the 1989 issue of ASHRAE 62.1–1989, which also overruled the 1981 requirements. Evidently, this put an end to the SBS pandemic. The potential for emissions in building materials changed over time. In accordance with ASHRAE guidelines, code-compliant ventilation and upkeep, the elimination of smoking, and notable advancements in ambient air quality have all contributed to the interior air environment's acceptability.

### 4.2.2. Epidemiology

According to certain research, women report experiencing SBS symptoms at a higher rate than do men. Whether this is caused by biological, social, or occupational factors is not entirely clear, though. In order to improve scientific understanding of gender differences under the Sick Building Syndrome phenomenon, 1464 office workers participated in a 2001 study that was published in the Journal Indoor Air (Brasche, Bullinger, Morfeld, Gebhardt, & Bischof, 2001). Through the use of surveys, ergonomic analyzes, building assessments, and physical, biological, and chemical variables, the researchers were able to obtain results that are comparable to previous studies on SBS and gender. The research team discovered that not only were prevalence rates varied in most test variables, but there was also a significant gender gap in the working conditions. Men's workplaces, for instance, are typically much larger and have better overall job characteristics. Second, there was a discernible difference in the reporting rates, with women reporting at a rate that is about 20% higher than that of men. There may be a variation in reporting willingness because this data was consistent with that discovered in earlier research.

The reporting rates of sick building syndrome may differ by gender because women tend to report more symptoms than men do. In addition, some research has shown that women are more likely to experience mucosal dryness and facial erythema, and they have a more responsive immune system. Additionally, some claim that women are more likely than men to work in clerical positions, where they are exposed to specialized office supplies and equipment (such as toner-based printers and blueprint machines), while men are more likely to work outside of offices (Godish Thad, 2001). This suggests that women are also more exposed to indoor environmental factors.

### 4.2.3. Signs and Symptoms

Numerous detrimental health effects have been linked to aerosol exposure in humans. Symptoms reported by building occupants include sensory irritation of the nose, eyes, or throat; general health issues or neurotoxic reactions; skin irritation; infectious diseases; and taste and odor sensations. General malaise has been linked to exposure to inadequate lighting conditions (Figure 4.1).

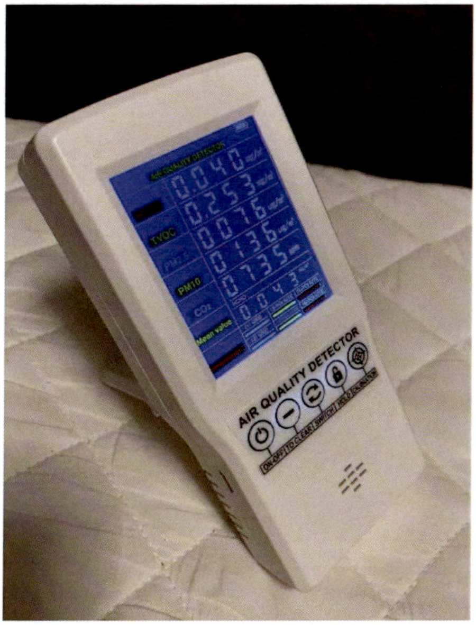


**Figure 4.1.** *An air quality monitor.*

*Source: By Raralu4440 – Own work, CC BY-SA 4.0,* https://commons.wikimedia.org/w/index.php?curid=121600720.

The presence of bacteria and fungi in the moist air of residential buildings and commercial offices has been linked to extrinsic allergic alveolitis. In 2017, a study found a correlation between a number of respiratory inflammatory diseases and concrete proof of damp-related property damage (Pind C. Ahlroth, 2017). The symptoms that have been reported have been categorized by the WHO into several broad categories, such as mucous membrane irritation, which includes irritation of the eyes, nose, and throat; neurotoxic effects, which include headaches, fatigue, and irritability; asthma and symptoms similar to asthma, such as tightness in the chest and wheezing); skin dryness and irritation; gastrointestinal complaints, and more (Apter, 1994). Many sick residents may report isolated symptoms that don't seem to be related. The increased incidence of illnesses in general that start or worsen within a short period of time, typically within a few weeks, is crucial to their discovery. Most of the time, SBS symptoms go away as soon as the occupants leave the specific room or area. Nevertheless, a variety of neurotoxins can have residual effects that might not go away once the occupant vacates the building. There may be long-term health effects in certain cases, especially in sensitive people (Joshi Sumedha, 2008).

## 4.2.4. Cause

According to the United States Environmental Protection Agency's (EPA) air quality ratings, contaminated urban air is deemed unacceptable. As a result, ASHRAE has recognized the need for treatment such as filtration, for which HVAC professionals typically use carbon-impregnated filters and similar materials. The human body will become irritated by various toxins in different ways. While some people are extremely sensitive to dust, others are more allergic to mold. Cooking fumes or deteriorating fiberglass insulation are examples of minor issues that can be exacerbated into much more serious indoor air quality issues by inadequate ventilation.

**Table 4.2.** *Types of Problems Found in 203 Indoor air Quality Investigations Carried out by National Institute for Occupational Safety and Hygiene*

| Problem | No. | % | Notes |
|---|---|---|---|
| Contamination (inside) | 36 | 18 | Exposure to chemical or other toxic agents generated within the office space, e.g., methyl alcohol from a spirit duplicator, methacrylate from a copier, sulfur dioxide from a heating system, amines used in a humidification system, chlordane used as a pesticide |
| Contamination (outside) | 21 | 10 | Exposure to a chemical or other toxic substance originating from a source outside the building, e.g., motor vehicle exhaust fumes, construction activity, underground petrol spillage |
| Contamination (building fabric) | 7 | 3 | Problems from the material used to construct the building (figure excludes asbestos), e.g., formaldehyde, fiberglass |
| Inadequate ventilation | 98 | 48 | Symptoms may be due to low levels of multiple contaminants and/or poor ventilation |
| Hypersensitivity pneumonitis | 6 | 3 | Problems due to a reaction to microorganisms in the building environment |
| Cigarette smoking | 4 | 2 | |

| | | | |
|---|---|---|---|
| Humidity | 0.9 | 4 | |
| Noise/illumination | 2 | 1 | |
| Scabies | 1 | 0.5 | |
| Unknown | 19 | 9 | |

*Source: Melius 1984,* https://www.ncbi.nlm.nih.gov/pmc/articles/PMC2796751/.

The HVAC AC system can trap common products like paint, insulation, rigid foam, particle board, plywood, duct liners, exhaust fumes, and other chemical contaminants from indoor or outdoor sources, as well as biological contaminants. When fan coils are used to recycle this air, the overall oxygenation ratio decreases and becomes hazardous. The weakened immune system of people living in buildings in contaminated urban areas can be exacerbated by additional stressors like bad lighting and loud traffic. All common building construction products contain adhesives that contain certain volatile organic compounds (VOCs), which are toxic to humans. These volatile organic compounds, or VOCs, can have both short-term and long-term negative health effects on building occupants, such as cancer, paralysis, lung failure, and other conditions. Biological contaminants include mold spores, bacteria, fungi, viruses, pollen, and pollen grains. All of these substances have the potential to induce SBS, or allergic reactions. Furthermore, outside pollution, like car exhaust, can seep into poorly constructed buildings and lead to high CO and $CO_2$ concentrations as well as poor indoor air quality. The symptoms of adult SBS were linked to a past medical history of asthma, eczema, and allergic rhinitis.

According to a 2015 study on the relationship between SBS and indoor air pollutants in Iranian office buildings, symptoms such as headaches, nausea, nasal irritation, dyspnea, and dry throat have been shown to worsen with rising carbon dioxide levels in the building. It has been discovered that some work environments are associated with particular symptoms. For instance, there was a significant correlation found between increased light intensity and malaise, eye pain, and dry skin. It has also been discovered that symptoms like sneezing, skin redness, itchy eyes, and headaches are correlated with higher temperatures, whereas sneezing, skin redness, and eye pain are correlated with lower relative humidity. A contributing factor to sick building syndrome was discovered when ASHRAE Standards 62–73 and 62–81 reduced the required ventilation from 10 cubic feet per minute (4.7 L/s) per person to 5 cubic feet per minute (2.4 L/s) per person in 1973 in response to the oil crisis and conservation concerns. In addition to ventilation based on the zone floor area delivered to the breathing zone, the 2016 revision of the ASHRAE ventilation standards stipulates that each occupant must receive 5 to 10 cubic feet per minute of ventilation, depending on the type of occupancy.

#### 4.2.4.1. Workplace

While poor interpersonal relationships, poor communication, and excessive work stress or dissatisfaction are frequently linked to sick building syndrome (SBS), recent research indicates that stress and environmental sensitivity can also play a significant role in the development of SBS. Higher effects were observed when low support and high job demands were present in the psycho-social work environment. The study came to the conclusion that aspects of the psycho-social work environment seem to have a greater bearing on differences in the prevalence of symptoms than does the physical environment of office buildings. Nevertheless, regardless of workplace stress, there is still a connection between sick building syndrome and worker symptoms.

Certain stressors at work are associated with particular symptoms of SBS. There is a significant correlation between general symptoms (such as headache, abnormal tiredness, feeling cold, or nausea) and workload and work conflict. While upper respiratory symptoms are linked to crowded workspaces and low work satisfaction. Ventilation rates have been linked to work productivity and have been found to be a contributing factor to SBS. Production increases significantly as ventilation rates rise, by 12.7% for every twofold increase in ventilation rate. Certain symptoms of SBS may result from printer effluent, which is released into the office air as ultra-fine particles (UFPs) when toner is burned during the printing process. A portion of office workers may be sensitive to various toxins found in printer effluent, which could result in SBS symptoms. Certain SBS symptoms are also linked to particular careers. Social workers, transportation, communication, and healthcare have the highest rates of general symptom prevalence. Technical work is linked to skin symptoms like eczema, itching, and rashes on the hands and face. The industries with the lowest rates of sick building syndrome symptoms are forestry, agriculture, and sales.

#### 4.2.4.2. Home

Home-related factors can also lead to sick building syndrome. Comparing laminate flooring to stone, tile, and cement flooring, there is a greater chance of chemical exposure and subsequent SBS symptoms. Increased symptoms were also linked to recent redecorating, new furnishings within the last year, dampness and related factors, pet ownership, and cockroach presence. Although it's unclear if this was because of repellent use or the presence of mosquitoes, their presence was also linked to an increased number of symptoms.

- **Mold:** Mycotoxin or mold contamination indoors has been linked in some studies to sick building syndrome. The connection between mold and sick building syndrome, however, is debatable and not well-supported.

#### *4.2.4.3. Indoor Temperature*

It has been demonstrated that indoor temperatures below 18°C (64°F) are linked to higher rates of respiratory and cardiovascular illnesses, elevated blood pressure, and higher hospitalization rates.

### 4.2.5. Diagnosis

While building-related illness (BRI) consists of specific, diagnosable symptoms brought on by specific agents (chemicals, bacteria, fungi, etc.), sick building syndrome (SBS) is characterized by a wide range of non-specific symptoms.). Usually, these are measurable, quantifiable, and identifiable. In BRi, there are typically four causative agents: irritant, toxic, infectious, and immunologic. For example, Legionella pneumophila, the typical causative agent of Legionnaire's disease, is a particular type of organism whose presence in a building can be determined by clinical observations.

### 4.2.6. Prevention

- Reduction of time spent in the building
- If living in the building, moving to a new place
- Fixing any deteriorated paint or concrete deterioration
- Regular inspections to indicate for presence of mold or other toxins
- Adequate maintenance of all building mechanical systems
- Toxin-absorbing plants, such as sansevieria
- Roof shingle non-pressure cleaning for removal of algae, mold, and *Gloeocapsa magma*
- Employing ozone to eradicate the various causes, including bacteria, viruses, molds, mildews, VOCs, and even smells. Despite its commercial success and widespread perception, high-ozone shock therapy has been shown in multiple studies to be ineffective.
- Replacement of water-stained ceiling tiles and carpeting
- Paints, adhesives, solvents, and pesticides should only be used in well-ventilated spaces or when the source of the pollution is not in use.

- Increasing the quantity of air exchanges—a minimum of 8.4 air exchanges per 24-hour period is advised by the American Society of Heating, Refrigeration, and Air-Conditioning Engineers.
- Increased ventilation rates that are above the minimum guidelines
- Proper and frequent maintenance of HVAC systems
- UV-C light in the HVAC plenum
- Installing HVAC air cleaning equipment to get rid of bioeffluents, or human odors, and volatile organic compounds (VOCs).
- All particles in the house, including ultrafine particles (UFPs) measuring less than 0.1 μm, are completely removed by central vacuums.
- Using a HEPA filter vacuum cleaner on a regular basis will help you collect and hold onto 99.97% of particles down to and including 0 micrometers.
- The practice of exposing bedding to sunlight is connected to a study conducted in a high-humidity region where damp bedding was prevalent and linked to SBS.
- When it comes to lighting, it should be natural and designed to give people control.
- Moving office printers—possibly to a different building—outside the air conditioning perimeter.
- Replacing current office printers with lower emission rate printers
- Identification and removal of products containing harmful ingredients

## 4.2.7. Management

As a general, non-specific term, SBS lacks a known cause or treatment. Any known treatment would be related to the particular disease that was ultimately brought on by exposure to recognized toxins. Removing the impacted individual from the related building is the only way to alleviate the situation in all of them. However, BRI uses a treatment that is suitable for the contaminant found inside the building (e.g., antibiotics to treat Legionnaires' illness).

Enhancing a building's indoor air quality (IAQ) can reduce or even completely eradicate ongoing exposure to pollutants. However, two independent authors' Cochrane review of 12 studies on mold and dampness remediation in private homes, workplaces, and schools found inconsistent results in children and very low to moderate quality of evidence in reducing adult asthma symptoms. As in the case of mold toxins, the patient's recovery may involve treating the acute symptoms of a particular illness. The treatment of different illnesses related to buildings is essential to the overall understanding of SBS. Thorough examination by licensed building specialists and medical professionals can assist in pinpointing the precise reason for the BRI and provide evidence of an infection's causal pathway. With this information, it is theoretically possible to remove contaminants from a building and reconstruct it using fresh materials. Three factors are most likely to account for office BRI: a broad range in a population's threshold of response (susceptibility); a spectrum of responses to a particular agent; or variability in exposure within large office buildings.

The three steps that people with office BRI should take are history, examinations, and interventions because it can be difficult to isolate any one of the three aspects of the condition. History explains the process of keeping track of and documenting the health of employees who are suffering from BRI. It also gathers information about prior building modifications or associated activities. Monitoring employee health and administering exams go hand in hand. This step involves physically inspecting the entire workspace and assessing any risks to the health of the staff. Interventions are made in accordance with the findings of the report on the examination and history.

**Table 4.3.** *Comfort Levels Recommended by the Chartered Institution of Building Services Engineers (CIBSE 1986)*

| **Parameter** | **Recommended Level** |
|---|---|
| Temperature (dry bulb) | 19–23°C |
| Relative humidity (RH) | 40–70% |
| – | More than 55% RH is needed in carpeted buildings with under floor heating to avoid electrostatic shocks |
| *Ventilation* | |
| Delivery of fresh air | 8 l/s/person (minimum) |
| – | 16 l/s/person where some smoking |
| – | 25 l/s/person where heavy smoking |
| Total air supply | 4–6 air changes/h |
| Air speed | 0.1–0.3 m/s. Less than 0.1 m/s causes stuffiness. More than 0.3 m/s causes drafts. For air speeds higher than 0.1 m/s, CIBSE recommends an increase in air temperature to take account of air movement |
| Sound | 46 dBA is the upper limit for general office work |

| | |
|---|---|
| Lighting | 500 lux for general office work, 750 lux for deep-plan offices and work at drawing boards, proofreading, etc. |

*Source: Joshi SM. The sick building syndrome. Indian J Occup Environ Med. 2008 Aug;12(2):61–4. doi: 10.4103/0019–5278.43262. PMID: 20040980; PMCID: PMC2796751.*

## 4.3. CURRENT STRATEGIES FOR MONITORING AND CONTROL OF IAQ

This section centers on the necessity of creative monitoring systems to balance energy efficiency and improve the quality of life for occupants. Together, the tactics give people and building managers the ability to track, evaluate, and improve indoor air quality, creating healthier interior spaces. Real-time data-driven decision-making is made possible by the application of cutting-edge materials, creative sensors, and intelligent technologies, which supports efficient and sustainable IAQ management.

### 4.3.1. Development of Materials for IAQ Sensors

In the last ten years, there has been a lot of interest in two-dimensional (2-D) nanostructured materials because of their special chemical and physical characteristics, which make them useful for designing and manufacturing gas-sensing devices. Outstanding characteristics of 2-D materials were shown to exist, including a high surface sensitivity, a large surface-to-volume ratio, and good semiconducting properties. Furthermore, combining two-dimensional nanostructured materials with three-dimensional materials is suggested as a viable method for creating high-performing IAQ-monitoring sensors. Four unique architectures for IAQ monitoring have been shown to be possible by combining 2-D nanostructured materials with 3-D bulk materials, 2-D nanostructures, 1-D nanostructures, and 0-D nanomaterials (Figure 4.2). Quantum dots (QDs), nanoparticles, nanocrystals, and nanoclusters are examples of 0-D nanomaterials of nanoscale (<100 nm) size that have been used in the creation of photonic, electronic, and chemical sensing devices because of their unique optical and electrochemical properties. Because of their unique morphology, high surface-to-volume ratio, and quick electron transport, 1-D nanostructures like nanotubes, nanowires (NWs), nanofibers, and nanorods have been shown to work well when combined with 2-D materials in a variety of IAQ-sensing applications. Simultaneously, different types of 2-D–2-D heterostructures, such as graphene/MoS2, graphene/black phosphorus, and graphene/WS2/graphene, have been applied in novel devices for IAQ monitoring due to the formation of Van der Waals heterostructures through direct stacking of different 2-D materials. Lastly, layered 2-D nanostructured semiconductors can be integrated with conventional bulk 3-D semiconductor substrates in the design of novel gas-sensing devices by growing or transferring 2-D materials onto bulk 3-D semiconductors. In conclusion, heterostructure-based devices have the potential to be applied to a variety of IAQ detection tasks due to the superior qualities of the materials that make them up.

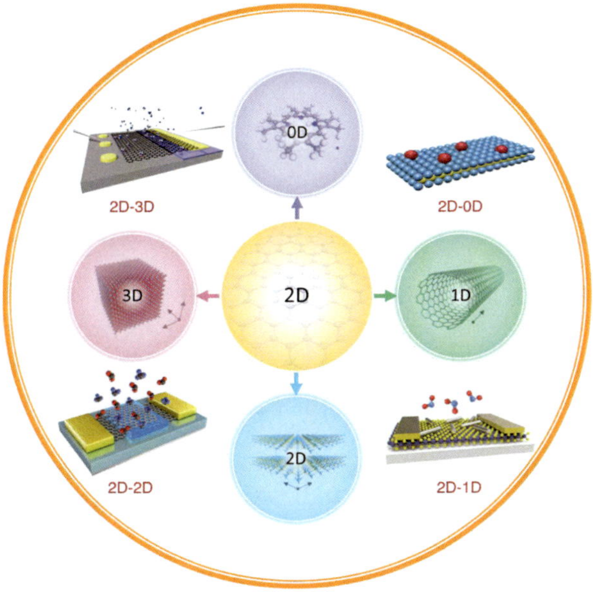


**Figure 4.2.** *Schematic overview of 2-D nanostructured materials and different types of heterostructures with 0-D, 1-D, 2D, and 3-D materials used for gas-sensing applications.*

*Source: Bag A., & Lee N.-E. (2019). Gas sensing with heterostructures based on two dimensional nanostructured materials: A review. J. Mater. Chem. C.; 7:13367. doi: 10.1039/C9TC04132J.*

One important tactic for IAQ monitoring has emerged: the creation of low-cost sensor networks and systems. Nowadays, metal oxide (MOx) sensors are thought to be among the best technologies because of their affordability, simplicity, high sensitivity, and compatibility with contemporary electronics. Their portability and affordability make them the perfect choice for remote and portable monitoring systems. Metal oxide semiconductor gas sensors have gained attention recently as a potential platform for the indoor air quality detection of CO and VOCs. Several metal oxides have reportedly been used in IAQ-sensing applications, including $SnO_2$, ZnO, $TiO_2$, $In_2O_3$, $Fe_2O_3$, $MoO_3$, $Co_3O_4$, CuO, NiO, and CdO.

## 4.3.2. Advanced Technologies for Monitoring of IAQ

Many methods have been developed and made available for real-time monitoring. However, because of their expanding application in the Industry 4.0 revolution, wireless sensor networks (WSN) and Internet of Things (IoT)-based systems are the most widely used technologies developed for IAQ monitoring.

### *4.3.2.1. Internet of Thing (IoT)-Based Systems*

With the growth of big data, the Internet of Things (IoT), and mobile technologies in recent years, machine-learning technologies have emerged as popular ones with great potential for real-time IAQ monitoring. The advent of portable IAQ-monitoring devices powered by the Internet of Things has made it possible to easily monitor and control air quality in real time these days. Numerous IoT-based IAQ-monitoring systems and gadgets, including open-source data processing and transmission technologies, have been developed and released.

"Electronic noses" (E-noses), which are biomimetic devices that mimic the functions of mammals' olfactory systems, are one of the major uses of IoT in IAQ monitoring. Generally, an E-nose system consists of four basic components: (i) A multi-sensor array; (ii) software with digital pattern recognition algorithms; (iii) an information-processing unit (i.e., an artificial neural network (ANN)), and (iv) reference library databases. Target gases can be readily monitored by E-noses through the detection and discrimination of types and concentrations through the use of chemical sensor arrays in conjunction with classification algorithms. As a result, due to their affordable and portable IAQ monitoring feature, E-noses have garnered a lot of attention worldwide. As an instance, Tastan et al. suggested the E-nose system, a low-cost, portable, Internet of Things (IoT)-based, real-time monitoring tool that uses a variety of sensors to measure various air pollutants (i.e., $NO_2$, PM10, $CO_2$, and $CO_2$) as well as air parameters (i.e., humidity, and temperature). In particular, the suggested E-nose is made with four detection units: the MH-Z14A ($CO_2$ sensor), the GP2Y1010AU (dust sensor), the MICS-4514 ($NO_2$ and CO sensor), and the DHT22 (temperature and humidity sensor). The software for these units is open-source, inexpensive, and simple to install. These sensors use the 32-bit ESP32 Wi-Fi controller to monitor, and data is received and stored in a cloud server by a mobile device interface built on the Blynk IoT platform (Figure 4.3A). It is expected that this idea will function as a model for upcoming IAQ-monitoring equipment. In additional studies, Chen et al. created the high-performance smart E-nose system, which consists of a wireless data transmission unit, a mobile phone receiver, a read-out electronic circuit, a multiplexed tin oxide ($SnO_2$) nanotube sensor array, and a data processing application. Because this nanotube sensor uses nanotube sensors with a large surface area for gas molecule interaction, it has higher sensitivity for gas detection and discrimination at room temperature when compared to conventional devices. This E-nose device can be considered to have state-of-the-art sensitivity for air pollutant detection at room temperature based on metal oxide sensors because it has been shown to detect indoor target air pollutants using only a basic vector-matching recognition algorithm. Consequently, it is significant that the smart E-nose system and device can be used for high-performance indoor environment quality monitoring in smart homes, smart buildings, or even smart cities in the present and the future (Figure 4.3B).

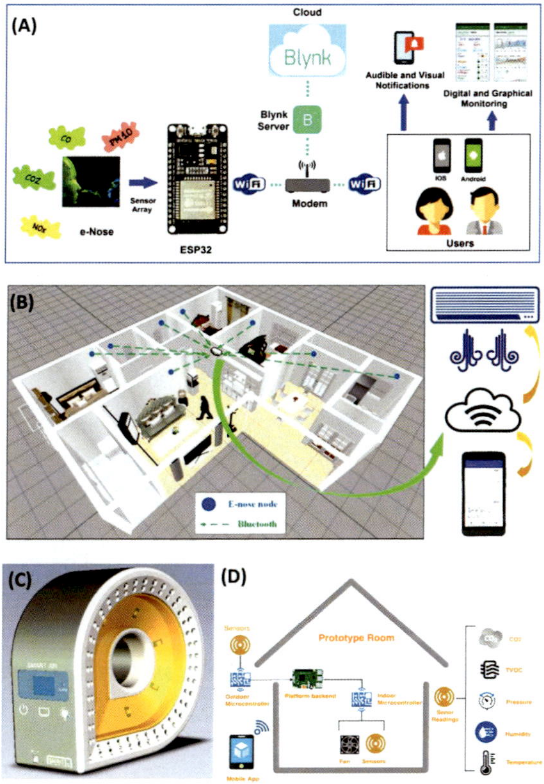


**Figure 4.3.** *Illustration of the proposed electronics nose (E-nose) system architecture.*

*Note: (A); the proposed E-nose application in future smart buildings (B); primitive concept design of Smart-Air (C); and a smart ventilation system for optimizing indoor air quality (IAQ) Levels in buildings (D).*

*Source: Tran VV, Park D, & Lee YC. (2020). Indoor air pollution, related human diseases, and recent trends in the control and improvement of indoor air quality. Int. J. Environ Res Public Health, 17(8), 2927. doi: 10.3390/ijerph17082927. PMID: 32340311; PMCID: PMC7215772.*

Recently, cloud computing and Internet of Things (IoT)-based smart devices have been developed to effectively transmit real-time data to an IoT sensor network and precisely monitor IAQ. Jo et al. (2020) introduced a Smart-Air device to collect reliable and accurate data for IAQ monitoring (Figure 4.3C). Three parts make up the Smart-Air device: an LTE modem, a microcontroller, and an array of sensors for detecting pollutants. The VOC, laser PM, $CO_2$, CO, and temperature/humidity sensors are among the sensors found in the Smart-Air device. In addition, the Smart-Air platform uses Internet of Things and cloud computing technologies to monitor and manage IAQ anywhere, at any time. Crucially, testing conducted by the Korean Ministry of Environment has shown how dependable the device is. All incoming data is kept in the cloud of the web server to offer resources for additional IAQ analysis. A recently developed strategy for IAP reduction by demand-controlled ventilation systems (DCVs) based on $CO_2$ detection is based on the concept of combining ventilation system activation with IAQ-variable-level detection. The IAQ index computation, mobile app, backend, and hardware make up the four main parts of this system. In order to precisely and automatically control the ventilation system and maintain an appropriate concentration of air pollutants by balancing indoor and outdoor parameters, the system's operation depends on a number of real-time sensors and algorithms. This sensing platform offers an IoT-based system for controlling the IAQ levels in a building based on recorded data, as illustrated in Figure 4.3D. A number of variables are measured simultaneously, including temperature, humidity, VOCs, and $CO_2$ in both the indoor and outdoor environments. This information is then used to calculate the IAQ index, which is a tool for algorithm-based ventilation control. The system also enables management and observation through a mobile app.

#### *4.3.2.2. Wireless Sensor Network (WSN)-Based Systems*

Over the last few years, wireless sensor networks, or WSNs, have drawn a lot of interest from various monitoring scenarios. WSN is typically made up of small units, or nodes, that play a crucial role in wirelessly interacting with other nodes in the system and gathering data from sensors in the surrounding area. ZigBee, a wireless standard based on the IEEE 802.15.4 specification, is regarded as the most reliable communication protocol because of its low cost, low consumption, and low data rate. Furthermore, networks operating on the ZigBee protocol can operate in power-saving mode. WSN systems are typically used in conjunction with a distributed sensor network that is linked to a cloud system. Through a gateway, ZigBee-based sensor nodes send field measurement data to the cloud system. The collected data from the sensor network has frequently been processed, stored, monitored, and visualized using the cloud computing systems that are optimized for this purpose. Furthermore, artificial intelligence techniques are employed to process and analyze the obtained data in order to optimize the detection of air pollutants. The huge number of nodes set up to offer pertinent data on the distribution of IAQ in various locations means that WSN systems have the potential to greatly improve air quality monitoring and measurement efficiency. The WSN is an automatic IAQ monitoring system that provides real-time data to the user or building manager. It measures IAQ in various areas autonomously, accurately, and simultaneously. It has been discovered that WSN systems are typically used for temperature, relative humidity, CO, $CO_2$, luminosity, and a range of other environmental parameters. Common VOCs monitored and detected include benzene, ethylbenzene, toluene, and xylene.

### 4.3.3. Air Purification Technologies for IAQ Improvement

The development of air purification technologies, optimization of ventilation systems, and control of emission sources are thought to be the three primary strategies for enhancing IAQ. The exploration of innovative technologies for the purpose of filtering and purifying indoor air pollutants has garnered significant attention in the past ten years. The most advanced methods for eliminating gas pollutants have long been recognized to be adsorption processes utilizing carbon-based filter media. There are numerous residential and commercial HVAC filters that have been developed and used in various buildings, i.e., household, shopping mall, school, and office building. Adsorptive HVAC filters have been shown to be able to efficiently remove a number of harmful pollutants, such as PM, aerosols, $O_3$, and VOCs. HVAC filters come in a variety of shapes, including cartridges filled with granulated activated carbon or bag filters, and filter cassettes composed of flat sheet filter media. However, due to a significant decrease in its equilibrium adsorption capacity at sub-ppm concentrations of pollutants, this technology has demonstrated a significant limitation with respect to adsorption efficiency. Furthermore, it has been shown that the presence of humidity significantly lowers the adsorption efficiency of carbon-based filters. Therefore, one possible strategy for the development of HVAC filter development strategies is to enhance the air filter system's long-term operation and performance.

Extended use can encourage the growth of germs and mold, which effectively reduces the effectiveness of the air filters. It has been shown that non-thermal plasma (NTP) technologies are an efficient way to deal with these issues. Radicals, ions, electrons, and UV photons are examples of the quasi-neutral environment that is created in order for NTP techniques to function. Numerous non-thermal discharge plasma methods, including surface discharge, dielectric barrier discharge, and corona discharge, have been developed for the purpose of eliminating indoor air contaminants like VOCs, $SO_x$, and $NO_x$. It has been demonstrated that the air filter can act as an efficient electric precipitator due to electrostatic force, and that the ozone produced by the electric discharge demonstrates a notable germicidal power. Nevertheless, NTP technologies also have certain drawbacks, including: (i) low energy efficiency; (ii) performance being greatly impacted by humidity; and (iii) secondary hazardous pollutants, like CO, HCHO, and $NO_x$, forming. The technique's practical applications for indoor air purification have been limited by these drawbacks.

Advanced oxidation processes (AOPs)-based technologies have garnered significant attention in development strategies for controlling indoor air pollution because they produce the least amount of secondary pollution. Among these, photocatalysis—the process of exposing semiconductor photocatalysts to sufficiently intense irradiation to cause them to disintegrate—has been regarded as a promising technology. When compared to traditional adsorption methods, photocatalysis shows a number of noteworthy characteristics. These include: (i) direct degradation of gaseous pollutants, especially volatile organic compounds (VOCs), into $CO_2$ and $H_2O$ in ambient conditions; and (ii) suitability for low-concentration pollutant removal (sub-ppm levels). In photocatalysis technology, semiconductors such as titanium dioxide ($TiO_2$), graphitic polymers, tungsten oxide ($WO_3$), carbon nitride (CN), Bi, Ag, cadmium sulfide (CdS), metal oxides MOx (M = Fe, Zn, V), perovskite, and others are utilized as photocatalysts. However, $TiO_2$ has drawn increased attention for IAP purification because of its inexpensive cost, potent oxidative power, and chemical stability.

### 4.3.4. Smart Home for IAQ Control

The "smart home" in particular, as well as home automation in general, have garnered a lot of attention lately. These days, a key component of smart cities is the smart home, which can be managed by wearable technology and smartphone applications. The suggested home energy management system served as its cornerstone (HEMS). Actually, the original purpose of the smart home's development was to maximize ventilation technology and increase energy efficiency. But these days, the phrase "smart home" is most frequently linked to a variety of home automation systems, including household appliances, communications, and entertainment electronics. Future advancements in smart home technology are probably going to focus on two main areas: (i) controlling appliances, blinds, and ventilation to save energy costs and improve comfort, safety, and health; and (ii) using robotic machines and smart sensor technologies to help the elderly and disabled. The smart home, however, also presents chances to lessen the detrimental effects of IAP on people's health. Undoubtedly,

IAQ and air hygiene will receive more attention as smart home technology is implemented in buildings more frequently. The three primary parts of a typical smart building system are the controller and devices, the power supply media, and the user gateway media. User gateway media, which combines a supervisory control and data acquisition (SCADA) interface with a smartphone application, enables residents to remotely monitor and control appliances from computers or smartphones via Wi-Fi or a wireless module. Power supply media is a cutting-edge technology that uses renewable energy sources to maximize energy efficiency. Lastly, the temperature, air pollutant concentrations, and fan speed are automatically controlled by the controller and devices. These days, sensor technology is the key to smart buildings. All meteorological parameters and pollutant concentrations are automatically and continuously recorded in the SCADA system through the use of multiple sensors. Specifically, smart sensor systems can record data and use that data to take appropriate actions (e.g., modifying the temperature and/or ventilation settings) to raise IAQ and lower IAP.

## 4.4. INDOOR AIR POLLUTION: SOLUTIONS

Recurring negative health effects that are accurately recorded may serve as a helpful marker for an issue with indoor air quality. Even though indoor air pollutants have been linked to a number of detrimental effects, there is still much discussion and misunderstanding regarding the exposure levels or times that are required to cause any given health problem. The science is still in its infancy, but measuring indoor air pollution levels is still essential. Three tactics have been used in general to address indoor air pollution.

### 4.4.1. Source Control

The best approach to enhance indoor air quality is to locate and remove possible sources of pollution or drastically lessen their influence by regulating their emissions. To reduce the quantity of pollutants released into the atmosphere, some pollutants, like asbestos, can be sealed or enclosed, while others, like gas stoves, need to be adjusted. Because source control rarely necessitates expensive adjustments or extra equipment, it is frequently a more economical method of preserving indoor air quality.

### 4.4.2. Ventilation

Increasing the quantity of outdoor air vented indoors is one strategy to reduce the concentration of indoor air pollution. Energy-efficient ventilators that facilitate the exchange of fresh air are becoming increasingly common in new homes and businesses, despite the fact that most heating and cooling systems are not built to mechanically vent fresh air indoors. The most effective way to control indoor air quality is through ventilation, especially in areas where there is a lot of exposure to pollutants, like paints, solvents, kerosene heat, welding, cooking, or soldering. To improve the exchange of fresh air in these conditions, it is essential to open windows, turn on attic fans, and adjust vent controls.

### 4.4.3. Air Cleaners

As a result of worries about environmental tobacco smoke, air cleaners of all shapes and sizes have started to hit the market (ETS). Air purifiers aren't very good at removing particles, but they can be very successful at removing gaseous pollutants. Furthermore, studies indicate that common houseplants can significantly lower dangerous chemical concentrations. While indoor air quality can benefit from plants, they can also act as a source of mold spores and indoor pollen. Furthermore, there isn't any proof as of yet that plants are able to filter appreciable amounts of ETS. Studies conducted by agencies such as the U.S. EPA have resulted in a greater recognition of the effects of indoor air pollution. In order to help define pollutant concentrations, additional indoor air quality monitoring is necessary due to the presence of particulate matter and organic and inorganic gases. Controlling excessive exposure to indoor air pollutants will require ongoing attention to indoor air quality.

## 4.5. EPA APPROACH AND PROGRESS

While most people are aware that air pollution outdoors can be harmful to their health, it's possible that they are unaware of the serious consequences indoor air pollution can have as well. U.S. EPA studies of human exposure to air pollutants indicate that indoor levels of many pollutants may be 2 to 5 times higher, and occasionally more than 100 times higher than outdoor levels. Indoor levels of many pollutants can range from 2 to 5 times higher. The majority of people are thought to spend up to 90% of their time indoors, making these levels of indoor air pollution especially concerning. Our exposure to indoor air pollutants is thought to have increased over the past few decades due to a number of factors, including the construction of more tightly sealed buildings, lower ventilation rates for energy efficiency, the use of synthetic furnishings and building materials, and the use of chemically formulated household cleaners, pesticides, and personal care products. The EPA and its Science Advisory Board have conducted comparative risk assessments in recent years, and the results consistently place indoor air pollution in the top five environmental hazards to public health. The EPA has started a concentrated effort to learn more about indoor air pollution and to lower people's exposure to air pollutants in their homes, workplaces, schools, and other indoor environments where people live, work, and play. This effort is being carried out in close collaboration with other Federal agencies and the private sector.

### 4.5.1. EPA Indoor Air Program

The EPA has developed a comprehensive program to better understand the problem of indoor air pollution and to take decisive action to reduce people's exposure to indoor air contaminants of all types. This is because indoor air pollution can have a serious impact on the health of those who may experience it, not to mention the financial costs to society if it is ignored. Prudent public policy requires that reasonable efforts be made to lower people's exposure to potentially harmful levels of indoor air pollutants by using the resources available to the Federal government under current laws, even in the absence of a complete scientific understanding of indoor air pollution. A regular component of the design, construction, upkeep, and operation of public and commercial buildings, residences, healthcare and childcare facilities, educational institutions, and other special use buildings must be pollution prevention and the effective resolution of indoor air quality issues of all kinds. To gain a deeper understanding of the variables influencing indoor air quality, exposure patterns, health effects, and control strategies for enhancing indoor air quality, an efficient research and development program is required. Utilizing both non-regulatory and regulatory tools made possible by a number of Federal laws, EPA is carrying out this program to protect manufacturers, architects, engineers, builders, building owners and managers, and building occupants by disseminating information and offering incentives for action.

### 4.5.2. Other Federal Agencies

Due to their status as significant property managers or as statutory authorities, more than 20 Federal agencies bear responsibility for indoor air quality. These agencies work together to coordinate their activities through a number of channels, such as the interagency Committee on Indoor Air Quality (CIAQ), which convenes quarterly to share information on indoor air-related matters. The National Institute for Occupational Safety and Health, the Occupational Safety and Health Administration, the Department of Energy, the Consumer Product Safety Commission, and the EPA are the five federal agencies that co-chair the CIAQ. In order to help ensure that federal actions are well-coordinated, EPA also collaborates closely with other agencies on regulatory and information development projects. These agencies also sponsor a large number of EPA's guidance and public information documents.

# REFERENCES

1. Apter, A. (1994). Epidemiology of the sick building syndrome. *Journal of Allergy and Clinical Immunology, 94*(2), 277–288. https://doi.org/10.1053/ai.1994.v94.a56006. PMID 8077580.
2. Baldacci, S., Maio, S., Cerrai, S., Sarno, G., Baiz, N., Simoni, M., Annesi-Maesano, I., & Viegi, G. (2015). Allergy and asthma: Effects of the exposure to particulate matter and biological allergens. *Respiratory Medicine, 109*, 1089–1104. https://doi.org/10.1016/j.rmed.2015.05.017.
3. Bernstein, J. A., Alexis, N., Bacchus, H., Bernstein, I. L., Fritz, P., Horner, E., Li, N., Mason, S., Nel, A., Oullette, J., et al. (2008). The health effects of nonindustrial indoor air pollution. *Journal of Allergy and Clinical Immunology, 121*, 585–591. https://doi.org/10.1016/j.jaci.2007.10.045.
4. Brasche, S., Bullinger, M., Morfeld, M., Gebhardt, H. J., & Bischof, W. (2001). Why do women suffer from sick building syndrome more often than men?--subjective higher sensitivity versus objective causes. *Indoor Air, 11*(4), 217–222. https://doi.org/10.1034/j.1600-0668.2001.110402.x. ISSN 0905-6947. PMID 11761596. S2CID 21579339.
5. Brickus, L. R., Cardoso, J., & Neto, F. R. D. A. (1998). Distributions of indoor and outdoor air pollutants in Rio de Janeiro, Brazil: Implications to indoor air quality in bayside offices. *Environmental Science & Technology, 32*, 3485–3490. https://doi.org/10.1021/es980336x.
6. Brook, R. D., Rajagopalan, S., Pope III, C. A., Brook, J. R., Bhatnagar, A., Diez-Roux, A. V., Holguin, F., Hong, Y., Luepker, R. V., Mittleman, M. A., et al. (2010). Particulate matter air pollution and cardiovascular disease: An update to the scientific statement from the American Heart Association. *Circulation, 121*, 2331–2378. https://doi.org/10.1161/CIR.0b013e3181dbece1.
7. Brooks, B. D., & Davis, W. F. (1992). *Understanding indoor air quality*. CRC Press. ISBN 0849388465.
8. Bruno, R. C. (1983). Sources of indoor radon in houses: A review. *Journal of the Air Pollution Control Association, 32*, 105–109. https://doi.org/10.1080/00022470.1983.10465550.
9. Chen, K., Gao, W., Emaminejad, S., Kiriya, D., Ota, H., Nyein, H. Y. Y., Takei, K., & Javey, A. (2016). Printed carbon nanotube electronics and sensor systems. *Advanced Materials, 28*, 4397–4414. https://doi.org/10.1002/adma.201504958.
10. Colt, J. S., Lubin, J., Camann, D., Davis, S., Cerhan, J., Severson, R. K., Cozen, W., & Hartge, P. (2004). Comparison of pesticide levels in carpet

dust and self-reported pest treatment practices in four U.S. sites. *Journal of Exposure Analysis and Environmental Epidemiology, 14*, 74–83. https://doi.org/10.1038/sj.jea.7500307.

11. Godish, T. (2001). *Indoor environmental quality*. CRC Press. ISBN 1-56670-402-2.
12. Hamanaka, R. B., & Mutlu, G. M. (2018). Particulate matter air pollution: Effects on the cardiovascular system. *Frontiers in Endocrinology, 9*, 680. https://doi.org/10.3389/fendo.2018.00680.
13. Holt, E., Audy, O., Booij, P., Melymuk, L., Prokes, R., & Klánová, J. (2017). Organochlorine pesticides in the indoor air of a theatre and museum in the Czech Republic: Inhalation exposure and cancer risk. *Science of the Total Environment, 609*, 598–606. https://doi.org/10.1016/j.scitotenv.2017.07.203.
14. Hospodsky, D., Qian, J., Nazaroff, W. W., Yamamoto, N., Bibby, K., Rismani-Yazdi, H., & Peccia, J. (2012). Human occupancy as a source of indoor airborne bacteria. *PLOS ONE, 7*(e34867). https://doi.org/10.1371/journal.pone.0034867.
15. Huang, Y., Yang, Z., & Gao, Z. (2019). Contributions of indoor and outdoor sources to ozone in residential buildings in Nanjing. *International Journal of Environmental Research and Public Health, 16*, 2587. https://doi.org/10.3390/ijerph16142587.
16. Hwang, H.-M., Park, E.-K., Young, T. M., & Hammock, B. D. (2008). Occurrence of endocrine-disrupting chemicals in indoor dust. *Science of the Total Environment, 404*, 26–35. https://doi.org/10.1016/j.scitotenv.2008.05.031.
17. Jo, J., Jo, B., Kim, J., Kim, S., & Han, W. (2020). Development of an IoT-based indoor air quality monitoring platform. *Journal of Sensors, 2020*(8749764). https://doi.org/10.1155/2020/8749764.
18. Joshi, S. M. (2008). The sick building syndrome. *Indian Journal of Occupational and Environmental Medicine, 12*(2), 61–64. https://doi.org/10.4103/0019-5278.43262. ISSN 0973-2284. PMC 2796751. PMID 20040980.
19. Koivisto, A. J., Kling, K. I., Hänninen, O., Jayjock, M., Löndahl, J., Wierzbicka, A., Fonseca, A. S., Uhrbrand, K., Boor, B. E., & Jiménez, A. S. (2019). Source specific exposure and risk assessment for indoor aerosols. *Science of the Total Environment, 668*, 13–24. https://doi.org/10.1016/j.scitotenv.2019.02.398.
20. Komarnicki, G. J. K. (2005). Lead and cadmium in indoor air and the

urban environment. *Environmental Pollution, 136*, 47–61. https://doi.org/10.1016/j.envpol.2004.12.006.

21. Kulmala, M., Asmi, A., & Pirjola, L. (1999). Indoor air aerosol model: The effect of outdoor air, filtration, and ventilation on indoor concentrations. *Atmospheric Environment, 33*, 2133–2144. https://doi.org/10.1016/S1352-2310(99)00070-9.
22. Madany, I. M. (1994). The correlations between heavy metals in residential indoor dust and outdoor street dust in Bahrain. *Environment International, 20*, 483–492. https://doi.org/10.1016/0160-4120(94)90197-X.
23. Miller, M. R., Shaw, C. A., & Langrish, J. P. (2012). From particles to patients: Oxidative stress and the cardiovascular effects of air pollution. *Future Cardiology, 8*, 577–602. https://doi.org/10.2217/fca.12.43.
24. Mudarri, D., & Fisk, W. J. (2007). Public health and economic impact of dampness and mold. *Indoor Air, 17*(3), 226–235. https://doi.org/10.1111/j.1600-0668.2007.00474.x. ISSN 0905-6947. PMID 17542835. S2CID 21709547.
25. Oh, H.-J., Jeong, N.-N., Sohn, J.-R., & Kim, J. (2019). Personal exposure to indoor aerosols as actual concern: Perceived indoor and outdoor air quality, and health performances. *Building and Environment, 165*(106403). https://doi.org/10.1016/j.buildenv.2019.106403.
26. Passarelli, G. R. (2009). Sick building syndrome: An overview to raise awareness. *Journal of Building Appraisal, 5*, 55–66. https://doi.org/10.1057/jba.2009.20.
27. Pind, C. A. (2017). Patient-reported signs of dampness at home may be a risk factor for chronic rhinosinusitis: A cross-sectional study. *Clinical & Experimental Allergy, 47*(11), 1383–1389. https://doi.org/10.1111/cea.12976. PMID 28695715. S2CID 40807627.
28. Rashed, M. N. (2008). Total and extractable heavy metals in indoor, outdoor, and street dust from Aswan city, Egypt. *Clean, 36*, 850–857. https://doi.org/10.1002/clen.200800062.
29. Raub, J. A., Mathieu-Nolf, M., Hampson, N. B., & Thom, S. R. (2000). Carbon monoxide poisoning–A public health perspective. *Toxicology, 145*, 1–14. https://doi.org/10.1016/S0300-483X(99)00217-6.
30. Sauni, R., Verbeek, J. H., Uitti, J., Jauhiainen, M., Kreiss, K., & Sigsgaard, T. (2015). Remediating buildings damaged by dampness and mold for preventing or reducing respiratory tract symptoms, infections and asthma. *Cochrane Database of Systematic Reviews, 2015*(2), CD007897. https://doi.org/10.1002/14651858.CD007897.pub3. PMC 6769180. PMID 25715323.

31. Salonen, H., Salthammer, T., & Morawska, L. (2018). Human exposure to ozone in school and office indoor environments. *Environment International, 119*, 503–514. https://doi.org/10.1016/j.envint.2018.07.012.
32. Seow, W. J., Downward, G. S., Wei, H., Rothman, N., Reiss, B., Xu, J., Bassig, B. A., Li, J., He, J., & Hosgood, H. D. (2016). Indoor concentrations of nitrogen dioxide and sulfur dioxide from burning solid fuels for cooking and heating in Yunnan province, China. *Indoor Air, 26*, 776–783. https://doi.org/10.1111/ina.12251.
33. Shahzad, S. S., Brennan, J., Theodossopoulos, D., Hughes, B., & Calautit, J. K. (2016). Building-related symptoms, energy, and thermal control in the workplace: Personal and open plan offices. *Sustainability, 8*(4), 331. https://doi.org/10.3390/su8040331.
34. Stolwijk, J. A. (1991). Sick-building syndrome. *Environmental Health Perspectives, 95*, 99–100. https://doi.org/10.1289/ehp.919599. ISSN 0091-6765. PMC 1568418. PMID 1821387.
35. Tang, X., Misztal, P. K., Nazaroff, W. W., & Goldstein, A. H. (2015). Siloxanes are the most abundant volatile organic compound emitted from engineering students in a classroom. *Environmental Science & Technology Letters, 2*, 303–307. https://doi.org/10.1021/acs.estlett.5b00256.
36. Tastan, M., & Gökozan, H. (2019). Real-time monitoring of indoor air quality with Internet of Things-based e-nose. *Applied Sciences, 9*, 3435. https://doi.org/10.3390/app9163435.
37. U.S. Environmental Protection Agency. (2003). EPA assessment of risks from radon in homes. U.S. Environmental Protection Agency. Retrieved from https://www.epa.gov/radon/risk-radon.
38. U.S. Environmental Protection Agency. (2020). Carbon monoxide's impact on indoor air quality. Retrieved January 28, 2020, from https://www.epa.gov/indoor-air-quality-iaq/carbon-monoxides-impact-indoor-air-quality.
39. U.S. Environmental Protection Agency. (2020). Indoor particulate matter. Retrieved January 28, 2020, from https://www.epa.gov/indoor-air-quality-iaq/indoor-particulate-matter.
40. U.S. Environmental Protection Agency. (2020). Pesticides' impact on indoor air quality. Retrieved January 28, 2020, from https://www.epa.gov/indoor-air-quality-iaq/pesticides-impact-indoor-air-quality.
41. U.S. Environmental Protection Agency. (2020). Volatile organic compounds' impact on indoor air quality. Retrieved January 28, 2020, from https://www.epa.gov/indoor-air-quality-iaq/volatile-organic-compounds-impact-indoor-air-quality.

42. Weschler, C. J., & Nazaroff, W. W. (2014). Dermal uptake of organic vapors commonly found in indoor air. *Environmental Science & Technology, 48*, 1230–1237. https://doi.org/10.1021/es405490a.
43. Weschler, C. J., & Nazaroff, W. W. (2012). SVOC exposure indoors: Fresh look at dermal pathways. *Indoor Air, 22*, 356–377. https://doi.org/10.1111/j.1600-0668.2012.00772.x.
44. World Health Organization. (1988). *Indoor air quality: Biological contaminants: Report on a WHO meeting, Rautavaara* (Vol. 31). World Health Organization Regional Office for Europe.
45. World Health Organization. (2009). *WHO handbook on indoor radon: A public health perspective*. World Health Organization.

Chapter 5

# Ambient Air Quality Monitoring

## Contents

# INTRODUCTION

An efficient air quality management system must include ambient air monitoring. There are several reasons to gather this kind of information.

- assess the extent of pollution;
- provide air pollution data to the general public in a timely manner;
- support implementation of air quality goals or standards;
- evaluate the effectiveness of emissions control strategies;
- provide information on air quality trends;
- provide data for the evaluation of air quality models; and
- support research (e.g., long-term studies of the health effects of air pollution).

Every given pollutant can be measured using a variety of techniques. A monitoring strategy developer should review the available options and decide which approaches are most suitable, considering factors such as the primary purposes of the data, equipment purchase costs up front, ongoing expenses, system dependability, and user-friendliness. The goal of the monitoring determines where monitoring stations should be placed. The majority of air quality monitoring networks are built with human health goals in mind, and their monitoring stations are located in densely populated areas. They could be in urban areas, close to busy roads, or in areas of special concern (e.g., a hospital, school, or specific sources of emissions). In order to measure background pollution levels, monitoring stations may also be set up far from cities and emissions sources. Systems are required to record and store data, analyze data and present findings, and guarantee that data are of a sufficient quality.

## 5.1. UNDERSTANDING AMBIENT AIR QUALITY

The quality of the air outside and occasionally inside is known as ambient air quality. The concentrations of pollutants in the air that are considered safe for human, natural, and animal health are known as ambient air quality standards. As a result, standards are not universally defined, and different organizations employ standards that are frequently dissimilar. Another name for ambient air quality monitoring is outdoor air quality testing.

### 5.1.1. Definition of Air Quality Monitoring

The process of determining the amount and kinds of pollutants in the atmosphere in accordance with air quality standards is known as air quality monitoring. By keeping an eye on the pollutants in the atmosphere, air quality monitoring enables us to take appropriate action to improve air quality.

## 5.1.2. Need for Ambient Air Quality Monitoring or Testing

The introduction of chemicals, particulates, or other hazardous materials into the atmosphere results in damage to living things or artificial organisms. This is known as air pollution. Numerous factors contribute to the increase of air pollutants in the atmosphere, which alters its composition and has an impact on the biotic environment. In addition to the amount released by air pollution sources, the atmosphere's capacity to either absorb or disperse these emissions determines the concentration of air pollutants. Air pollution originates from a variety of sources, including domestic, industrial, and vehicular sources. The health of the populace and properties is being negatively impacted by the high concentration of air pollutants in the ambient air. Government action was taken to stop the decline in air quality. in 1981 with the passage of the Air (Prevention and Control of Pollution) Act. The Environment (Protection) Act of 1986 has further highlighted the obligation. Thus, from 1984 to 1985, the Central Pollution Control Board launched the National Ambient Air Quality Monitoring (NAAQM) Network on a nationwide scale. Later on, the initiative was rebranded as the National Air Quality Monitoring Program (NAMP).

## 5.1.3. Ambient Air Quality Monitoring or Testing Parameters

In N. A. M. P. twelve air contaminants, namely. The following substances have been identified for routine monitoring at the necessary locations/industrial units: sulfur dioxide ($SO_2$), nitrogen oxides such as $NO_2$, particulate matter (PM10), particulate matter (PM2.5), ozone ($O_3$), lead (Pb), carbon monoxide (CO), ammonia ($NH_3$), benzene ($C_6H_6$), benzo-pyrene (BaP), arsenic (As), and Nickel (Ni). Particulate matter is the most significant pollutant in the atmosphere when considering all of these factors. It is also known as dust monitoring or particulate matter monitoring. According to the National Ambient Air Quality Testing Standards (NAAQS) and CPCB Notification. These results can be explained in a non-technical manner so that the extent of pollution is understood by all. The term for it is the Air Quality Index (AQI).

## 5.1.4. Ambient Air Quality Monitoring or Testing Standards

In ambient air testing, the National Ambient or Outdoor Air Quality Standards (NAAQS Standards) are used. These are listed below.

| **Parameters** | **Unit** | **Limit** |
|---|---|---|
| Sulphur Dioxide ($SO_2$) | $mg/M^3$ | $\leq 80$ |
| Oxides of Nitrogen | $mg/M^3$ | $\leq 80$ |
| Particulate Matter PM10 | $mg/M^3$ | $\leq 100$ |

| | | |
|---|---|---|
| Particulate Matter PM2.5 | mg/M$^3$ | ≤ 60 |
| Ozone ($O_3$) | mg/M$^3$ | ≤ 180 |
| Lead (Pb) | mg/M$^3$ | ≤ 1.0 |
| Carbon Monoxide (CO) | mg/M$^3$ | ≤ 04 |
| Ammonia (NH3) | mg/M$^3$ | ≤ 400 |
| Benzene (C6H6) | mg/M$^3$ | ≤ 05 |
| Benzo(a)Pyrene (BaP) | mg/M$^3$ | ≤ 01 |
| Arsenic (As) | mg/M$^3$ | ≤ 06 |
| Nickel (Ni) | mg/M$^3$ | ≤ 20 |

*Source:* https://www.ppsthane.com/ambient-air-quality-monitoring-testing.

## 5.2. PARTICULATE MATTER COLLECTION

All across the world, particulate matter (PM) is a very common and significant air pollution. The main source of particulate matter pollution is combustion, which includes wildfires, automobiles, and industrial processes. Although particulate matter's concentration and composition vary widely across the globe, most nations find that it significantly affects public health. It is categorized based on the size of its particles; the majority of the pollutant's harmful effects on human and environmental health are attributed to the smaller category of PM2.5. Despite extensive research on the effects of PM, the majority of the world's population still experiences levels of PM that are above benchmarks, necessitating action to produce cleaner air.

### 5.2.1. Particulate Matter Composition

Particulate matter, which includes dust, pollen, ash, soot, dirt, and smoke, is made up of both solid and liquid particles rather than a single substance. While some of these particles are too small to be seen with the human eye, others may be apparent, such as when haze appears. Based on the size of the particles, particulate matter is commonly classified into four categories: coarse, PM10, PM2.5, and PM1. PM 10 and PM2.5 are composed of particles with a diameter of 10 or 2.5 μm, respectively, whereas coarse PM particles are 10 μm or larger. Since they are small enough to enter the lungs deeply, PM10 and PM2.5 are both regarded as inhalable particles. The smallest particles, with a diameter of less than one micrometer, make up PM1. Due to a lack of widespread monitoring and regulation, PM1 is not as frequently monitored as PM2.5 or PM10 because it is not covered by most air quality standards, despite the possibility that its small size makes it more hazardous to human health. However, because of its small particle size, PM1 is thought to be responsible for a large amount of PM2.5's health effects (Figure 5.1).

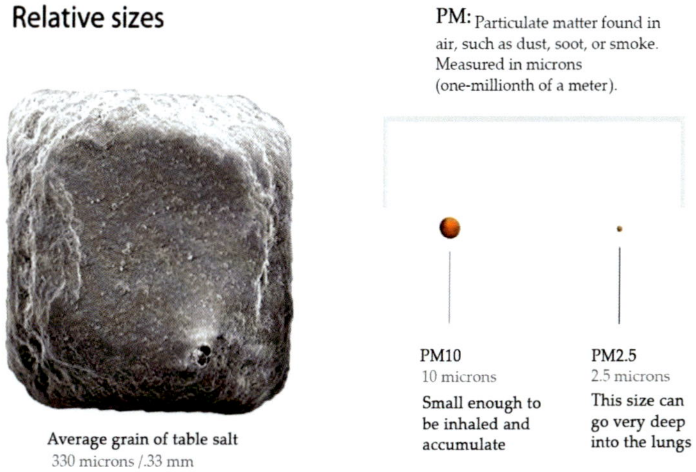


**Figure 5.1.** *A size comparison between PM10, PM2.5, and an average grain of table salt illustrates the minuteness of particulate matter. Because it is composed of such small particles, particulate matter can be inhaled into the lungs, aggravating both respiratory symptoms as well as other negative health effects.*

*Source: Minnesota Pollution Control Agency,* https://www.clarity.io/blog/air-quality-measurements-series-particulate-matter.

## 5.2.2. Pollution sources and Distribution for Particulate Matter

The main sources of particulate matter emissions are combustion sources like automobiles, diesel engines, and industrial buildings. Both primary and secondary sources can produce particulate matter (PM). Primary sources include wood stoves, construction sites, unpaved roads, fires, and smokestacks. Additionally, some soil-disturbing activities, like driving or agricultural work, may stir up coarser particulate matter, which raises the overall PM levels. Secondary sources, on the other hand, release gases into the atmosphere that combine to form PM. Particulate matter is created by complex chemical reactions between certain pollutants from other pollution sources, such as sulfur dioxide and nitrogen oxides, which can be released by power plants or coal fires. These interactions occur in the atmosphere. Additionally, as they cool, gaseous particles in the air can become solid or liquid. Compared to rural areas, urban areas typically have higher concentrations of particulate matter. According to one study, large central metropolitan counties have a mean of 11.21 "PM days," or days when PM2.5 levels are higher than the target 24-hour average set by the National Ambient Air Quality Standards (NAAQS), while rural counties only have 0.95 days. The central counties, which have higher densities of people and activities that contribute to particulate matter pollution, were also found to have higher concentrations of PM2.5.

### 5.2.3. Particulate Matter Impacts on Human Health

There are substantial detrimental impacts on one's health that come with particulate matter exposure. A Harvard study published in 2021 discovered that 10.2 million excess deaths worldwide in 2012 were caused by particulate matter air pollution from burning fossil fuels. With the use of the study's methodology, scientists were able to connect the burning of fossil fuels to PM2.5 exposure-related premature deaths, which they attribute to the cause of one in five premature deaths globally. According to the same study, 350,000 premature deaths in the US were attributed to fossil fuel pollution in 2018. Due to the tiny size of PM particles, which can enter the body, travel through the lungs and airways, and even enter the bloodstream, many of the harmful health effects of PM are caused by these particles (Figure 5.2).

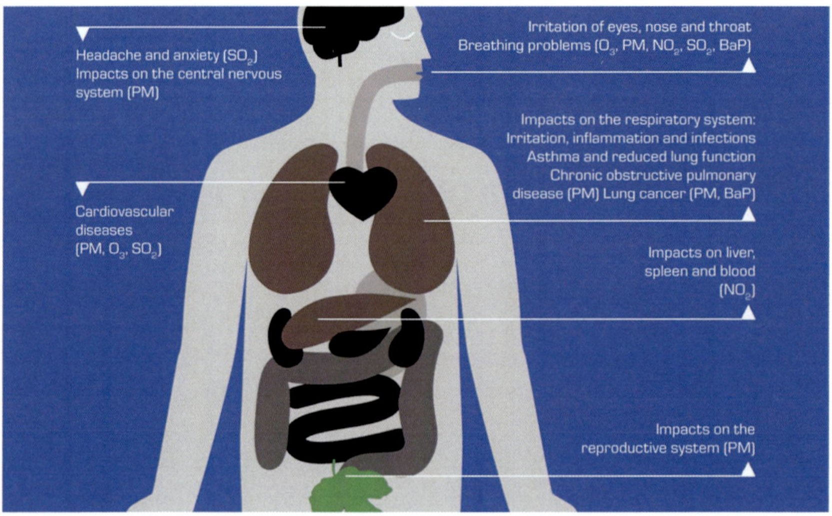


**Figure 5.2.** *Infographic shows the range of health impacts associated with PM and other air pollutants.*

*Source: European Environment Agency.*

The biggest health risk is from PM2.5, which has particles even smaller than PM10; PM10 particles tend to gather in the upper respiratory tract, but PM2.5 can enter the lungs, damage lung tissue, and spread throughout the body. Acute bronchitis, asthma episodes, and increased susceptibility to respiratory infections can all result from brief exposure to PM, which can last for hours or days. Living in an environment with high PM concentrations for an extended period of time is known to cause chronic bronchitis, reduced lung function, and early death. Children are particularly vulnerable to PM exposure because of a number of factors, including the fact that their respiratory systems are still developing, they breathe in more air per pound of body weight than adults do, and they are more likely to be active outside. A higher risk of negative health effects also exists for those who already have lung or heart disease.

Additionally, studies have shown that the brain can be impacted by particulate matter pollution exposure. Exposure to PM during childhood can change the size of the developing brain, potentially increasing the risk of emotional and cognitive issues in later life. Additionally, studies have shown a connection between dementia and particulate matter exposure. A 2017 USC study discovered that the risk of dementia in older women nearly doubled for those who lived in areas where PM2.5 exposure exceeded the USEPA's standard of 12 $\mu g/m^3$. Extrapolating these results to the broader population, exposure to PM pollution would account for roughly 21% of dementia cases globally. According to a 2019 USC study that focused on the relationship between PM exposure and Alzheimer's disease, women in their 70s and 80s who had previously been exposed to higher levels of PM2.5 showed more signs of memory decline and Alzheimer-like brain atrophy than those who had breathed cleaner air.

## 5.2.4. Environmental Impacts of Particulate Matter Pollution

Wind has the ability to carry particulate matter great distances, settling it far from the source of the emission. This implies that a broad geographic area's air quality may be significantly impacted by a PM-emitting event, such as a massive wildfire. These days, it is not unusual for there to be haze in the air, especially in big cities, which reduces visibility due to particulate matter. By turning lakes and streams acidic, altering the nutrient balance in coastal waters and large river basins, depleting soil nutrients, harming crops and forests, causing acid rain, and compromising ecosystem diversity, high levels of particulate matter can also change the overall health of the environment.

## 5.2.5. Regulation of Particulate Matter, Globally and in the U.S.

The World Health Organization (WHO) establishes guidelines and standards for permissible PM levels worldwide. The 24-hour mean value for PM2.5 is 25 μg/m³, while the annual mean value is 10 μg/m³, according to WHO benchmarks. These figures are 20 and 50 μg/m³, respectively, for PM10. The WHO sets interim targets for nations whose PM pollution levels are higher than these advised maximum levels. These targets are meant to gradually reduce pollution levels until the country's air is deemed clean by the WHO. However, numerous scientific studies conducted since the publication of the current WHO Air quality guidelines in 2005 have confirmed that there is, in fact, no safe level of PM for human health. National and regional organizations are in charge of regulating PM pollution in various regions on a smaller scale. These organizations, which set goals in their respective regions, are the EU and the USEPA. The USEPA controls the amount of "inhalable" particles that are present outside, such as PM2.5 and PM10, but excludes particles larger than 10 μm in diameter. One of the six criteria pollutants covered by the National Ambient Air Quality Standards (NAAQS) is particulate matter. USEPA is required by the Clean Air Act to establish standards for these six categories of pollutants, which are thought to be hazardous to the environment and public health.

The USEPA establishes primary and secondary standards for PM regulation. The main goal of setting primary standards is to safeguard the general public's health, especially the health of vulnerable groups like children, the elderly, and people who have asthma. Secondary standards are designed to safeguard the environment, the welfare of the general public, and animals, crops, vegetation, and buildings from harm. They also aim to minimize instances of reduced visibility. These regulations cover PM2.5 and PM10, with 24-hour and annual standards included. The current primary and secondary standards for PM2.5 set by the EPA are even stricter than those recommended by the WHO, with levels of 12.0 μg/m³ and 15.0 μg/m³ (annual average), respectively. We are starting to see PM regulated at a more local level, in the areas where people live and work, in addition to global and regional targets. Employers who employ outdoor workers are now required by law in California, Washington, and Oregon to monitor air quality and take protective measures for their workers. Specifically, they must limit or eliminate their exposure to wildfire smoke when the air quality index (AQI) rises to 151 or higher.

## 5.2.6. PM Prevalence Around the World

Globally, there are variations in the concentration and distribution of PM pollution in terms of both space and time. According to a 2017 study, PM2.5

concentrations rose by 38% between 1960 and 2009, with China and India experiencing the majority of this increase. The number of deaths worldwide attributable to PM pollution increased by 89% to 124% during the same time period, with increases primarily occurring in China and India. This increase in PM2.5 concentration is correlated with this increase in deaths. The concentration of PM2.5 has also dropped in Europe and the US at the same time. A startling 91% of people on the planet reside in regions where PM2.5 levels are higher than what the World Health Organization considers to be safe air. Around 4 billion people reside in regions where the WHO's least stringent target of 35 μg/m³ for mean annual concentrations of PM2.5 is not met. According to the WHO, people who live above an annual average PM2.5 concentration of 10 μg/m³ are more likely to die from lung cancer and cardiovascular problems if they are exposed to pollution over an extended period of time.

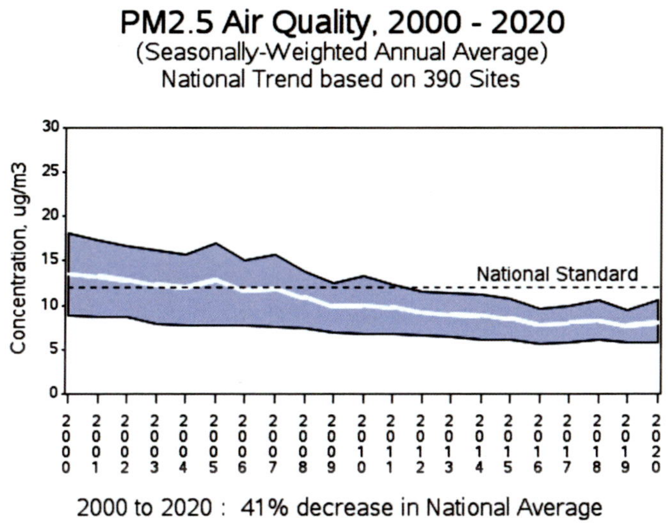


**Figure 5.3.** *Over the course of 2000 to 2020, the annual average levels of PM2.5 have decreased and now sit below the national standard for this pollution under NAAQS. This 20 year period reflects a 41% decrease in PM2.5 levels when looking at the United States as a whole. Despite these gains on the national level, certain regions and communities within the country still experience heightened levels of PM2.5 exposure, especially during certain pollution-emitting events such as wildfires.*

*Source:* https://www.clarity.io/blog/air-quality-measurements-series-particulate-matter.

By this measure, more than 90% of people on the planet are exposed to contaminated air. In contrast, there have been marginal drops in PM2.5 levels in the US. These figures, however, only take into consideration the average PM2.5 concentration across the country; they do not take into consideration regional variations or elevated levels within particular communities. Unhealthy levels of PM pollution are still present in many parts of the United States (Figure 5.3).

## 5.2.7. How PM is Measured

The most popular instruments for measuring particulate matter either measure the particle size distribution or the PM concentration. The gravimetric method, which draws air onto a filter where particles can gather, yields the most accurate measurement. After that, the filter is weighed and its original weight prior to the air being drawn in is compared. This method also makes it possible to analyze the particles chemically to find out more about their composition, since they are collected directly. Because of this measurement method's high standard for data accuracy, regulatory bodies like the EU and USEPA use it. It can only provide an average value for the time period it was deployed, and real-time data is not possible due to the procedure used to gather and weigh the filters. Due to the requirement for skilled personnel to operate the instruments, this method also has significant operating costs.

By examining various light characteristics, such as light scattering, light absorption, and light extinction, and how they respond to the presence of particles, optical instruments can also be used to measure particulate matter. The most widely used of these techniques is the optical particle counter (OPC), which measures the amount of light scattered by the particles after they are illuminated by a light source like a laser diode and detected by a photodetector. To improve an OPC's accuracy, it can be verified and calibrated against FRM and FEM instruments, like a USEPA reference-grade monitor. To indicate its concentration in the air, particulate matter is commonly expressed in micrograms per cubic meter ($\mu g/m^3$). This concentration is commonly expressed using the Air Quality Index (AQI) for PM, which categorizes air quality into six groups according to the degree of health risk posed by the current pollution.

## 5.2.8. Use Cases for Measuring PM

Particulate matter is commonly measured and reported, particularly through the PM2.5 AQI, to act as a stand-in for overall air quality. PM can also be used to measure the amount of air pollution linked to specific industries, like mining and construction. Measuring particulate matter can assist in estimating the total amount of air pollutants emitted from construction sites, as these locations often produce high concentrations of air pollutants. Determining the amount of air pollution caused by mining operations can be aided by measuring PM. When vegetation and soil are removed, surface mining releases particulate matter into the atmosphere and exposes the soil to weather that can disperse the particles. PM measurement tools can be used to measure the effects of mining on the general quality of the air. Measuring the amount of particulate matter released by wildfires can help categorize the impact of the fire on the quality of the air and also help people understand the danger that the air poses at any given time. PM is the primary pollutant that accounts for the health risks posed by wildfire smoke.

## 5.3. GASEOUS POLLUTANT COLLECTION

Gases can be collected by condensation, adsorption, or absorption—in an evacuated container using grab sampling. The physical characteristics of the gas of interest will determine the best approach. In the analysis conducted using wet chemical methods, absorption plays a crucial role. The most modern and affordable methods available before continuous monitoring instrumentation was developed were those that used absorption. Due to the wide range of techniques for analyzing the resultant solution, atmospheric sampling has made considerable use of the absorption of gaseous pollutants. The process of drawing gases to a boundary surface, where they are concentrated, is known as adsorption. Adsorption techniques are widely employed in air pollution monitoring to gather a particular gas or combination of gases. Usually, the procedure involves extracting a gas by passing it through a container that is filled with an adsorbent, like silica gel, activated carbon, or a molecular sieve. After being collected, the gas can be chemically treated, heated, or passed through a system containing inert carrier gases in order to remove it from the adsorbent for analysis. Gases can be collected using static or passive sampling techniques such as diffusion onto a collection medium, large particle settling into a container, or impaction of particles on sticky paper. The benefits of simplicity and affordability make static sampling a popular choice. However, the usefulness of static sampling data is limited due to long averaging times (up to 30 days) and semi-quantitative methodology.

Non-condensable gases can also be collected using evacuated glass, stainless steel, or other non-reactive evacuated containers for grab sampling; the methods of collection vary depending on the particular gas properties involved. Condensation and cryogenic trapping are further methods for gathering gases. This technique basically involves forcing the air stream through a series of progressively colder collection chambers. The gaseous components will condense to a liquid phase if the temperatures inside the chambers are roughly equal to or lower than the boiling point of the air inside or the gaseous components themselves. The designated EPA Federal Reference Methods (FRM) for sampling and analysis of pollutants measured from the ambient air for Criteria Pollutants (i.e., Particulate Matter – PM2.5 and PM10 – Lead (Pb), Sulfur Dioxide ($SO_2$), Nitrogen Dioxide ($NO_2$), Ozone ($O_3$), and Carbon Monoxide (CO), can be found in 40 CFR part 50 Appendices A through M. These methods outline the measurements that must be taken in order to determine compliance with the primary and secondary National Ambient Air Quality Standards.

Equivalent methods that have been designated to measure equivalency to the reference method may be used in addition to reference methods. The designation of an equivalent method can be done automatically or manually. When both methods are applied to the measurement of pollutant concentrations in an actual atmosphere, the manual method must show a consistent relationship to the reference method in order to be considered for equivalency. Each instrument needs to meet certain performance requirements in order to be used with automated methods. When it comes to tracking gaseous pollutants in the atmosphere, automated or instrumental techniques have taken center stage. The ability to transfer data directly into computer systems, increased sensitivity in fulfilling the needs of particular applications, and real-time data output are some of the main benefits of instrumental methods over manual ones. Chemiluminescence, nondispersive infrared, and ultraviolet florescence are a few of the instrument techniques currently used for ambient air monitoring. An additional crucial factor in ambient air monitoring is the average duration of the pollutants sampled. The amount of time needed to gather a representative sample and the intended use of the data are usually what dictate the average times used to calculate pollutant concentrations. 24-hour averages, for instance, would be suitable for long-term trends in air quality. For pollutants like $O_3$, which only reach their peak levels for brief periods of time, an 8- or 1-hour averaging time would be preferable. Continuous or real-time air monitoring devices offer an ongoing log of concentrations that are subject to change over time. Therefore, hourly average concentrations are provided for comparison to respective air quality standards through the integration of data from continuous air monitoring instruments.

### 5.3.1. General Characteristic of Instruments for Air Monitoring

The earliest manual methods for measuring air quality were quite basic. A sample was frequently taken at a location and then brought back to a lab for examination. As a result of these measurements' great reliability and relatively low cost, this is actually still a common practice today. Since the 70th century, automatic analyzers have been incorporated into monitoring networks. These offered the advantage of cheap labor costs and continuous, highly resolved measurements. As a result of these ongoing measurements becoming necessary for regulatory purposes, urban monitoring networks were set up in numerous cities to keep an eye on adherence to the newly emerging EC Directive limit values for air quality. The networks were then extended as a result of promises made by various governments to increase urban monitoring in specific nations and enhance the accessibility of air quality data for the general public. High measurement sensitivity is a fundamental requirement for analyzers used in air monitoring, i.e., low limits for quantitation (LOQ) and detection (LOD). It provides an opportunity to detect pollutants at the necessary concentrations. The following extra features must be included in instruments that are intended to obtain analytical data either instantly or very slowly:

- provide high resolution data (characterized by a low response time),
- provide automatic calibration and zeroing,
- extended operation without service or maintenance.

Measurement methods based on physical (or physicochemical) principles have become increasingly common in air quality assessments in recent years. These techniques entail determining a pollutant's physical attribute directly, sometimes following its interaction with another substance. Improved stability, sensitivity, and dependability may be attained with ease using this method. In addition, the practical application needs less upkeep. This principle allows for the easy automation of instruments, making them useful for continuous measurements required for current air quality assessments. It is particularly pertinent to environmental monitoring since many current standards make reference to a given time frame, i.e., One hour, two days, or a year. Instruments can be stationary or on-site, depending on where measurements are taken. In the first instance, analysis is carried out in a lab setting using advanced equipment like X-ray fluorescence spectrometers, mass, and electron mobility meters. Pollution levels can be measured in the field thanks to on-site systems. The devices, which are typically simple, reasonably priced, and portable, have a lot of potential for use in remote areas since access to an advanced laboratory is not necessary. The capacity for quick assessment and problem-solving is the primary benefit of on-site analysis.

There are two categories for monitoring systems: mobile and stationary. The majority of currently in use automated and manual systems that track ambient aerosols and gaseous pollutants in air typically carry out stationary measurements, i.e., They have a direct connection to a particular location or area around that location. The assessment of spatial and temporal variations of air pollutants is not feasible with the data collected from a single monitoring site. A continuous-monitoring device that is portable or transportable is referred to as mobile. Typically, they are made to carry out analytical measurements devoid of any preprocessing. A self-contained, battery-operated gadget is referred to as portable when it can be carried or worn by the user, or it can need to be placed in a particular location that needs to be watched over using specialized vehicles. Transportable gas monitors are suitable for mounting on automobiles, aircraft, balloons, ships, and space shuttles, but not on industrial trucks or mining equipment. The registered pollution values for mobile systems need to be correlated with site-specific data and current meteorological conditions (temperature and humidity).

Over the past ten years, a number of publications have demonstrated the benefits of mobile systems for obtaining data on the temporal and spatial distribution of atmospheric trace gases without requiring a dense network of fixed stations. Most of the proposed systems are based on the application of mobile laboratories (Bukowiecki et al., 2002; Gouriou et al., 2004; Pirjola et al., 2004), equipped with appropriate monitors. Pollutants in a stream of vehicles can also be measured using other systems; however, these require the installation of a measuring unit on any vehicle rather than one that is specific to a van. In the field of developing instruments for assessing air quality, the general trend is to integrate multiple instruments into a single system, creating what are known as hybrid multisensor systems. These systems are managed by a microprocessor that can transmit the collected data to a central station, usually through a wireless system. Data is gathered in the central station from both larger areas and individual objects, such as houses and plants. Numerous systems come with components that allow you to diagnose the entire instrument as well as test the sensors. They usually feature alarms that alert the user to any potentially hazardous situation resulting from the breach of a value limit. These battery-operated systems have a prolonged operational lifespan of multiple days or months.

The relatively quiet atmosphere of the laboratory is not the same as the environments in which analyzers are used. Wide variations in the surrounding temperature as well as vibrations must be tolerated by analyzers. As a result, many systems are totally sealed to function regardless of external circumstances and to withstand the assault of monkey-wrench mechanics.

### 5.3.2. Mobile System for On-road Measurements of Air Pollutants

The initial monitoring system prototype is made up of original equipment that measures atmospheric air gas pollution in motion. This equipment is also referred to as ARPOL (abbreviated from ARMAAG, Agency of Regional Monitoring of Atmosphere of Gdańsk Agglomeration and Gdańsk POLytechnic), and it is intended to test atmospheric air pollution with the following chemical compounds: $C_6H_6$, $NO_2$, NOx, CO, and $CO_2$. Any mobile vehicle, such as a bus or taxicab, can have this equipment installed. It communicates the results of airborne concentration measurements of the substances mentioned above to a server that functions as a database. The server also receives data on the vehicle's location and speed, as well as the temperature and humidity of the surrounding air. A radio modem that operates in GSM/GPRS systems is used to transmit data between the equipment and the server.

In the laboratory of Gdańsk University of Technology's Department of Radiocommunication Systems and Networks, an ARPOL monitoring station has been designed and constructed. Researchers from the ARMAAG Foundation and the Chemical Faculty at Gdańsk University of Technology have chosen the collection of commercially available gas concentration sensors in air. The system is built in accordance with the specifications of the sensors and the intended use. Using the most recent generation of semiconductor (thick-film) sensors, which are composed of nanostructures with a grain-size distribution ranging from 30 to 50 nm, has been determined as a consequence of the decisions made. After preliminary investigations the following sensor (all from Figaro Engineering Inc from Osaka, Japan) were chosen: TGS2442 for CO, TGS4161 for $CO_2$, TGS823 for benzene, TGS2106 for $NO_2$ and TGS2201 for $NO_x$. Some of these sensors were designed for measurement of gases in ppm (parts per million, $10^{-6}$) concentration, but the concentration of $NO_x$, CO or $C_6H_6$ in typical air condition is of the order of ppb (parts per billion, $10^{-9}$). Nonetheless, a comparison of the data from several months' worth of measurements demonstrated that, following accurate calibration and compensation of humidity and air temperature, they were sufficiently stable and accurate to identify the regions and sources of the highest levels of road traffic-related pollution. Internal heaters are used to heat each gas concentration sensor, which serves as a substrate for structural sensing. Because there is no temperature sensor integrated into the substrate of the gas-sensing element, the substrate's temperature cannot be regulated. By measuring the air temperature using an additional semiconductor sensor and the appropriate calibration equations, the substrate temperature change caused by the change in air temperature surrounding the sensor is corrected. It can take several tens of minutes for gas concentration sensors to heat up properly. Measurement results obtained during this time frame cannot be regarded as reliable. Moreover, a power outage that causes results to become unstable for a variable amount of time—up to several minutes—means that a monitoring device equipped with these sensors must continuously heat the sensors throughout the duration of the measurement campaign. Measurements with high sensitivity and accuracy are guaranteed by special electronic circuits. In the design of ARPOL a microcontroller of the AtMega series (supplier: Atmel Corporation from San Jose, USA) with built-in analog-to-digital converters has been used; the microcontroller circuit has been connected to a ready-made GSM/GPRS modem board (supplier: Enfora from Richardson, USA) and a GPS signal receiver (supplier: Trimble Navigation Ltd. from Sunnyvale, USA), as well as a series RS232 interface (RS232 signal converter based on Maxim chip, supplier: Maxim Integrated Products Inc. from Sunnyvale, USA) and the required additional supply circuits. Additionally, a connector for attaching a board with potential extra sensors not included in the original design as well as transistor circuits and actuators that enable the application of control signals have been added to the main board.

The entire electronic circuitry of the main ARPOL component will be assembled on a single printed board in order to maximize the equipment's immunity to shocks. It has been decided to place all chemical sensors, as well as external temperature and air humidity sensors, at one edge of the main board due to the possibility of heat buildup from some equipment devices (power supply circuits in particular). This will allow the microprocessor part with power supplies to be separated from the sensors in the housing by an appropriate dividing partition. Enclosure of the equipment in a housing that allows atmospheric air to reach the temperature, humidity, and gas concentration sensors while protecting the system from the effects of precipitation. The housing enables the device to be stationary at specific locations throughout the town and mounted in an automobile. A portion of the sensor-equipped electronic board has been covered with a plastic channel that ends in an air intake and outlet and has extra "roofs" to keep the rain out. The device features a forced sampling system opening in the direction of driving to lessen the impact of engine exhaust gas. A microcontroller registers an alarm state when there is a blower failure or reduction in efficiency that results in a decrease in air flow in the measurement channel. This is done to ensure that there is a uniform air exchange in the channel with sensors.

The mobile monitoring system (MMS) for measuring and analyzing atmospheric air pollution along road traffic arteries is described schematically in Figure 5.4. This system is made up of an ARPOL unit that is mounted on a car and a WWW server that has a database set up for the gathering and processing of measurement data. The values of concentration of chemical compounds polluting the atmosphere are transmitted through radio, through the mobile GSM/GPRS network and the Internet network to a WWW server located anywhere, e.g., at Gdańsk University of Technology (Gdańsk Polytechnic). These data are processed there, and their veracity is checked. They are then arranged by measurement date and, if required, shown by a standard WWW page browser on an authorized user's monitor.

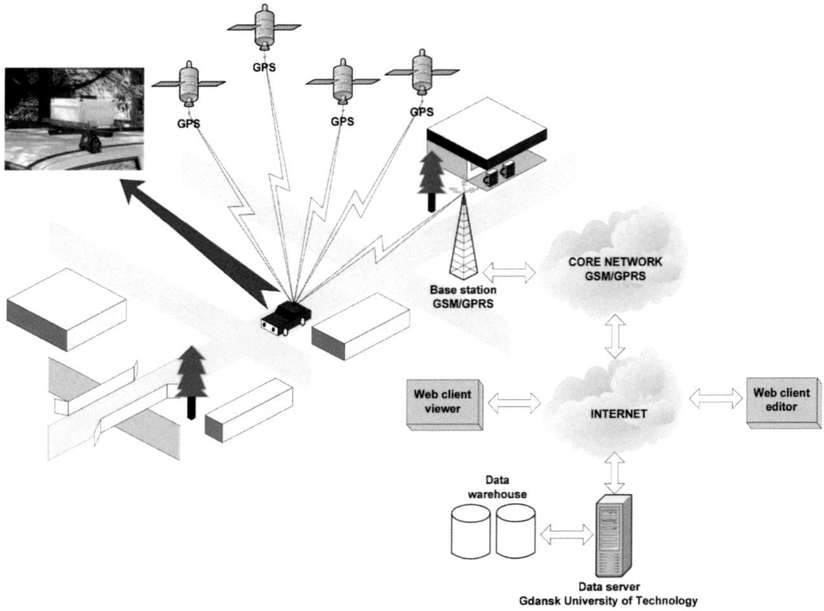


**Figure 5.4.** *The concept of the mobile monitoring system for chemical agents control in the air.*

*Source: J. R, Namiesnik J, Sadowski J, Stefanski J, Wardencki W. Monitoring of Gaseous Air Pollution [Internet]. The Impact of Air Pollution on Health, Economy, Environment and Agricultural Sources. InTech; 2011. Available from:* http://dx.doi.org/10.5772/20771.

The WWW server and ARPOL unit configuration is kept to a minimum in order to simplify the monitoring system's operation. This implies that a variety of servers can be used to implement this system. One could characterize the suggested monitoring unit as maintenance-free. A monitoring station that has been properly configured and set up can run independently for several months. Short messages are used to communicate to the appropriate person details regarding the condition of a device, potential malfunctions in the operation of electronic components, or issues with data collection (SMS).

## 5.3.3. Sensors Description

Many air pollutants are released into the atmosphere by road traffic, but the most significant ones are nitrogen oxides (NO and $NO_2$, together known as NOx), which can cause lung damage and asthma. Other air pollutants include sulfur dioxide ($SO_2$), particulate matter (PM), carbon monoxide (CO), carbon dioxide ($CO_2$), and volatile organic compounds (VOCs), which can also form secondary pollutants with varying degrees of efficiency. Particulate matter in the air can only be measured by mass and air filtration over a predetermined period of time. This can't be done with semiconductor sensors, and it doesn't work with mobile monitoring devices. Thick film semiconductor sensors can be used to measure additional air pollutants. Monitoring unit prototypes for the suggested systems make use of particular Figaro sensors.

- **TGS2442:** CO sensor, response: resistance of the $SnO_2$ layer on the layer of electrical insulation. Time of response: a few seconds.
- **TGS4161:** $CO_2$ sensor, response: electromotive force, which is produced by the electrochemical reaction between solid cation ($Na^+$) electrolyte and gold and lithum carbonate electrodes. Response time is one minute, and recovery time is two and a half minutes.
- **TGS823:** VOC sensor. Several volatile organic compounds, including benzene ($C_6H_6$), ethanol, n-hexane, acetone, isobutane, and others, can cause this sensor to react. However, it is primarily used to detect benzene because it is the compound with the highest concentration. Response: $SnO_2$ layer resistance, tens of seconds for response.
- **TGS2106:** $NO_2$ sensor, response: resistance of the semiconductor metal-oxide layer. Reaction time: a few seconds.
- **TGS2201:** Only $NO_x$ is measured using the double CO and $NO_x$ sensor; the metal-oxide semiconductor layer's resistance is the response. Time of response: a few seconds.

While the concentration of $NO_x$, CO, or $C_6H_6$ in typical air conditions is of the order of ppbv (parts per billion, $10^{-9}$), some of these sensors were designed for measurement of gases in ppmv (parts per million, $10^{-6}$). Comparing data from several months of measurements, however, showed that air temperature and humidity, after correct calibration and compensation, are sufficiently stable and accurate to identify the sources and areas with the highest levels of road traffic-related pollution.

## 5.3.4. Air Monitoring Unit

Block diagram of mobile monitoring unit is presented in Figure 5.5. This version of the measurement device converts the electric response of the sensor to gas concentrations using a microcontroller, A/D converters, and sensor unit. Along with linking measurement results to GPS coordinates, the microcontroller is also in charge of using a GSM/GPRS link to transmit these data to a server. The interior of one of monitoring units built in Gdansk University of Technology laboratory is shown in Figure 5.6.

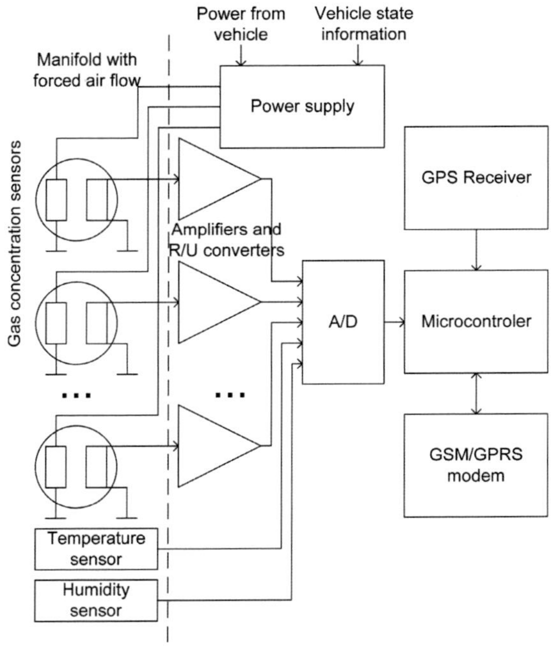


**Figure 5.5.** *Block diagram of mobile monitoring unit.*

*Source: J. R, Namiesnik J, Sadowski J, Stefanski J, Wardencki W. Monitoring of Gaseous Air Pollution [Internet]. The Impact of Air Pollution on Health, Economy, Environment and Agricultural Sources. InTech; 2011. Available from:* http://dx.doi.org/10.5772/20771.

The temperature, humidity, and gas concentration sensors are installed inside the manifold (top of Figure 5.6). Because air is cooling the sensor casings, a change in flow rate will affect the substrate temperature of the gas sensors, forcing air flow inside the manifold through a fan whose rotation speed is controlled by a microprocessor.

Gas sensors' electrical responses are transformed into voltage signals either by resistive voltage dividers for resistive output or by very high impedance operational amplifiers for EMF output. The signals are then passing through low-pass filters with a cutoff frequency of roughly 10 Hz for all sensors, with the exception of TGS2442, which needs to measure pulsed resistance, so the filter cutoff frequency needs to be higher than 100 Hz. A microprocessor that averages digital values following A/D conversion over a 10-second period performs additional noise-filtering.

**Figure 5.6.** *Mobile monitoring unit prototype.*

*Source: J. R, Namiesnik J, Sadowski J, Stefanski J, Wardencki W. Monitoring of Gaseous Air Pollution [Internet]. The Impact of Air Pollution on Health, Economy, Environment and Agricultural Sources. InTech; 2011. Available from:* http://dx.doi.org/10.5772/20771.

The average digital values are sent directly to the server, requiring no further conversion. The server uses nonlinear equations that are generated during the calibration process to calculate the gas concentration. Within each packet, additional data is sent to the server.

- monitoring unit identification number;
- geographical coordinates of measurement point (from GPS receiver);
- speed and direction of movement (from GPS receiver);
- date/time of measurement (from GPS receiver);
- temperature and humidity of air in manifold;
- temperature of electronic circuits outside manifold;
- external or internal power supply voltage;
- state of vehicle engine in case of mounting the device on car/bus;
- error flags which indicate condition of monitoring unit.

TCP/IP protocol stack and GPRS/EDGE radio link are used for data transmission. Data records are encoded as unique HTTP requests in order to prevent the need to write special software on the server side. PHP scripts can easily handle these kinds of requests, and once the raw data has been decoded, records are written to MySQL databases. Air pollution has a detrimental effect on health and can even be fatal in certain situations. As a result, many nations are now very concerned about the issue of air quality and are trying to control emissions of harmful substances into the atmosphere, enhance fuel quality, and incorporate environmental protection regulations into the energy and transportation sectors. Even though the quality of the air has improved recently, air pollution is still an issue. As a result, more work needs to be done on a local, national, European, and worldwide scale. Air quality control requires the monitoring of air pollution, which is accomplished through a range of analytical techniques using various measurement instruments with varying sensitivities and specificities. Monitoring is essential to managing and safeguarding ambient air quality, and it is a fundamental component of all environmental protection initiatives.

Member states of the European Union are required by law to measure the cleanliness of the air in their atmosphere and to implement preventive measures against excessive pollution that surpasses permissible threshold limits. In order to fulfill these responsibilities, the member nations must create inventories of the zones and agglomerates where limiting norms are broken as well as develop and implement plans and programs that explain how to follow the legal restrictions. As a result, the mobile monitoring system referred to as ARPOL—which is exceptional on a national level—can improve the current atmospheric air cleanliness measurement system and make it easier to make the right decisions by analyzing different emission scenarios. The devised apparatus is a component of an expert system that surely enhances ecological safety in a particular area. Rather than being used to evaluate air quality, it might be a cheap and practical tool (given the low investment and exploitation costs) for evaluating the ecological effects of new initiatives to limit pollution emissions. By adding data from non-stationary sources to the measurement basis, it helps to promote more meticulous environmental planning.

## 5.4. HAZARDOUS AIR POLLUTANTS (HAPS)

Under the National Emissions Standards for Hazardous Air Pollutants (NESHAPs) or Hazardous Air Pollutants (HAPs) regulations, hazardous substances that are not formally designated as criteria pollutants are classified as non-criteria pollutants. Pollutants that are known to cause cancer or other major health effects in humans or that are suspected of causing harm to ecosystems are referred to as hazardous or toxic air pollutants. Countless stationary industrial facilities, major point sources (like electric power plants or utilities), smaller area sources (like dry cleaners), and mobile sources (like cars) all release harmful air pollutants into the atmosphere. Congress amended the Clean Air Act of 1990 in response, realizing the possible health risk posed by these pollutants. A new strategy for regulating hazardous air pollutants (HAPs) was established by these amendments. Congress specifically designated 189 compounds as hazardous air pollutants when it revised the Clean Air Act. The U.S. EPA has been directed to develop technologically based emissions standards for all these pollutants. All pollutants that are known or believed to cause cancer and other harmful health effects are included in this list.

The CAA Amendments classified pollutants and designated 189 substances that needed to be regulated. Volatile organic compounds make up the majority of the 188 HAPs on this list, which also includes compounds like carbon tetrachloride, chlorine, ethylene oxide, cadmium, and manganese. There are two stages to the new strategy for controlling HAPs. For stationary sources of the 188 HAPs, the first mandate calls for the creation of technology-based emissions standards. The assessment of lingering issues or hazards and the necessary development of new regulations to address the sources of those issues constitute the second phase. The National Air Toxics Program will move away from a technologically-based approach that is primarily focused on regulation and toward a risk-based approach. More and better data regarding ambient HAP levels, HAP emission sources, and the exposure of humans and the environment to HAPs will be needed for this. The National Air Toxics Program's ability to measure the effects of air toxics emissions on the environment and public health will be crucial to its success. In order to help identify areas of concern, characterize risks, and track progress, the EPA has launched a number of National Air Toxics Assessment (NATA) activities. Monitoring the ambient air is yet another crucial NATA component. The characterization of ambient concentrations and deposition in representative areas, establishment of trends, assessment of the efficacy of HAP reduction strategies, and provision of data to support and evaluate dispersion and deposition models are all made possible by ambient measurements. Currently, ambient data on hazardous air pollutants are being produced by about 300 monitoring sites.

In order to create a monitoring network that is indicative of issues with air toxics on a national level and offers a way to gather data on a more localized basis, the EPA is collaborating with state and local air monitoring organizations. Eventually, the network will be an amalgam of data from numerous monitoring programs, such as rural IMPROVE networks that offer data on HAP trace metals, the new urban PM2.5 chemical speciation, and PAMS, which offers data on VOCs and aldehydes. Long-running air toxics monitoring programs in several states yield measurements on a wide range of HAPs, including significant urban HAPs. The Urban Air Toxics Monitoring Program (UATMP), which is funded by an EPA contractor, and the Non-methane Organic Compound (NMOC) and Speciated Non-methane Organic Compound (SNMOC) monitoring program both support some of these state initiatives. Thirteen carbonyl compounds and 39 HAP volatile organic compounds are measured as part of the UATMP's toxics monitoring program. Among the numerous pollutants that act as ozone precursors, PAMS also measures HAPs. The PAMS program mandates the regular, year-round measurement of volatile organic compounds (VOCs), comprising nine HAPs: 2,2,4-trimethylpentane, m/p-xylene, o-xylene, n-hexane, styrene, toluene, and xylene. The variety of organic substances of potential concern and the lack of standardized and documented methods make it often difficult to measure organic pollutants, which include many HAPs, in ambient air. Assessing and reducing the effects of hazardous air pollutants (HAPs) on the environment and public health depend heavily on monitoring these pollutants. These are important things to remember:

1. **Identification of HAPs**: Pollutants known or suspected of causing cancer or other major health effects, like damage to the nervous system or reproduction, are known as HAPs. Benzene, formaldehyde, and mercury are a few examples.
2. **Monitoring Methods**:
   - **Direct Measurement**: Particle counters, spectrometers, and gas chromatographs are a few examples of instruments that can measure particular air pollutants directly.
   - **Sampling**: To measure pollution concentrations, air samples can be gathered on filters or sorbents and examined in labs.
   - **Remote Sensing**: Technologies that can monitor large areas for specific pollutants include satellite imagery and remote sensors based on the ground.
3. **Regulatory Framework**: Standards and regulations for HAPs monitoring are in place in many nations. For example, the Clean Air Act in the United States establishes emission standards and mandates compliance monitoring.
4. **Health and Environmental Impacts**: Depending on exposure duration and concentration, HAPs can have both short- and long-term health effects. Targeted interventions are made possible by monitoring, which assists in identifying exposure sources and trends.
5. **Community and Environmental Justice**: Monitoring contributes to the protection of communities from disproportionate exposure to HAPs, especially those that are disadvantaged or vulnerable.
6. **Data Analysis and Reporting**: In order to determine trends,

pinpoint pollution hotspots, and gauge how well pollution control strategies are working, monitoring data is analyzed. It is imperative for transparency and accountability that this information be reported to the public and policymakers.

7. **Emerging Technologies**: The precision and effectiveness of HAPs monitoring are being improved by developments in sensor technology and data analytics, which is strengthening our understanding of and capacity to reduce air pollution.

In summary, maintaining regulatory compliance, advancing environmental justice, and protecting public health all depend on efficient hazardous air pollution monitoring. In order to effectively address the challenges presented by HAPs in our atmosphere, it is imperative that monitoring technologies and procedures constantly improve.

## 5.5. CLEAN AIR ACT

The main federal law governing air quality in the United States is the Clean Air Act (CAA), which aims to lower and regulate air pollution on a national level. It is among the first and most significant modern environmental laws in the United States, having been passed in 1963 and revised numerous times thereafter. As with many other major U.S. federal environmental statutes, the Clean Air Act is administered by the U.S. Environmental Protection Agency (EPA), in coordination with state, local, and tribal governments. The EPA creates comprehensive administrative regulations to implement the requirements of the law. These regulations are carried out by related regulatory programs, which are frequently intricate and technical. The National Ambient Air Quality Standards program, which establishes limits for specific pollutant concentrations in outdoor air, and the National Emissions Standards for Hazardous Air Pollutants program, which establishes limits for specific hazardous pollutant emissions from specific sources, are two of the most significant. Requirements are set by other programs for industrial facilities, car fuels, and other technologies and activities that affect air quality. More recent initiatives focus on particular issues like climate change, ozone layer preservation, and acid rain. The California Air Quality Authority (CAA) has faced numerous legal challenges from environmental organizations requesting stricter enforcement as well as from states and utilities requesting more regulatory flexibility. The Clean Air Act has significantly decreased air pollution and improved US air quality, though the precise benefits depend on what is counted. The EPA attributes these improvements to the Act, which has the potential to save trillions of dollars and countless lives annually.

## 5.5.1. History

The United States saw a sharp rise in air pollution between the Second Industrial Revolution and the 1960s. In the wake of the 1948 Donora smog incident, states passed a number of laws aimed at reducing air pollution, the public started to discuss air pollution as a serious issue, and Congress started debating whether or not to act further. The United States Bureau of Mines, which was interested in smoke abatement (reducing smoke from burning coal), and the United States Public Health Service, which handled industrial hygiene and was worried about the causes of lung health issues, were at the time the main federal agencies interested in air pollution. The first federal legislation addressing air pollution was passed by Congress in 1955, following several years of hearings and proposals. A research and training program was authorized by the Air Pollution Control Act of 1955, which sent $3 million annually to the U.S. For five years, the Public Health Service did not directly control the sources of pollution. The research program under the 1955 Act was continued in 1959, 1960, and 1962 while Congress deliberated whether to enact additional regulations.

Congress started enlarging federal air pollution control laws in 1963 in an effort to hasten the nation's complete eradication of air pollution. The new law's programs were initially administered by the U.S. Secretary of Health, Education, and Welfare, and the Air Pollution Office of the U.S. Public Health Service, until they were transferred to the newly created EPA immediately before major amendments in 1970. Since then, the Clean Air Act has been implemented by the EPA, and in 1977 and 1990, Congress added significant regulatory programs. Most recently, the U.S. Supreme Court's ruling in *Massachusetts v. EPA* resulted in an expansion of EPA's CAA regulatory activities to cover greenhouse gases.

### *5.5.1.1. Clean Air Act of 1963 and Early Amendments*

The Clean Air Act of 1963 was the first federal legislation to permit the U.S. federal government to take direct action to control air pollution. It authorized the HEW Secretary to hold conferences and take direct action against interstate air pollution when state action was deemed insufficient. It also extended the 1955 research program and encouraged cooperative state, local, and federal action to reduce air pollution. Appropriated funds totaling $95 million over three years were allocated to support the development of state pollution control programs. First federal vehicle emissions standards were established by the Motor Vehicle Air Pollution Control Act of 1963, which went into effect for 1968 models. These regulations represented reductions from the emissions levels of 1963 by 72% for hydrocarbons, 56% for carbon monoxide, and 100% for hydrocarbons found in crankcases. Additionally, a new section authorizing the reduction of international air pollution was added by the law.

The Air Quality Act of 1967 allowed HEW to define air quality regions, develop technical documentation that would allow states to set ambient air quality and pollution control technology standards, require states to submit implementation plans for improving the quality of the air, and authorize HEW to take direct abatement action in air pollution emergencies. It also permitted the creation of interstate air pollution control agencies and authorized planning grants to state agencies responsible for controlling air pollution. Additionally, it approved the continuation of research on ambient monitoring methods, control strategies, and inventories of air pollutant emissions. This made it possible for the federal government to look into enforcing interstate air pollution transport laws and to conduct extensive ambient monitoring studies and stationary source inspections for the first time. The 1967 act also approved more research on ambient monitoring methods, control strategies, and inventories of air pollutant emissions. Although there were only six states with air pollution programs in 1960, the federal funding and legislation of the 1960s caused all 50 states to have air pollution programs by 1970.

### *5.5.1.2. 1970 and 1977. Amendments*

Congress significantly increased the federal mandate by imposing extensive federal and state regulations for both industrial and mobile sources in the Clean Air Amendments of 1970. In order to achieve aggressive air pollution reduction goals, the law established the National Ambient Air Quality Standards (NAAQS), New Source Performance Standards (NSPS), and National Emissions Standards for Hazardous Air Pollutants (NESHAPs). It also significantly strengthened federal enforcement authority. During his first term, EPA Administrator William Ruckelshaus focused 60% of his time to enact the strict amendments, which required the automobile industry to reduce its emissions by 90%. Senators had grown impatient with the industry for failing to reduce emissions under earlier, laxer air regulations.

In 1977, significant modifications were made to the Clean Air Act (1977 CAAA). Provisions for the Prevention of Significant Deterioration (PSD) of air quality in areas meeting the NAAQS were the main focus of the 1977 Amendments. Requirements concerning sources in non-attainment areas for NAAQS were also included in the 1977 CAAA. A geographical area that does not satisfy one or more federal air quality standards is known as a non-attainment area. Major permit review requirements were established by both of these 1977 CAAAs to guarantee the achievement and upkeep of the NAAQS. A policy of offset trading, which was first implemented in Los Angeles in 1974, was also adopted as part of these amendments. This policy permits new sources to purchase additional reductions from existing sources in order to offset their emissions. The 1977 Clean Air Act Amendments mandated additional standards for non-attainment areas and Prevention of Significant Deterioration (PSD) of air quality for areas meeting the NAAQS.

### *5.5.1.3. 1990 Amendments*

In 1990, the Clean Air Act underwent yet another significant set of changes (1990 CAAA). The 1990 CAAA significantly expanded the federal government's power and jurisdiction. Authorization was given to new regulatory initiatives for the issuance of stationary source operating permits and the management of acid rain. A significantly enlarged program for limiting toxic air pollutants included the NESHAPs. The NAAQS attainment and maintenance requirements were greatly expanded and changed. Provisions for protecting the stratosphere's ozone layer, more enforcement power, and bigger research initiatives were among the other changes.

Regulations pertaining to stationary source operating permits and acid deposition (acid rain) control were introduced by the 1990 Clean Air Act. A cap-and-trade program was one of the measures intended to reduce sulfur dioxide emissions; in contrast to previous Clean Air Act versions, this program allowed power companies greater latitude in fulfilling the law's objectives. The amendments expanded the NESHAP program with a list of 189 hazardous air pollutants to be controlled within hundreds of source categories, in accordance with a specific schedule, going far beyond the original criteria pollutants. Also, the NAAQS program was made larger. Additional new provisions addressed research program expansion, enhanced enforcement power, and stratospheric ozone protection. In 1990, more changes were made to address the issues of toxic air pollution, acid rain, and ozone depletion. They also established a national permit program for stationary sources and strengthened enforcement powers. The amendments also established new auto gasoline reformulation requirements, set Reid vapor pressure (RVP) standards to control evaporative emissions from gasoline, and mandated new gasoline formulations sold from May to September in many states. Reviewing his tenure as EPA Administrator under President George H. W. Bush, William K. Reilly characterized passage of the 1990 amendments to the Clean Air Act as his most notable accomplishment.

### *5.5.1.4. 2022 Amendments*

Signed by President Joe Biden in August 2022, the Inflation Reduction Act amends the Clean Air Act to permit the EPA to oversee $27 billion in grants to green banks across the country through a competitive funding mechanism to be named the Greenhouse Gas Reduction Fund. For the $14 billion National Clean

Investment Fund, the recipients are the consumer-focused Climate United Fund ($6.97 billion to a consortium of Calvert Impact, Self-Help Ventures Fund and Community Preservation Corporation), the Coalition for Green Capital ($5 billion), and Power Forward Communities ($2 billion to a consortium of Enterprise Community Partners, Local Initiatives Support Corporation, United Way, Habitat for Humanity and Rewiring America); collectively they have pledged 60% of funds would go to low-income and marginalized communities, well above the 40% required by Biden. Together, they have pledged that 60% of the funds will go to low-income and marginalized communities, well above the 40% required by Biden. Four CDFI Intermediaries—Opportunity Finance Network, Inclusive, Native CDFI Network, and Appalachian Community Capital—will receive a combined total of approximately $5.1 billion from the $6 billion Clean Communities Investment Accelerator program to distribute funds exclusively and deeply into these communities. A coalition of community organizations known as the Justice Climate Fund will receive $940 million. The Biden administration estimates that through a wide range of initiatives, they will reduce emissions by up to 40 million metric tons by 2032 and leverage $7 from the private sector for every dollar of public investment. By the time the IRA expires, the remaining $7 billion will fund 60 distributed generation projects in low-income neighborhoods, which are expected to save 900,000 households $8 billion in energy costs.

The government anticipates receiving high rates of return from the Act on investments made by the private sector from the Fund. The Act also designates carbon dioxide and other greenhouse gases as substances to be regulated by the EPA, in reaction to the Supreme Court case *West Virginia v. EPA*, which limited the EPA's authority to institute a program such as the Obama-era Clean Power Plan. The EPA is also given more freedom to promote renewable energy thanks to the IRA.

## 5.5.2. Regulatory Programs

The following are the principal Clean Air Act regulatory initiatives as of right now.

### *5.5.2.1. National Ambient Air Quality Standards*

The amount of ground-level ozone ($O_3$), carbon monoxide (CO), lead (Pb), sulfur dioxide ($SO_2$), nitrogen dioxide ($NO_2$), particulate matter ($PM_{10}$, $PM_{2.5}$), and other air pollutants that are permitted is regulated by the National Ambient Air Quality Standards (NAAQS). The permissible limits of a number of air pollutants in the ambient air in the US are established by the NAAQS. The federal government did not have primary responsibility for developing ambient air quality standards prior to 1970, and there was no national program for their development prior to 1965.

In order to establish NAAQS and air quality standards for the air pollutants that most threatened public health and welfare, the EPA was mandated by the 1970 CAA amendments to identify these pollutants. Primary NAAQS refers to the health-based standards, whereas standards designed to safeguard public welfare other than health (e.g., agricultural values) were referred to as NAAQS secondary. The EPA published regulations for sulfur oxides, nitrogen dioxide, hydrocarbons, carbon monoxide, sulfur dioxide particles, and photochemical oxidants in 1971 (36 FR 22384). Lead was initially controlled by mobile source authorities rather than being listed as a criteria pollutant by the EPA. However, following the Natural Resources Defense Council's (NRDC) successful lawsuit in 1976, the EPA was forced to do so (43 FR 46258). The 1977 CAA Amendments created a process for regular review of the NAAQS list, and created a permanent independent scientific review committee to provide technical input on the NAAQS to EPA. EPA added regulations for PM2.5 in 1997 (62 FR 38652), and updates the NAAQS from time to time based on emerging environmental and health science.

### *5.5.2.2. National Emissions Standards for Hazardous Air Pollutants*

The amount of 187 hazardous air pollutants that can be released from industrial facilities and other sources is regulated by the National Emissions Standards for Hazardous Air Pollutants (NESHAPs). Hazardous air pollutants, also known as air toxics or HAPs, are air pollutants that endanger human health and welfare and are not covered by NAAQS. The guidelines used to regulate, minimize, and completely get rid of HAP emissions from stationary sources like industrial buildings are called NESHAPs. Following the creation of a list of HAPs, the 1970 CAA mandated that the EPA create national emissions standards for each of them. The first NESHAPs were standards based on health. The EPA's list was formalized in the 1990 CAA Amendments, which also mandated the development of technology-based standards based on maximum achievable control technology (MACT). Numerous NESHAP regulations have been released by the EPA over the years, creating NESHAPs according to industrial process, category of industry source, and pollutant. For mobile sources (transportation), there are also NESHAPs, but these are mainly managed by the mobile source authorities. The 1990 amendments also established a procedure requiring EPA to evaluate and revise its NESHAPs every eight years, identify any risks that persisted after MACT was applied, and develop any new regulations required to safeguard public health.

### *5.5.2.3. New Source Performance Standards*

The equipment that must be installed in both newly constructed and renovated industrial facilities, as well as the guidelines for identifying which facilities are new, are outlined in the New Source Performance Standards (NSPS). The 1970 CAA mandated that the EPA create standards for newly built and renovated stationary sources, or industrial facilities, utilizing the most effective emission reduction system that the [EPA] determines has been sufficiently demonstrated, while also accounting for the associated costs. The following year, the EPA published its first NSPS rule, which covered nitric and sulfuric acid plants, Portland cement plants, incinerators, and steam generators (36 FR 24876). Numerous NSPS regulations have been released by the EPA since then, mostly categorized by source type. The regulations encourage the use of available pollution control technologies across the industry. However, these standards encourage prolonging the lives of existing facilities because they only apply to new and modified sources.

### *5.5.2.4. Acid Rain Program*

An emissions trading program called the Acid Rain Program (ARP) helps power plants reduce the pollutants that contribute to acid rain. In order to address the problem of acid rain, specifically the emissions of sulfur dioxide ($SO_2$) and nitrogen oxides (NOx) from fossil fuel-powered electric power plants as well as other industrial sources, a new title was created by the 1990 CAA Amendments. The United States' first emissions trading program, known as the Acid Rain Program, capped overall emissions and gradually decreased them through the trading of emissions credits as an alternative to imposing direct emission controls. The program was implemented in two phases. The first phase called for more than 100 power plants with a capacity of more than 100 megawatts to reduce $SO_2$ emissions by 30.5 million tons by January 1995. Facilities surpassing 75 megawatts were given a deadline of January 2000 under the second stage. All of the program's legally mandated objectives have been met.

### *5.5.2.5. Ozone Layer Protection*

The goal of the CAA ozone program is to gradually phase out the use of chemicals that deplete the ozone layer through technological transition. CAA Title VI, added by the 1990 CAA Amendments, mandated regulations regarding the use and production of chemicals that harm Earth's stratospheric ozone layer, in line with the US commitments made in the Montreal Protocol. Programs to phase out ozone-destroying substances, track their import and export, determine exemptions for their continued use, and define practices for their destruction, equipment maintenance and servicing, identification of new alternatives to those still in use, and technician licensing are all carried out by the EPA under Title VI.

### *5.5.2.6. Mobile Source Programs*

regulations governing the pollutants that automobiles' internal combustion engines can emit. Congress has imposed progressively stricter regulations on vehicle engine technology and tailpipe emission reductions since 1965. These days, the law mandates that the EPA create and update regulations on pollutants that could endanger public health from a wide range of motor vehicle classes. These regulations use technology to achieve the maximum amount of emission reduction possible while taking availability, cost, energy, and safety into account.

- **On-Road Vehicles Regulations:** For a number of vehicle classes that operate on roads, the EPA establishes regulations regarding exhaust gases, evaporative emissions, air toxics, refueling vapor recovery, and vehicle inspection and maintenance. SUVs, pickup trucks, passenger cars, minivans, and passenger vans are all covered under the EPA's light-duty vehicle regulations. Big trucks and buses are covered by the regulations for heavy-duty vehicles.
- **Vehicle Testing Program:** When the air pollution testing system for automobiles was first created in 1972, it made use of driving cycles that were intended to mimic Los Angeles rush hour at that time. The EPA published the precise fuel economy results derived from the test up until 1984. The EPA started modifying city (also known as the Urban Dynamometer Driving Schedule, or UDDS) results by 10% and highway (also known as the High Way Fuel Economy Test, or HWFET) results by 22% in 1984. This was done to account for changes in driving conditions that had occurred since 1972 and to improve the correlation between EPA test results and actual driving. In order to further enhance the correlation between fuel economy and emission estimates and actual reports, the EPA proposed in 1996 to update the Federal Testing Procedures to include a new higher-speed test (US06) and an air conditioner-on test (SC03). A 12-year review cycle for the test procedures was established when the revised testing methodology was finalized in December 2006 and intended to be used in vehicles of the 2008 model year.

Drivers can enter real-world fuel economy data into a database on the EPA's fuel economy website and compare it with other drivers' and the original EPA test results through a program called Your MPG, which was introduced in February 2005. The EPA tests a very small number of cars for fuel efficiency. Of the cars the EPA tests, two thirds are chosen at random, and the other third are tested for particular purposes. Driving cycles have been used to estimate how far an electric vehicle can travel on a single charge, despite being developed originally as a benchmark for fossil-fueled cars.

- **Non-road Vehicles Regulations:** Regulation of aircraft emissions was made possible by the 1970 CAA amendments (42 U. S. C. § 7571), and in 1973 the EPA started to regulate. In order to comply with international standards, the EPA finalized its most recent limits on NOx emissions from gas turbine aircraft engines with rated thrusts above 26.7 kiloNewton (3 short ton-force) in 2012. These engines are primarily commercial jet aircraft engines. Since 2010, the EPA has been looking into the possibility of regulating lead in fuels for small aircraft, but it hasn't taken any action yet. The 1990 CAA Amendments expanded the scope of EPA regulation to include marine transport, heavy equipment, small engines powered by diesel (compression-ignition), gas and other fuels (spark-ignition), and locomotives. They also added rules for a nonroad engine program.
- **Voluntary Programs:** To encourage and support a decrease in transportation-related air pollution, EPA has created a number of voluntary programs. These include components of the Clean Diesel Campaign, the Ports Initiative, the SmartWay program (which is targeted at the freight transportation industry), and other initiatives.

### *5.5.2.7. Fuel Controls*

Since 1967, the federal government has controlled the chemical makeup of fuels used in transportation; in 1970, substantial new authority was added to further safeguard public health. One of EPA's earliest actions was the elimination of lead in U.S. gasoline beginning in 1971, a project that has been described as "one of the great public health achievements of the 20th century." The chemical makeup of gasoline, avgas, and diesel fuel in the US is still regulated by the EPA.

### *5.5.2.8. Stationary Source Operating Permits*

A nationwide operating permit program, known as the Title V Program and encompassing thousands of major industrial and commercial sources, was authorized by the 1990 amendments. Large companies had to deal with air pollutants, measure their amount, create a plan to reduce and control them, and report on a regular basis. This reduced the number of facility requirements to one document. Depending on how severe the non-attainment status was in the area, permits were needed for sources emitting as little as 50, 25, or 10 tons of

VOCs annually in non-attainment areas. State and local governments are the ones who issue most permits. The EPA may take over if the state is not keeping a close enough eye on the requirements. By contacting the EPA, members of the public can request to see the permits. The permission needs to be renewed and has a maximum duration of five years.

### *5.5.2.9. Monitoring and Enforcement*

One of the most well-known features of the Clean Air Act is the authority granted to the EPA to enforce the law's numerous regulations, track down infractions, and force regulated entities to comply. The majority of enforcement cases end in settlement, with fines much below the highest statutory limits. Lately, a significant number of the biggest Clean Air Act settlements have been achieved with automakers who were allegedly manipulating the Act's fuel and vehicle requirements (e.g., the Dieselgate scandal of 2015).

### *5.5.2.10. Greenhouse Gas Regulation*

The EPA regulates greenhouse gas (GHG) emissions primarily through the above-discussed programs. EPA began regulating GHG emissions following the 2007 Supreme Court ruling in *Massachusetts v. EPA*, the EPA's subsequent endangerment finding, and development of specific regulations for various sources. The EPA's authority to regulate carbon dioxide emissions was questioned by the court in *West Virginia v. EPA* (2022) but restored by Congress with the Inflation Reduction Act of 2022, which clarified that carbon dioxide is one of the pollutants covered by the Clean Air Act.

GHGs from stationary sources are regulated by Part C of Title I of the Act, and standards for mobile sources have been set in accordance with Section 202 of the CAA. The 2010 and 2012 EPA auto emission standards for greenhouse gas emissions aim to save over $4 billion in oil barrels and $1.17 trillion for consumers by halving emissions from specific vehicles and doubling the fuel efficiency of passenger cars and light-duty trucks by 2025. A two-phase program to lower greenhouse gas emissions from medium- and heavy-duty trucks and buses has also been proposed by the agency. The national greenhouse gas inventory reporting program is also under the control of the EPA. Following the Supreme Court decision in *West Virginia v. EPA*, which ruled that Congress did not grant EPA the authority to require "outside the fence" options for limiting carbon dioxide at power plants, the Inflation Reduction Act of 2022 specifically defined carbon dioxide, hydrofluorocarbons, methane, nitrous oxide, perfluorocarbons, and sulfur hexafluoride as greenhouse gases to be regulated by the EPA, as well as giving the EPA the ability to regulate the inclusion of renewable sources, notably, through a $27 billion green bank, among other methods.

#### 5.5.2.11. Others

The aforementioned programs are typically built upon or intersected by other significant but less fundamental Clean Air Act regulatory initiatives.

- **Risk Assessment**. Many EPA regulatory programs involve risk assessment and management, even though they are not regulatory programs in and of themselves. The EPA has worked to streamline and arrange its numerous risk assessment procedures over time. The Commission on Risk Assessment and Management was established by the 1990 CAA Amendments with the responsibility of developing recommendations for a risk assessment framework. This work has been the foundation for numerous subsequent reports.
- **Visibility and Regional Haze**. The EPA requires states to create plans to increase visibility by lowering pollutants that contribute to haze, and it monitors visibility and air clarity (haze) at 156 protected parks and wilderness areas.
- **Interstate pollution control**. States must regulate emissions that will materially contribute to NAAQS nonattainment or maintenance in a downwind state under the Clean Air Act's good neighbor provision. For many years, EPA has battled to pass regulations that carry out this mandate. Between 2003 and 2005, it created the Clean Air Interstate Rule, which the courts invalidated in 2008. Then, between 2009 and 2011, EPA created the Cross-State Air Pollution Rule, which is still being challenged in court as it is updated.
- **Startup, Shutdown, & Malfunction**. In order to address excess emissions during startup, shutdown, and malfunction periods—when facility emissions may momentarily be significantly higher than standard regulatory limits—the EPA promulgates rules for the states.

### 5.5.3. The Clean Air Act and States

#### *5.5.3.1. State Implementation Plans*

State Implementation Plans (SIPs) had to be created as part of a cooperative federalist program for creating pollution control standards and initiatives, according to the 1963 act. Rather than create a solely national program, the CAA imposes responsibilities on the U.S. states to create plans to implement the Act's requirements. SIP regulations were first published by the EPA in 1971 and 1972. Eight requirements were imposed by the 1970 Amendments on an implementation plan. Even if the Agency determines the plan is not practical, the EPA must approve plans that follow the Senate's three-year mandate for primary air quality standards. In *Union Electric Co. v. Environmental Protection Agency* the Supreme Court considered whether the Agency was required to reject plans that were not technologically or economically feasible. The court decided that states could adopt emission standards stricter than national standards and "force" technology (i.e., require installation of more advanced technologies).

### *5.5.3.2. Non-Attainment Areas*

SIP requirements were added by the 1977 CAA Amendments for nonattainment areas, or areas that had not met the applicable NAAQS. Until all reasonably possible control measures could be implemented, states were obliged to adopt plans that made reasonable additional progress toward attainment in these areas. Significant modifications to SIP requirements in nonattainment areas were included in the 1990 CAA Amendments due to the fact that attainment progress was substantially slower than Congress had initially directed.

### *5.5.3.3. Prevention of Significant Deterioration*

By introducing requirements for the Prevention of Significant Deterioration (PSD), the 1977 CAA Amendments changed the SIP requirements. These regulations safeguard places that have already complied with the NAAQS, especially national parks and wilderness areas. In addition to purifying contaminated air, the PSD provision mandates that SIPs maintain clean, high-quality air. In order to determine whether PSD requirements would be met, the new law also mandated New Source Reviews, or investigations into proposed construction of new polluting facilities.

### *5.5.3.4. Federalism*

The Clean Air Act gave the EPA enforcement powers and mandated that by 1977, states create State Implementation Plans outlining how they would comply with new federal ambient air quality standards. This federal cooperative model is still in use today. Since addressing pollution control issues frequently necessitates a unique understanding of local conditions like geography, industrial activity, transportation, and housing patterns, the law acknowledges that states should take the lead in implementing the Clean Air Act. States cannot, however, impose controls that are less stringent than the federal minimum standards established by the EPA. Every SIP needs to be approved by EPA, and if one is rejected, EPA may continue to enforce the CAA in that state. For instance, a lawsuit and a federal state implementation plan were created for California after the state was unable to comply with the new standards imposed by the 1970 amendments.

In order to support clean air programs, the federal government also provides funding, expert studies, engineering designs, and scientific research to the states.

The law also forbids states from establishing standards that are stricter than those set by the federal government, but it grants California a special exemption because of its history of problems with smog pollution in urban areas. In actuality, the EPA is consulted when California's environmental agencies determine new vehicle emission standards; the most recent approval was granted in 2009. In order to avoid having to develop separate emission systems for different states, the California standard was adopted by twelve other states and established the de facto standard that automakers later accepted. President Donald Trump, however, made an attempt to revoke this waiver in September 2019, claiming that the elimination of the stricter emissions regulations would make cars safer because they had made them too costly. Andrew Wheeler of the EPA added that although the organization values federalism, it is unable to permit any one state to set national standards. Governor Gavin Newsom of California declared his intention to sue the federal government, viewing the action as part of Trump's political vendetta against the state. Twenty-three states, along with the District of Columbia and the cities of New York City and Los Angeles, joined California in a federal lawsuit challenging the administration's decision. The Trump-era rule was overturned by the Biden administration in March 2022, enabling California to reimpose stricter auto emissions regulations.

### 5.5.4. Effects

The Journal of Economic Literature published a review study in 2022 that found strong causal evidence that the CAA enhanced air quality (Aldy Joseph, Auffhammer Maximilian, & Morgenstern Richard 2022); Cropper, Maureen; Fraas, Arthur. In contrast to the baseline of the 1970 and 1977 regulatory programs, the most recent study conducted by the EPA estimates that by 2020, the United States would lose roughly $60 billion annually due to the updates brought about by the 1990 Clean Air Act Amendments, while reaping about $2 trillion in benefits (in the form of monetized health and lives saved). A 2020 study conducted for the Natural Resources Defense Council projected net economic benefits of up to $3.8 trillion (32 times the cost of the regulations) and 370,000 avoided premature deaths and 189,000 fewer hospital admissions annually. Similar findings have been drawn by other research (Ross Kristie, Chmiel James, & Ferkol Thomas, 2012).

Over the past five decades, vehicles such as automobiles, trains, and boats, along with their engines, have significantly reduced their emission of pollutants like hydrocarbons, carbon monoxide, nitrogen oxides, and particulate matter by approximately 99%. Similarly, the discharge of volatile organic compounds, carbon monoxide, nitrogen oxides, and lead from individual vehicles has decreased by over 90%, leading to a significant drop in national pollutant emissions. Despite a considerable rise of over 400% in annual miles driven, this decrease has been substantial. Since the 1980s, we've seen a reduction of 25% in ground-level ozone, an 80% decrease in mercury emissions, and a nearly complete eradication of 90% of atmospheric lead pollution due to the switch from leaded to unleaded gas. A 2018 study revealed that the Clean Air Act played a crucial role in the 60% reduction in pollution emissions by the manufacturing sector between 1990 and 2008 (Shapiro Joseph & Walker Reed, 2018).

## 5.5.5. Legal Challenges

Numerous lawsuits have been filed against the EPA since its founding regarding the authority granted by Congress and the EPA's rulemaking under the Clean Air Act. Some of the major suits where the Clean Air Act has been focal point of litigation include the following:

- **Train v. Natural Resources Defense Council, Inc., 421. U.S. 60 (1975):** States had nine months from the time the EPA published the new standards to submit their implementation plans under the Clean Air Act. The Natural Resources Defense Council appealed the EPA's approval of multiple state plans that permitted deviations from their emissions limits. The Supreme Court ruled that the EPA's approval was lawful and that a state is "at liberty to adopt whatever mix of emission limitations it deems best suited to its particular situation," provided that the "ultimate effect of a State's choice of emission limitations is compliance with the national standards for ambient air."
- **Chevron U.S.A., Inc. v. Natural Resources Defense Council, Inc., 467. U.S. 837 (1984):** The Clean Air Act required the EPA to control emissions from sources of air pollution, but it did not specify what constituted a source for the purposes of regulation. As a result, the EPA interpreted what a source was in accordance with the law. The Supreme Court ruled in a 7–0 decision that the EPA had judicial deference to establish their own interpretation of the law when the law is ambiguous and the interpretation is reasonable and consistent, despite the EPA's interpretation being contested. This idea—known as the Chevron deference—applies to any executive agency that receives authority from Congress.
- **Whitman v. American Trucking Ass'ns, Inc., 531. U.S. 457 (2001):** The trucking industry filed a challenge against the EPA's

rule in lower courts after the agency's rulemaking to set NAAQS standards for diesel truck emissions. The industry claimed the rule violated the nondelegation doctrine and did not provide sufficient justification for the new levels. The D.C. The EPA's rule did not take into account the costs of putting emissions regulations and controls in place, according to the Circuit Court, which ruled in favor of the trucking industry. The D.C was overturned by the Supreme Court. District Court's decision, which upheld the constitutionality of the Clean Air Act's power transfer from Congress to the EPA and said that the law did not oblige the EPA to take costs into account when deciding what air quality controls to implement.

- **Massachusetts v. Environmental Protection Agency, 549. U.S. 497 (2007):** The EPA concluded in 2003 that the Clean Air Act's language did not permit them to regulate emissions from motor vehicles, and even if they could, they would not be motivated to do so in response to pressure from states and environmental organizations to control carbon dioxide and other greenhouse gas emissions from motor vehicles. The EPA was sued by several states and agencies for neglecting to take action against what they deemed to be dangerous air pollutants. The Supreme Court decided 5–4 that the EPA was required by the Clean Air Act to regulate carbon dioxide and other greenhouse gases as air pollutants, and that if the EPA did not regulate these emissions, it would be subject to additional legal action. Although there has been ongoing disagreement over the ruling, the Court's ruling in Massachusetts v. The EPA was regarded as historic because it allowed additional environmental lawsuits to be filed in court to compel businesses to address climate change.
- **American Electric Power Co. v. Connecticut, 564. U.S. 410 (2011):** Electric utility companies were sued by a number of states and cities, claiming that their emissions were a public nuisance, and that the companies should use a cap-and-trade system to reduce their emissions. In an 8–0 ruling, the Supreme Court held that private companies cannot be sued by third parties for emissions-related matters because the EPA is expressly granted this authority under federal common law through the Clean Air Act.
- **Utility Air Regulatory Group v. Environmental Protection Agency, 573. U.S. 302 (2014):** In 2010, the EPA released new regulations that extended the scope of the regulated air pollutants and greenhouse gas emissions regulations to include light and heavy engines as well as smaller stationary sources. Numerous regulatory

organizations and power companies contested these new regulations because they significantly broadened the range of facilities that required environmental permits before construction could begin. Although it invalidated some of the EPA's new regulations impacting smaller sources, the Supreme Court largely maintained the EPA's authority under the Clean Air Act.

- **Michigan v. EPA, 576. U.S. 743 (2015):** The EPA published new regulations in 2012 designating new pollutants—like mercury—as hazardous materials whose emissions from power plants need to be controlled. Implementing mercury pollution controls on new and existing plants can be costly. To that end, the EPA was sued by a number of states, businesses, and other organizations alleging that the cost factor was overlooked in the analysis that resulted in the new regulations. The Supreme Court decided, 5–4, that it was improper for the EPA to ignore the costs associated with these pollution controls and that these costs must be taken into account for any determination that the EPA makes that is required and appropriate under the Clean Air Act.
- **West Virginia v. EPA, 597. U.S. (2022):** The Clean Power Plan (CPP), which sought to mitigate climate change by lowering carbon dioxide emissions from coal-burning power plants, was proposed by the EPA during the Obama administration in 2014. Using its authority under 42 U.S.C. § 7411(d), the agency sought to regulate emissions from existing power plants. Several states were successful in their requests to federal courts for an emergency stay to prevent the plan's implementation after the final version was unveiled in 2015. Since the CPP was never put into effect, the Trump administration's EPA passed the Affordable Clean Energy (ACE) rule in 2017, repealing the CPP. On January 19, 2021, the D.C began a federal court challenge to the ACE rule brought by a different group of states and environmental advocacy organizations. A different set of states and environmental advocacy groups challenged the ACE rule in federal court, and on January 19, 2021, the D.C. Circuit Court of Appeals vacated the rule because it relied on a flawed interpretation of § 7411(d). A number of states petitioned the Supreme Court, arguing that the D.C. Circuit's interpretation of § 7411(d) was too broad, and four petitions were consolidated as *West Virginia v. EPA*. In a 6–3 ruling on June 30, 2022, the Supreme Court declared the CPP unconstitutional on the grounds that it was unconstitutional under the major questions doctrine, requiring more specific Congressional approval than was reasonably argued to be contained in the statute.

### 5.5.6. Current Challenges

It has been observed that ethnic minorities are 1.5 times more likely than ethnic majority people to live in an area affected by air pollution, as poor air quality is still a problem in the US. The COVID-19 pandemic is still affecting the United States, but research indicates that residents who live in areas with air pollution are 8% more likely to have long-term COVID effects, which could be fatal. This leads to a discrepancy between American communities of color and the majority, despite the fact that the air pollution these communities experience is frequently caused by redlining, which locates minority communities near heavily polluting industries and busy roads. The Environmental Protection Agency (EPA) and the Sierra Club have suggested a number of solutions, such as decreasing permissible emission levels, strengthening enforcement of the Clean Air Act, and providing funding for clean energy transportation services. Although the Clean Air Act has been in place since 1970, the fossil fuel industry has attempted to override it, which has resulted in increased air pollution in many areas.

The San Joaquin Valley, which suffers from poor air quality due to harmful agricultural practices, heavy traffic on major roadways, and the oil industry, is one area where the Clean Air Act has not been very effective. According to research, PM 2.5, or fine particulate matter, is the most common air pollutant in the area. Pregnant women who are exposed to it may experience health problems like worsening asthma, a reduction in FEV1, weakened immunity, and an increased risk of preterm birth. Additional symptoms of PM 2.5 exposure include chronic bronchitis, decreased lung function in children, and hospitalizations for heart and lung issues, which may result in early death, particularly if the patient has a history of health issues. Although there was a significant drop in PM 2.5 pollution in 2004 due to a decrease in biomass burning and agriculture, PM 2.5 pollution is still a common occurrence today and levels have risen. In order to lower toxic emissions into the atmosphere, groups like the California Air Resources Board (CARB) have suggested that the San Joaquin Valley implement stronger regulations. The San Joaquin Valley Air Pollution Control District, the EPA, and the CARB are currently working to enforce Clean Air Act regulations and provide funding for additional sustainable living initiatives.

### 5.5.7. Future Challenges

Even in 2017, certain American cities failed to meet federal requirements for ambient air quality. Tens of thousands of premature deaths may still be brought on by ground-level ozone pollution and fine-particle pollution. The United States' efforts to manage conventional air pollutants are complicated by climate change because of the possibility of more frequent air stagnation episodes brought on by warmer, drier summers. Extended dry spells that could fuel wildfires would also cause elevated air particle concentrations in specific areas. The Clean Air Act does not directly regulate transboundary air pollution, which means that ongoing agreements and international negotiations are necessary to control pollution entering and leaving the United States, especially with Canada and Mexico.

The Clean Air Act is still facing challenges with environmental justice. The Clean Air Act lessens the increased exposure of low-income and minority communities to air pollution by encouraging pollution reduction. However, there is a persistent overrepresentation of African Americans in areas with the worst air quality (Miranda Marie Lynn, Edwards Sharon, Keating Martha, & Paul Christopher, 2017). The most polluted areas in the US are home to dense populations of low-income and minority communities, which is thought to aggravate health issues in these populations. High exposure to air pollution is associated with a number of health issues, such as infant mortality, asthma, cancer, and premature death, all of which disproportionately affect communities of color and those with lower incomes. Each of these conditions has decreased as a result of the Clean Air Act's pollution reduction, which can also advance environmental justice for communities that are disproportionately affected by air pollution and worsening health outcomes.

## REFERENCES


1. Al Ahasan, M. A., Roy, S., Saim, A. H. M., Akter, R., & Hossain, M. Z. (2018). Arduino-based real time air quality and pollution monitoring system. *International Journal of Innovative Research in Computer Science & Technology, 6*(4), 81–86. https://doi.org/10.21276/ijircst.2018.6.4.8.
2. Aldy, J. E., Auffhammer, M., Cropper, M., Fraas, A., & Morgenstern, R. (2022). Looking back at 50 years of the Clean Air Act. *Journal of Economic Literature, 60*(1), 179–232. https://doi.org/10.1257/jel.20201626.
3. Bukowiecki, N., Dommen, J., Prévôt, A. S. H., Richter, R., Weingartner, E., & Baltensperger, U. (2002). A mobile pollutant measurement laboratory—measuring gas phase and aerosol ambient concentrations with high spatial and temporal resolution. *Atmospheric Environment, 36*(37), 5569–5579. https://doi.org/10.1016/S1352-2310(02)00694-5.
4. Clarke, A. G., & Watson, I. (1998). Electrochemical and other non-optical techniques. In *Industrial Air Pollution Monitoring* (pp. 192–211). Springer Netherlands. https://doi.org/10.1007/978-94-009-1435-3_11.
5. Denzer, R., Schimak, G., & Humer, H. (1992). Air pollution monitoring. In *Computer Techniques in Environmental Studies IV* (pp. 637–651). Springer Netherlands. https://doi.org/10.1007/978-94-011-1874-3_45.
6. Dovland, H. (1987). Monitoring European transboundary air pollution. *Environment: Science and Policy for Sustainable Development, 29*(10), 10–28. https://doi.org/10.1080/00139157.1987.9930854.
7. Gee, I. (2001). Monitoring indoor air pollution. *Indoor and Built Environment, 10*(3–4), 123–124. https://doi.org/10.1177/1420326x0101000301.
8. Gouriou, F., Morin, J.-P., & Weill, M.-E. (2004). On-road measurements of particle number concentrations and size distributions in urban and tunnel environments. *Atmospheric Environment, 38*(18), 2831–2840. https://doi.org/10.1016/j.atmosenv.2004.02.039.
9. Joans, S. M. (2017). Air pollution monitoring through image processing. *International Journal for Research in Applied Science and Engineering Technology, 5*(X), 1377–1382. https://doi.org/10.22214/ijraset.2017.10200.
10. Kasar, A. R., Khemnar, D. S., & Tembhurnikar, N. P. (2013). WSN based air pollution monitoring system. *International Journal of Science and Engineering Applications, 2*(4), 55–59. https://doi.org/10.7753/ijsea0204.1001.
11. Kasar, A. R., Khemnar, D. S., & Tembhurnikar, N. P. (2013). WSN based air pollution monitoring system. *International Journal of Science*

*and Engineering Applications, 2*(4), 55–59. https://doi.org/10.7753/ijsea0204.1001.

12. Lambor, S., Mahajan, A., Bodhankar, A., Prasad, D., Mahajan, S., Pujari, A., & Dhakalkar, R. (2023). Indoor air pollution monitoring. In *12th International Conference on Information Systems and Advanced Technologies "ICISAT 2022"* (pp. 434–441). Springer International Publishing. https://doi.org/10.1007/978-3-031-25344-7_40.
13. Lv, L., Liu, W., Fan, G., Zhang, T., Dong, Y., Chen, Z., Liu, Y., Huang, H., & Zhou, Y. (2016). Application of mobile vehicle lidar for urban air pollution monitoring. *Chinese Optics Letters, 14*(6), 060101–060106. https://doi.org/10.3788/col201614.060101.
14. Miranda, M. L., Edwards, S. E., Keating, M. H., & Paul, C. J. (2011). Making the environmental justice grade: The relative burden of air pollution exposure in the United States. *International Journal of Environmental Research and Public Health, 8*(6), 1755–1771. https://doi.org/10.3390/ijerph8061755. PMID 21776200. PMC 3137995.
15. Pathak, G., & Bhadrashetti, V. M. (2018). Air pollution monitoring system. *International Journal of Trend in Scientific Research and Development, 2*(5), 1848–1852. https://doi.org/10.31142/ijtsrd18191.
16. Pillai, P. G., & Saravanan, P. (2018). A study on air pollution monitoring system using wireless sensor network. *International Journal of Trend in Scientific Research and Development, 2*(2), 343–347. https://doi.org/10.31142/ijtsrd8374.
17. Pirjola, L., Parviainen, H., Hussein, T., Valli, A., Hämeri, K., Aalto, P., Virtanen, A., Keskinen, J., Pakkanen, T. A., Mäkelä, T., & Hillamo, R. E. (2004). Sniffer—a novel tool for chasing vehicles and measuring traffic pollutants. *Atmospheric Environment, 38*, 3625–3635. https://doi.org/10.1016/j.atmosenv.2004.03.047.
18. Ross, K., Chmiel, J. F., & Ferkol, T. (2012). The impact of the Clean Air Act. *The Journal of Pediatrics, 161*(5), 781–786. https://doi.org/10.1016/j.jpeds.2012.06.064. PMC 4133758. PMID 22920509.
19. Shapiro, J. S., & Walker, R. (2018). Why is pollution from U.S. manufacturing declining? The roles of environmental regulation, productivity, and trade. *American Economic Review, 108*(12), 3814–3854. https://doi.org/10.1257/aer.20151272.
20. Walker, D. S. (1998). Calibration of continuous gaseous emission measuring systems. In *Industrial Air Pollution Monitoring* (pp. 212–228). Springer Netherlands. https://doi.org/10.1007/978-94-009-1435-3_12.

21. Wardencki, W., Katulski, R., Stefański, J., & Namieśnik, J. (2008). The state of the art in the field of non-stationary instruments for the determination and monitoring of atmospheric pollutants. *Critical Reviews in Analytical Chemistry, 38*(4), 259–268. https://doi.org/10.1080/10408340802378254.

Chapter 6

# Measurement of Pollutant Emissions

## Contents

# INTRODUCTION

The process of calculating how much pollution, either in gaseous or particulate form, is released into the atmosphere from a particular source, like an industrial process, is known as emissions measurement. Emissions measurements are useful for creating emissions inventories as well as for determining the relative importance of a particular source in relation to other sources. Emissions measurements are used by government or industry personnel to evaluate how well control strategies are working. To find out how well a pollution control device captures pollutants, measurements of the gas stream can be taken both before and after the device. The amount of pollution that a source is permitted to emit is regulated, and emission measurements are used to assess compliance with those regulations. It is possible to take measurements quickly (e.g., hours), frequently known as a source test, or using techniques that measure continuously, frequently known as continuous emissions monitoring. In either case, it's critical to collect data using trustworthy methods from samples that accurately reflect the emission stream.

## 6.1. IMPORTANCE OF MEASUREMENT

There are numerous causes of air pollution. Urban environments can contain a wide range of materials, including various gases (most commonly carbon monoxide, nitrogen oxides, and sulfur dioxide, all related to fuel combustion), as well as solid and liquid particulates (such as soot from engines and fly ash escaping from incinerators). The built environment, the natural world (water, soil, crops, trees, and other vegetation), and human health are all affected differently by these various types of pollution (Jacobson Mark, 2012). In order to keep the quality of the air within legally mandated limits (imposed by regulators like the Environmental Protection Agency in the United States) or advisory guidelines proposed by organizations like the World Health Organization (WHO), one must first identify the causes of the pollution and then reduce or regulate them (Bower Jon, 1999). The WHO reports that more than 6,000 cities across 117 countries now regularly check the quality of their air.

## 6.2. TYPES OF MEASUREMENT

There are two ways to measure air pollution (in general) passively and actively.

### 6.2.1. Passive Measurement

It is relatively easy and inexpensive to use passive devices (Fan, & Zih-Hua Tina, 2011). They function by passively absorbing or gathering a sample of the surrounding air, which must subsequently be examined in a lab. The diffusion tube, which resembles a lab test tube and is attached to an object like a lamp post to absorb one or more particular polluting gases of interest, is one of the most widely used passive measurement methods. The tube is removed and sent to a lab for examination after some time. Another sort of passive device is the deposit gauge, which is among the earliest methods of measuring pollution. These are big funnels that gather particulates, such as soot, and pour them into sampling bottles for further analysis in a lab (Brimblecombe Peter, 1987) (Figure 6.1).

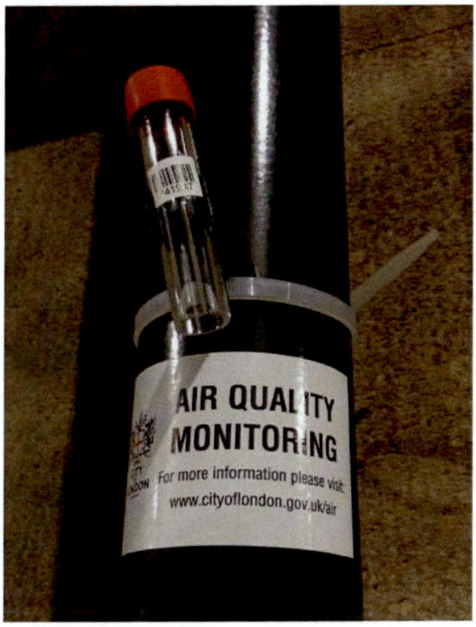


**Figure 6.1.** *A diffusion tube is an example of a passive air pollution monitor.*

*Source: By Etan J. Tal – Own work, CC BY-SA 4.0,* https://commons.wikimedia.org/w/index.php?curid=49918888.

## 6.2.2. Active Measurement

Although they are not always more sensitive or dependable than passive devices, active measurement devices are automated or semi-automated and have a tendency to be more complex and sophisticated (Fan & Zih-Hua Tina, 2011). They collect and store the air for later analysis in a laboratory, or they use fans to draw in air, filter it, and analyze it automatically there and then. Active sensors work in two ways: chemically or physically. Physical methods measure the amount of a certain wavelength of light that an air sample absorbs, for example, without altering the sample. Chemical techniques measure how a chemical reaction alters the sample in some way. The majority of automated sensors for air quality are instances of active measurement.

## 6.3. AIR QUALITY SENSORS

Small handheld devices, large-scale static monitoring stations in cities, and remote monitoring equipment used on airplanes and space satellites are all examples of air quality sensors.

### 6.3.1. Personal Air Quality Sensors

On one extreme of the spectrum, there exist diminutive, reasonably priced, transportable (and occasionally wearable), Internet-linked air pollution detectors, like PurpleAir and the Air Quality Egg. They generate measurements that are almost real-time and somewhat accurate, sampling gases and particulates continuously. Smartphone apps can analyze these measurements. Additionally, their data can be crowdsourced to create maps of pollution over large areas, either by itself or in conjunction with other pollution data. Most of them concentrate on measuring the five common types of air pollution—ozone, particulate matter, carbon monoxide, sulfur dioxide, and nitrogen dioxide—and can be used in both indoor and outdoor settings (Jiao Wan & Hagler Gayle, et al., 2016). Some monitor less prevalent contaminants like formaldehyde and radon gas.

This type of sensor was once expensive, but in the 2010s, there was a shift toward less expensive wearable devices that people could wear to track the quality of the air in their local area. These devices are now sometimes referred to as low-cost sensors (LCS) (Lewis, Lee James, Edwards Peter, Shaw Marvin, Evans Mat, et al., 2016). A recent review by the European Commission's Joint Research Center identified 112 examples, made by 77 different manufacturers (Karagulian, Gerboles, Barbiere, Kotsev, Lagler, et al., 2019).

With the use of personal sensors, people and communities can gain a better understanding of their exposure conditions and air pollution risks. For instance, a study team at UCSD under the direction of William Griswold distributed portable air pollution sensors to sixteen commuters and discovered urban valleys where structures trapped air pollution. Additionally, the group discovered that bus passengers are more exposed than car passengers.

### 6.3.2. Small-scale Static Pollution Monitoring

Static monitors continuously sample and measure the air quality in a specific urban location, in contrast to portable, low-cost monitors. In order to measure levels of nitrogen dioxide and other pollutants, active air quality monitors are occasionally permanently fixed alongside platforms in public areas like busy railroad stations (Hickman, Baker, Cai, Delgado-Saborit, & Thornes, 2018). Certain static monitors are made to provide real-time information on the state of the local air. EkoSłupek air monitors in Poland are equipped with tiny lamps on top that change color from red to green to indicate the quality of the air around them. The monitors measure a variety of polluting gases and particulates (Figure 6.2).

**Figure 6.2.** *An EkoSłupek air pollution sensor in Poland. The green light indicates good nearby air quality.*

*Source: By MichalPL – Own work, CC BY-SA 4.0,* https://commons.wikimedia.org/w/index.php?curid=115034190.

## 6.3.3. Large-scale Pollution Monitoring

On the other end of the spectrum from inexpensive sensors are the large, very expensive, static street-side monitoring stations that are part of metropolitan monitoring systems like the London Air Quality Network and the larger British network known as the Automatic Urban and Rural Network (AURN). These stations continuously sample a variety of different pollutants that are frequently found in urban air for local authorities. Data from more than 10,000 monitors are stored in the Air Quality System (AQS), a repository for air quality data maintained by the US EPA. Data on air quality is gathered by the European Environment Agency from 3,500 monitoring stations located throughout the continent (Figure 6.3).

**Figure 6.3.** *An air pollution monitoring station in Shaftesbury Avenue, London.*

*Source: By N19±, CC BY-SA 3.0,* https://commons.wikimedia.org/w/index.php?curid=56940719.

These kinds of sensors provide much more accurate measurements that are produced in almost real-time, and the results are used to create air quality indexes (AQIs). Medium-sized, portable monitors, which are sometimes installed in sizable wheeled cases or even integrated into smog-mobile sampling trucks, fall in between the two extremes of large-scale static and small-scale wearable sensors. Drive-by air pollution sensing systems, which use sensors installed on taxis, buses, trams, and other vehicles, have recently come to light as a viable method for monitoring the quality of the air (Zarrar Hassan & Dyo Vladimir, 2023). Because of their wide availability and broad geographic coverage, buses in particular have attracted a lot of attention as a mobile sensing platform (Caminha Cruz, Couto Souza, Kosmalski Maciel, Fladenmuller Anna, & Amorim Dias, 2018).

## 6.3.4. Remote Monitoring

Lidar (Richter, 1994), drones, and satellites can also measure air quality remotely from the air using techniques like gas filter correlation. The Global Ozone Monitoring Experiment (GOME) and NASA's MAPS (Mapping Pollution with Satellites) were two of the first satellite pollution monitoring projects. GOME measured global (tropospheric) ozone levels from the ESA European Remote Sensing Satellite (ERS-2) in 1995, and MAPS also measured the distribution of carbon monoxide in Earth's lower atmosphere in the 1990s.

## 6.4. SOURCE SAMPLING METHODOLOGY

The main forms of emissions from stationary sources are gases and solids. The effectiveness of the measurement is primarily dependent on the ability to gather a representative sample of gaseous or particulate pollutants. Understanding the characteristics of particulate and gaseous pollutants as well as the techniques used to measure them is crucial because no single method of measurement can accurately analyze every type of pollutant released from an exhaust stack. The U.S. E.P.A. Reference Methods specify the procedures and equipment necessary for measuring the various constituents of effluent gas from stationary sources. Complete explanations of the underlying ideas, their applicability, and the precise formulas required to compute the test results in standard units are also included in the methods. Every source test should be conducted carefully, just like any other scientific investigation, since it can be viewed as an original scientific investigation (Nicklin & Darabkhani, 2021).

The following Federal Reference Methods specify the protocols, tools, and source emission sampling.

- FRM 1 – Point Locations
- FRM 2 – Gas Flow Rate
- FRM 3 – Gas Molecular Weight
- FRM 4 – Gas Moisture Content
- FRM 5 – Particulate Matter
- FRM 6 – Sulfur Dioxide
- FRM 7 – Nitrogen Oxides
- FRM 8 – Sulfuric Acid Mist

Every single source test, though, varies slightly. Sampling situations can range from studies that are hot, wet, and dirty to those at ambient temperatures with very little moisture and particulate matter. These kinds of realities led to the development of source test methods that allow samples to be obtained at reasonable costs in a timely manner without compromising technical viability. The outcomes of stationary source emissions have been affirmed by the courts on multiple occasions. As a matter of fact, they are now the gold standard for assessing whether industries are adhering to or exceeding stationary source emission limitations. The techniques not only shown themselves to be trustworthy when compared to other sources, but they have also stood the test of time and scrutiny and are still in use today. For instance, the majority of states use the U.S. EPA Reference Methods or other comparable, stricter procedures that have the U.S. EPA's approval for conducting compliance testing.

Methods 1, 2, 3, and 4 deal with characterizing other aspects of the gas stream, whereas Methods 5, 6, 7, and 8 deal directly with the capture and measurement of particular pollutants in a gas stream. The first four methods are helpful in this regard, providing the data required to carry out additional methods or compute their outcomes. Among these approaches, Method 1 is distinct since it focuses on figuring out the array of sampling points and evaluating the suitability of the sampling site. Given that the first eight methods serve as the basis for all other methods, it is crucial to examine some of their key components in more detail.

### 6.4.1. Distributed Sampling Points

First, the velocity is measured, and then gas is extracted from a range of sampling points spaced evenly across the stack or duct's cross section to obtain a representative sample of the effluent. Less sampling sites are needed if only gaseous pollutants need to be sampled because stack gases are thought to be sufficiently mixed.

### 6.4.2. Sampling Distributed Over Time

A typical sampling time or run is approximately one hour, and each point should be sampled for the same amount of time. Usually, three runs are carried out, and the average of the results follows.

### 6.4.3. Isokinetic Sampling

Field investigators spend a lot of time and money trying to make sure that particulate matter sampling is done isokinetically, but before you conduct your next stack test, it might be worth considering how important isokinetic sampling is, especially with today's stricter emission standards and better control equipment. Because of the inertial effects of particles as they enter a sampling nozzle, skewed particulate concentration measurements can be avoided with the development of isokinetic sampling techniques. In order to achieve straight streamlines near the nozzle, the basic idea is to sample at the same speed as the gas stream. Heavy particles may miss the nozzle due to turns in the streamlines close to it, which would skew the concentration value that is being measured (Figure 6.4).

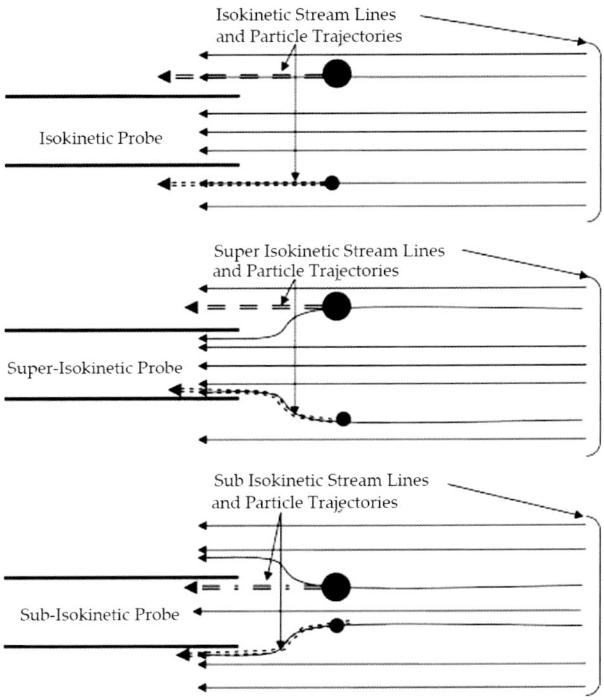


**Figure 6.4.** *Isokinetic sampling.*

*Source:* https://www.ldxsolutions.com/particulate-matter-isokinetic-sampling/.

Although the theory supporting isokinetic sampling makes sense, there are numerous situations in which non-isokinetic sampling has little impact. According to calculations, at normal duct velocities, a particle with an aerodynamic diameter of less than 2 microns has almost no inertia, and non-isokinetic sampling will not affect the concentration that is measured. Measuring 4-micron diameter particles at 50% higher or lower velocities than the gas stream will only result in a 2–3% measurement error. If you are sampling after a control device and are not required by contract or legislation to conduct formal emissions or EPA-type testing, you might want to think about doing away with the limitations of conventional isokinetic sampling. Time, money, and effort will all be saved.

## 6.4.4. Separation of Gas Constituents

The next step is to extract a predetermined amount of stack gas from the stack or duct using a sampling train, which is essentially a set of leak-proof parts designed to collect contaminants. Condensation, or bubbling the sample gas through chilled impingers (sealed glass vessels), some of which may contain liquid reagents to absorb gases, is the best method for capturing water vapor and gaseous pollutants. Filter paper is heated to capture solids, and glassware that connects to the probe is used to do so. A heated probe, or a hollow glass tube that is put into the stack or duct, is the first part of the sampling train. Gases are drawn through the other equipment components (the probe, filters, and impingers) by the final component, a pump. A gas meter measures the volume of the sampled gas, which is then exhausted into the atmosphere through an orifice at the by-pass valve. This orifice is used to adjust the sample train flow.

### 6.4.5. Sample Recovery and Analysis

Before use, filters are pre-weighed and reagent volumes are pre-measured. Each sampling train component's contents are meticulously recovered to a sealed vessel following each run, where they are weighed, measured, or subjected to other laboratory tests.

### 6.4.6. Data Recording

Appropriate measurement data are captured during every run. The sampling time, dry gas meter readings, orifice meter readings, differential pressures across the gas temperature, and sample train vacuum are all recorded point by point. The test results are then computed using the averages of these values and the total sample gas volume.

### 6.4.7. Calculation of Results

The ratio of the mass of the captured pollutant to the volume of gas sampled is used to determine the concentration of the pollutant in the sampled gas. These findings are expressed on a dry basis and adjusted to standard pressure and temperature. With the use of this standardized measure, test results conducted under particular circumstances can be compared to those of other tests or regulatory standards.

Equal or uniform sampling of gas in motion is referred to as isokinetic sampling. When the stack gases' fluid streamlines are not disrupted, this is achieved. When the gas velocity into the sample nozzle ($v_n$) equals the velocity of the approaching gas stream velocity ($v_s$), a 100% isokinetic source sample is obtained. To ascertain whether the results fall within the acceptable range of $90\% \geq I \leq 110\%$, the isokinetic variation (I) is computed. This is the equation for isokinetic variation.

$$\%\ \text{Isokinetic variation} = \frac{v_n}{v_s} \times 100$$

### 6.4.8. Expression of Results

The pollutant concentration can be used to calculate other test result expressions for broader applicability across sources and states. For example, an emission rate for the pollutant, or the amount emitted per unit of time, can be calculated using the concentration and the gas flow rate.

A single sampling train can be used to conduct multiple methods simultaneously by adding impingers, filters, or other suitable capture devices due to the similarity of the sampling methods and equipment, which allows for some degree of interchangeability and interrelation. For instance, it is simple to combine Methods 5 (which measures particulate emissions) and 8 (which measures sulfuric acid mist and sulfur dioxide) by incorporating extra impingers and an in-stack filter into the probe of the Method 5 sampling train. By simply strapping the velocity probe to the sampling probe, Method 2 can be combined with any other method when gas velocity measurements are needed to calculate test results. A manometer that is attached to the velocity probe is used to read velocity measurements. However, Method 2 (for gas velocity), Method 3 (for molecular weight), and Method 4 (for gas moisture content) are usually combined with Method 5 (for particulate emission). To compute a particulate emission rate, these supportive measurements—moisture content, molecular weight, and velocity—are necessary.

Instrument methods are a specific type of source sampling where a continuously recording instrument is used to analyze the gas. After being removed from a stack or duct, wastewater gas is routed to one or more continuously analyzing devices. A data acquisition system or a strip chart recorder receives signals containing the analytical results. Automated line graphs known as "strip charts" need to be interpreted by a technician as numerical values and then reduced to averages over suitable time intervals. Computers are the most common means of acquiring data, though, as they can store, reduce, and generate relevant reports from the data. The 1980s saw the introduction of instrumental methods, which are still being developed as technology develops (Figure 6.5).

Furthermore, it is imperative to offer measurements in "real-time," or instantaneously, as opposed to delaying the receipt of wet chemistry test results for several days or weeks. Scientists studying air pollution are moving toward using continuous emission monitoring (CEM) in place of periodic measurements thanks to developments in instrumental analyzers. Many metals and organic compounds at extremely low concentrations can be swiftly identified using modern instrumental test methods and analyzers. These kinds of instruments are the subject of significant current research.

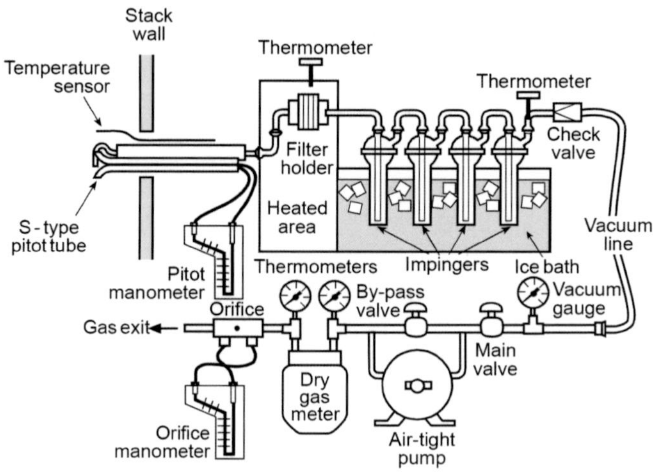


**Figure 6.5.** *Sampling train for particulate emissions.*

*Source:* https://www.4cleanair.org/wp-content/uploads/APTI/452combined.pdf.

## 6.5. CONTINUOUS EMISSION MONITORING

Systems for continuously monitoring emissions (CEMS) are a useful tool for keeping an eye on the streams of waste gases that are produced during industrial combustion. In industrial settings, CEMS can measure flue gas for carbon dioxide, carbon monoxide, and oxygen to provide information for combustion control (The Babcock and Wilcox Company, 2005). They are also employed as a way to adhere to state-permitted emission standards, other federal emission programs run by the US government, and air pollution regulations like the US Environmental Protection Agency's (EPA) Acid Rain Program. Analyzers to measure the concentrations of gases in the stream; devices to route a sample of that stream to the analyzers if they are located remotely; devices to condition the sample gas by eliminating water and other elements that could skew the reading; pneumatic plumbing with PLC-controlled valves to route the sample gas to and from the analyzers; a calibration and maintenance system that permits the introduction of calibration gases into the sample line; and a Data Acquisition and Handling System (DAHS) that gathers, stores, and processes each data point in order to calculate the total mass emissions. Even when the process it monitors is not in use, a CEMS is always in operation. They have the ability to consistently gather, document, and submit emissions data for process monitoring and/or legal requirements.

A sample probe, filter, umbilical sample line, gas conditioning system, calibration gas system, and several gas analyzers that represent the parameters under observation make up the typical CEM system. Sulfur dioxide, nitrogen oxides, carbon monoxide, carbon dioxide, hydrogen chloride, airborne particulate matter, mercury, volatile organic compounds, and oxygen are examples of typical emissions that are monitored. CEM systems are also capable of measuring moisture, flue gas opacity, and air flow. A PEMS is a type of monitoring system that measures particulate matter.

Within the U.S., in order to gather and submit the data, the EPA needs a system for data handling and acquisition. By adding measurements of flow rate, concentration measurements can be converted to mass/hour. Each type of source has its own subpart under 40 CFR parts 60 and 75, and these subparts determine the types of gases being measured and the calculations needed. Both a volumetric flow monitor and a $SO_2$ pollutant concentration monitor are used to calculate $SO_2$ emissions in pounds per hour. A diluent gas monitor and a NOx pollutant concentration monitor are both used to calculate the emissions rate for NOx in terms of weight per volume or weight per heat value (e.g., lbs/million Btu, lbs/ft$^3$, kg/kWh, cm$^3$/kg). It is occasionally necessary to measure opacity, depending on the kind of source. While $CO_2$ monitoring is occasionally necessary, it is best to use both a flow monitor and a $CO_2$ or oxygen monitor. The DAHS needs to be able to gather, log, and store data—typically every minute. Even when no process is running, a DAHS needs to be operational continuously around-the-clock for compliance reasons. The DAHS must take at least one reading every fifteen minutes for three quarters in order for the measurement to be considered valid. After that, the readings are averaged every hour.

## 6.5.1. Operation

Using a sample probe and a pump, a small sample of flue gas is drawn into the CEM system. A dilution-extractive probe is frequently used in fossil fuel combustion facilities to dilute the sample with clean, dry air to a ratio of usually 100:1, but it can range from 50:1 to 200:1. Pure flue gas can be hot, humid, and sticky when it contains certain pollutants, which is why dilution is used. After being diluted to the proper ratio, the sample is sent to a manifold from which individual analyzers may extract samples via a sample line, also known as an umbilical. Gas analyzers use a variety of methods to measure concentrations precisely. Among the often employed methods are fluorescence, chemiluminescence, infrared and ultraviolet adsorption, and beta-ray absorption. Following analysis, the gas leaves the analyzer and is vented outdoors via a common manifold shared by all analyzers. Each analyzer's signal output is received by a Data Acquisition and Handling System (DAHS), which is used to gather and store emissions data. Another sample extraction technique known as a dry extractive, hot dry extractive, or direct CEMS is frequently used in utility sources and industrial sources with low emission rates. The sample is transported into a sample conditioning unit at a high temperature via a heated sample line without being diluted. The sample is dried, usually using a chiller, to remove moisture after being filtered to remove particulate matter. After conditioning, the sample goes through a sampling manifold where it is measured by different gas analyzers; for combustion turbines and engines powered by natural gas or diesel, these analyzers usually measure $NO_x$ and $O_2$ (and occasionally CO). Most $NO_x$ analyzers use chemiluminescence to function. $O_2$ is drawn to a magnetic field in $O_2$ analyzers in order to determine the concentration. A suspended mirror inside the analyzer moves as a result of the $O_2$, and this movement modifies the amount of light that the mirror reflects onto a photocell. The $O_2$ concentration determines how much current is needed to bring the mirror back to center. To complete the necessary calculations, one must be able to calculate the sample's percentage of oxygen.

## 6.5.2. Quality Assurance

The system's accuracy is shown in a number of ways. Every day, a certified concentration of gas is introduced to the sample probe to conduct an internal quality assurance check. The Calibration Error percentage is then calculated by comparing the CEMS measurement with the known concentration. Moreover, a zero gas reading is obtained and contrasted. A calibration error percentage of more than twice the performance specification for five days in a row or more than four times the performance specification in a 24-hour period renders the CEMS out of control, making it impossible to rely on the data for accuracy until it is brought back under control. During out-of-control periods, data substitution will be applied. Regaining control of the CEMS is crucial because the data substitution method is typically not beneficial.

Continuous Emissions Monitoring Calibration Systems, which dilute gases to create calibration standards, are another tool permitted by the EPA. A specific percentage of accuracy must be met by the analyzer reading. Although the accuracy percentage can vary, most fall between 2.5 and 5%. An independent company must perform the annual (or bi-annual) certification of the system in power plants covered by the Acid Rain Program. To gather emissions data in tandem with the plant CEMS, the company will temporarily install a separate CEM system. We call this type of testing a Relative Accuracy Test Audit (RATA). In the U.S., periodic evaluations of the equipment must be reported and recorded. This includes semi-annual (or annual) RATA and bias tests, daily interference tests for flow monitors, and daily calibration error tests. Because CEMS equipment is costly, facilities may not always be able to afford it. In these situations, a facility will install analysis equipment at the emissions point that is not compliant with EPA regulations. A mobile CEMS company uses compliant equipment to measure emissions once a year for the equipment evaluation. Next, the outcomes are contrasted with the analyzer system that isn't compliant.

## 6.6. COMPLIANCE ASSURANCE MONITORING

For large emission units that depend on pollution control device equipment to achieve compliance assurance monitoring (CAM) is meant to offer a reasonable assurance of compliance with applicable requirements under the Clean Air Act (CAA). Monitoring is done to make sure that controls, whether installed or used in another way, are appropriately maintained and operated to maintain a degree of control that conforms with relevant regulations. In order to provide a reasonable assurance of compliance with applicable requirements, the CAM approach establishes monitoring with the following goals in mind: (1) documenting continued operation of the control measures within ranges of specified indicators of performance (such as emissions, control device parameters, and process parameters); (2) indicating any excursions from these ranges; and (3) responding to the data so that the cause or causes of the excursions are corrected.

### 6.6.1. Background

Title V of the Clean Air Act contains provisions that outline the specifications for permit applications, permit programs, and permit requirements and conditions. These clauses also cover other facets of the permits program, like application submission, compliance, enforcement, and permit approval. The EPA mandates that establishments that release pollutants into the atmosphere secure an operating permit. This permit, sometimes referred to as a "operating permit," details how the facility will adhere to rules and standards for emissions. Operating permits give specific information about the air pollution regulations that apply to each facility to facility owners, State inspectors, and the general public. The program for operating permits is expected to enhance adherence to current regulations and guarantee the achievement and sustenance of targeted reductions in emissions. Title VII of the Clean Air Act Amendments of 1990 also gives the EPA the authority to create regulations mandating that facilities keep an eye on the efficiency of their emission control equipment. The EPA proposed a "enhanced monitoring" rule in September 1993, which set general monitoring standards that facilities needed to meet in order to show ongoing compliance. The proposed rule was sharply criticized by a number of state and local agencies, business representatives, and other interested parties. In their opinion, the proposed rule was unduly restrictive and would have placed an undue burden on businesses to set up and run continuous emission monitoring systems, as well as on municipal and state agencies to carry out their operating permit programs. EPA has convened multiple meetings with key stakeholders since April 1995 in an effort to devise a novel, adaptable strategy for enhanced monitoring. EPA revised the enhanced monitoring rule through this stakeholder process, and in September 1995, they published a new draft rule that shifted the emphasis to compliance assurance. On August 2, 1996, the EPA revised the first draft of the CAM rule and released a second draft for public comment in response to feedback received. October 8, 1997 saw the publication of the final regulations.

### 6.6.2. CAM Rule Main Components

In order to reasonably assure compliance with emission limits and standards, the source owner or operator must undertake certain types of monitoring of currently installed control devices, as defined by the CAM rule. Through this monitoring, the owner or operator will be able to attest to compliance with the Title V operating permits program. The CAM rule includes Title V compliance certification language that allows the source owner or operator to use compliance assurance monitoring data to establish their compliance status with permit terms or conditions. They can then use this data to attest, as required by the Clean Air Act, that their facilities adhere to air pollution control regulations.

The rule exempts the owner or operator from additional CAM rule-related monitoring requirements in cases where continuous compliance monitoring is already specified in an operating permit. Instead, the owner or operator is instructed to use the data from continuous compliance monitoring to satisfy the CAM rule's certification and monitoring requirements. The rule mandates that the owner or operator of emission units with control equipment plan and carry out monitoring. An acceptable range—referred to as the "indicator range"—within which to operate the control device will be part of the monitoring. Typically, facility owners will identify the indicator ranges that, if the equipment is operated within those ranges, will reasonably assure compliance with emission limitations. They will do this by combining the results of performance tests with other information or equipment design. Limiting emissions and giving a reasonable guarantee that the facility is adhering to permit terms and conditions can be achieved by operating control devices within permissible ranges, as intended. The CAM rule offers optional tools for the State or local (or Federal if necessary) permitting authority to mandate more thorough review and enhancement of control practices if it is discovered that the control equipment is operating outside the indicator range for extended periods of time. Owners and operators of control equipment must promptly take corrective action to make the necessary adjustments to the equipment and notify State and local authorities of any potential compliance issues if the equipment is found to be operating outside of acceptable ranges.

## REFERENCES


1. Bell, M. L., Davis, D. L., & Fletcher, T. (2004). A retrospective assessment of mortality from the London smog episode of 1952: The role of influenza and pollution. *Environmental Health Perspectives, 112*(1), 6–8. https://doi.org/10.1289/ehp.6539. PMC 1241789. PMID 14698923.
2. Bower, J. (1999). *Monitoring ambient air quality for health impact assessment*. World Health Organization, Regional Office for Europe. ISBN 9789289013512. Retrieved March 29, 2022.
3. Breuer, D. (Ed.). (1999). *Monitoring ambient air quality for health impact assessment*. World Health Organization Regional Office for Europe. ISBN 9789289013512. Retrieved June 30, 2023.
4. Brimblecombe, P. (1987). *The big smoke: A history of air pollution in London since medieval times*. Routledge. ISBN 9781136703294.
5. Brimblecombe, P. (1987). *The big smoke: A history of air pollution in London since medieval times*. Routledge. ISBN 9781136703294.
6. Caminha, C., Couto, S., Kosmalski, M., Fladenmuller, A., & Amorim, D. (2018). On the coverage of bus-based mobile sensing. *Sensors, 18*(6), 1976. https://doi.org/10.3390/s18061976. PMC 6022044.
7. Conti, M., & Cecchetti, G. (2001). Biological monitoring: Lichens as bioindicators of air pollution assessment – a review. *Environmental Pollution, 114*(3), 471–492. https://doi.org/10.1016/S0269-7491(00)00224-4. PMID 11584645. Retrieved March 30, 2022.
8. Fan, Z.-H. T. (2011). Passive air sampling: Advantages, limitations, and challenges. *Epidemiology, 22*(1), S132. https://doi.org/10.1097/01.ede.0000392075.06031.d9. S2CID 75942106. Retrieved March 27, 2022.
9. Fine, G., Cavanagh, L., Afonja, A., & Binions, R. (2010). Metal oxide semi-conductor gas sensors in environmental monitoring. *Sensors, 10*(6), 5469–5502. https://doi.org/10.3390/s100605469. PMC 3247717. PMID 22219672.
10. Gilfrich, J., Burkhalter, P., & Birks, L. (1973). X-ray spectrometry for particulate air pollution—a quantitative comparison of techniques. *Analytical Chemistry, 45*(12), 2002–2009. https://doi.org/10.1021/ac60334a033. PMID 4762375.
11. Hickman, A., Baker, C., Cai, X., Delgado-Saborit, J., & Thornes, J. (2018). Evaluation of air quality at the Birmingham New Street Railway Station. *Proceedings of the Institution of Mechanical Engineers, Part F: Journal of Rail and Rapid Transit, 232*(6), 1864–1878. https://doi.org/10.1177/0954409717752180. PMC 6319510. PMID 30662169.

12. Jacobson, M. Z. (2012). *Air pollution and global warming: History, science, and solutions*. Cambridge University Press. ISBN 9781107691155. Retrieved March 29, 2022.
13. Jha, R. K. (2021). Non-dispersive infrared gas sensing technology: A review. *IEEE Sensors Journal, 22*(1), 6–15. https://doi.org/10.1109/JSEN.2021.3130034. S2CID 244564847. Retrieved March 29, 2022.
14. Jiao, W., Hagler, G., et al. (2016). Community Air Sensor Network (CAIRSENSE) project: Evaluation of low-cost sensor performance in a suburban environment in the southeastern United States. *Atmospheric Measurement Techniques, 9*(11), 5281–5292. https://doi.org/10.5194/amt-9-5281-2016. PMC 7425750. PMID 32802212.
15. Karagulian, F., Gerboles, M., Barbiere, M., Kotsev, A., Lagler, F., et al. (2019). Review of sensors for air quality monitoring: EUR 29826 EN. Luxembourg: Publications Office of the European Union. ISBN 978-92-76-09255-1. Retrieved March 28, 2022.
16. Lewis, A., Lee, J., Edwards, P., Shaw, M., Evans, M., et al. (2016). Evaluating the performance of low-cost chemical sensors for air pollution research. *Faraday Discussions, 189*, 85–103. https://doi.org/10.1039/C5FD00201J. PMID 27104223. Retrieved March 28, 2022.
17. Nicklin, D., & Darabkhani, H. G. (2021). Techniques to measure particulate matter emissions from stationary sources: A critical technology review using multi criteria decision analysis (MCDA). *Journal of Environmental Management, 296*, 113167. https://doi.org/10.1016/j.jenvman.2021.113167.
18. Ohlwein, S., Kappeler, R., Kutlar Joss, M., Künzli, N., & Hoffmann, B. (2019). Health effects of ultrafine particles: A systematic literature review update of epidemiological evidence. *International Journal of Public Health, 64*(4), 547–559. https://doi.org/10.1007/s00038-019-01202-7. PMID 30790006. S2CID 67791011.
19. Richter, P. (1994). Air pollution monitoring with LIDAR. *Trends in Analytical Chemistry, 13*(7), 263–266. https://doi.org/10.1016/0165-9936(94)87062-4. PMID 4762375. Retrieved March 28, 2022.
20. Saravanan, D., & Kumar, K. S. (2023). Improving air pollution detection accuracy and quality monitoring based on bidirectional RNN and the Internet of Things. *Materials Today: Proceedings, 81*, 791–796. https://doi.org/10.1016/j.matpr.2021.04.239.
21. Sorek-Hamer, M., Chatfield, R., & Liu, Y. (2020). Strategies for using satellite-based products in modeling PM2.5 and short-term pollution

episodes. *Environment International, 144*, 106057. https://doi.org/10.1016/j.envint.2020.106057.

22. Srivastava, A., & Majumdar, D. (2011). Monitoring and reporting VOCs in ambient air. In N. Mazzeo (Ed.), *Air Quality Monitoring, Assessment and Management* (pp. 137–148). InTech Open. ISBN 978-9533073170. Retrieved March 30, 2022.
23. The Babcock & Wilcox Company. (2005). *Steam: Its generation and use*. The Babcock & Wilcox Company. ISBN 0-9634570-1-2.
24. World Health Organization. (2020). *Methods for sampling and analysis of chemical pollutants in indoor air*. Copenhagen, Denmark: WHO Regional Office for Europe. ISBN 9789289055239. Retrieved July 29, 2023.
25. Zarrar, H., & Dyo, V. (2023). Drive-by air pollution sensing systems: Challenges and future directions. *IEEE Sensors Journal*. https://doi.org/10.1109/JSEN.2023.3305779. hdl:10547/625961. ISSN 1530-437X.

Chapter 7

# Emissions Inventory

## Contents

# INTRODUCTION

A pollution inventory, also known as an emissions inventory, is a list of all the pollutants released into the atmosphere. The total emissions for one or more particular greenhouse gases or air pollutants, originating from all source categories in a given geographic area and within a specified time span, usually a specific year, are typically included in an emission inventory.

The following features typically define an emission inventory:

- **Why**: The types of activities that cause emissions
- **What**: The chemical or physical identity of the pollutants included, and the quantity thereof
- **Where**: The geographic area covered
- **When**: The time period over which emissions are estimated
- **How**: The methodology to use

Emission inventories are gathered for use in policy-making processes as well as scientific applications.

## 7.1. PURPOSE OF EMISSIONS INVENTORIES

While not the only motivation, regulations are frequently the driving force behind creating an emissions inventory. Air emission inventories are specifically required by statutes or regulations governing emissions in order to ascertain the quantity of pollutants released into the atmosphere. For instance, base year inventories required for State Implementation Plans (SIPs) are outlined in the Clean Air Act (CAA) as amended in 1990. Furthermore, a national inventory of emissions from a specific industry or category of emission source is frequently required for the development of regulations. Smaller-scale uses for facility inventories include evaluating compliance with current emission regulations or permit conditions, conducting environmental impact assessments for proposed new sources, and providing information for human health risk assessment studies under various air quality management programs.

For many years, state and local organizations have used emissions inventories to measure the amount of pollutants emitted within their jurisdictions. Comprehending the diverse aims and objectives of an emissions inventory is crucial for devising an efficient planning process. In order to take into account as many factors as possible that could affect emissions in future years, agencies must also develop projections for their emission inventories. When future emissions are predicted using predicted activity levels and emissions controls, baseline emission estimates are extrapolated. This process is known as emission projections. Planning, assessing possible control measures, analyzing the effects of new sources, modeling future air quality, and evaluating the efficacy of air pollution control strategies for State Implementation Plans (SIPs) are all frequently done using projected emissions estimates. Because it provides a snapshot of emissions for a specific baseline year, a baseline emissions inventory is significant. However, baseline emissions might not accurately reflect emissions for a subsequent year due to the dynamic nature of sources and the air emissions they are associated with. The goal of emission projections is to take future growth and emissions controls into consideration. Typically, projections are not regarded as a step in the modeling inventory development process. Generally speaking, these processes are thought to be restricted to speciation, temporal allocation, and spatial allocation. In actuality, future projections of inventory emissions that have been gridded, speciated, and temporally allocated are not typically made. Rather, projections are made using the annual baseline emissions data. After that, these anticipated emissions can be speciated, temporally, and spatially allocated to produce modeling inventories of the anticipated emissions.

### 7.1.1. Historical Records

To find out what kinds of pollutants are present and in what amounts, a control agency may decide to perform an emissions inventory. Prior to the development of a control strategy, it is imperative to quantify the emissions of pollutants produced within a designated geographic area and comprehend historical emission patterns. In general, a thorough list of source emissions will be included in even the most basic inventory.

### 7.1.2. Temporal and Spatial Distribution

Determining the temporal and spatial distribution of pollutants in an area is the second use of the inventory. It is frequently insufficient to merely be aware of the kinds and total quantity of source emissions present in a sizable geographic area. When and where hazardous pollutants are emitted can either exacerbate or, on the other hand, lessen their effects. A pollutant's effects could get worse in the presence of moisture or sunlight, or they might not be sufficiently dispersed from a concentrated area by the atmospheric conditions.

### 7.1.3. Emission Reactivity

Agencies can also take into account the behavior of pollution in relation to other pollutants or natural elements in the surrounding environment by using an emissions inventory. Nearly every organic substance released into the atmosphere experiences a photochemical reaction. On the other hand, certain substances react more quickly or locally than others, making their effects on air quality more concentrated. Furthermore, contaminants released into the atmosphere might not have a significant impact on human health or welfare. However, they might become extremely toxic when mixed with other local emissions. Consequently, a thorough examination of every organic emission in a region is necessary for the development of a successful control plan.

### 7.1.4. Compliance

To ascertain the potential impact a pollutant may have on its surroundings, an emissions inventory is required. In addition to being able to recognize and control elevated pollutant levels, air pollution specialists need to be able to drastically cut down on or completely remove any pollutants that endanger public health or the ecosystem. One step in that direction is the area emissions inventory. This kind of inventory better reveals excess exposure during regular monitoring periods and efficiently accounts for all known types and amounts of pollutants. An agency that is mandated to use pollution control strategies to lessen or eradicate toxic substances within an area needs to have a precise accounting of all emissions. In order to establish a pattern of compliance with National Ambient Air Quality Standards, an accurate emissions inventory is frequently the first piece of information needed.

### 7.1.5. Regulatory Development

The emissions inventory is also used to inform the creation of public policy. An extensive analysis of emissions yields important insights into which pollution sources need more stringent regulation and how the new emission standard may affect air quality. Important data from the inventory is also contributed to public discussions evaluating the social and economic effects on the region. In conclusion, the inventory offers a helpful way to decide whether to grant or renew source permits in a certain area.

### 7.1.6. Ambient Air Monitoring

The emissions inventory is a valuable tool for helping choose locations for efficient ambient air quality monitoring stations. An agency's emissions data is combined with available ambient air measurements, meteorological data, and existing data to determine which locations would be best for further air monitoring and sampling. One of the best ways to assess violations of the National Ambient Air Quality Standard (NAAQS) and track source performance and progress is to use emissions data to identify the best locations for ambient air monitoring. Other requirements from the agency. The emissions inventory is also used to fulfill an air pollution control agency's general information needs.

The capacity to keep up-to-date emission data on current sources, monitor and report industrial growth in a region, and supply permit information for new source applications as mandated by public law are a few examples of such needs. An emissions inventory design should, therefore, be as adaptable as possible to accommodate the provision of any extra data that may be needed.

These seven applications for the emissions inventory only scratch the surface of possible uses for one. When resources are adequate, a well-built inventory can be as complex or detailed as time and resources allow. But ultimately, the emissions inventory's goal is to support the creation of a successful local strategy for reducing air pollution.

## 7.2. ELEMENTS OF AN EMISSIONS INVENTORY

The planning and design of a regional inventory can start as soon as there is some agreement on its goals or purposes. All things considered, the emissions inventory gives the pollution control executive the opportunity to research how an emission affects ambient air quality. A region's future needs and requirements can be reliably determined through effective inventories, even if the final design is more comprehensive and detailed than what is mandated by law. A few key components need to be included in order to create an inventory that is effective.

### 7.2.1. Geographic Area

The establishment of precise boundaries for a suitable geographic area to be researched is the first step in the emissions inventory process. An emissions inventory's scope and detail may be inflated by a very large area. Determining a boundary that will not contain more sources than can be sufficiently inventoried and monitored is crucial for this reason. The accuracy of the emissions inventory is contingent upon the agency's capacity to precisely identify and sample every point, area, and mobile source of pollution. For instance, a petroleum refinery might have a large number of emission points, from individual pumps and seals to process heaters. An area's need for additional staff and resources to manage its emissions inventory rises in tandem with the number of industrial profiles in the area. By focusing on the sources that are most concerning, agencies with limited resources could reduce the inventory.

### 7.2.2. Spatial and Temporal Characteristics

The location and timing of emissions are described by a source's spatial and temporal features. The features that define the position of point sources using UTM (Universal Transverse Mercator) grid coordinates are known as spatial characteristics. It is preferable to find the eight- or ten-digit grid coordinate location for each source, even though these coordinates can be roughly expressed in latitude and longitude. Emission points can be located more precisely when using UTM coordinates (usually within 5 meters of their exact location), grids are continuous and unaffected by political subdivisions, and UTM coordinates are a universal standard of measurement. The type and amount of pollutants released over time are reported by a source's temporal characteristics, which can range from annual to hourly observations and show how a source changes over time. Because the majority of activity happens once a year, emissions data are traditionally reported for a calendar year. However, in order to appropriately reflect the site's performance patterns, the agency might occasionally need to change the annual emission rates to seasonal, daily, or hourly rates. For instance, certain sources might only be available at specific times of day or year. Additionally, the time of day and season of the year have a significant impact on the activity levels of some mobile sources, surface coating, and pesticide operations. Because hot tar materials are used, road resurfacing projects are typically best finished in the warmer months. The pollution control agency must take these and other temporal and spatial variations of emissions into account (Figure 7.1).

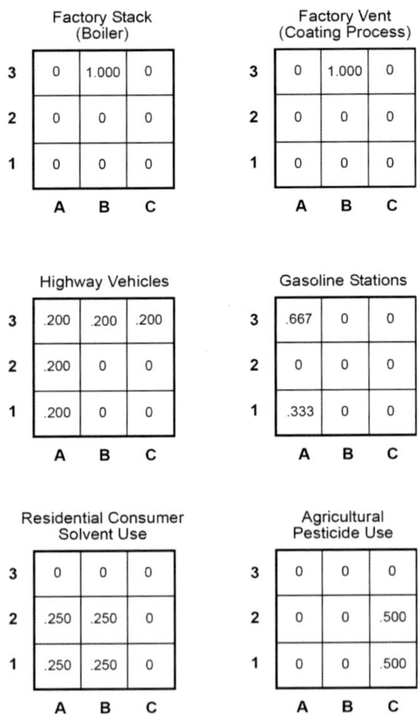


**Figure 7.1.** *Spatial allocation of emissions to grid cells.*

*Source:* https://www.4cleanair.org/wp-content/uploads/APTI/452combined.pdf.

### 7.2.3. Source Specific Data

The agency must next decide what source-specific data will be needed once the geographic boundaries and the kinds of data that will be recorded for a given area have been correctly defined. The location of an exhaust stack parameters, operational control devices, and additional physical attributes of the emission source are examples of this data. The creation of a useful emissions inventory depends on the accuracy of plant or facility data.

### 7.2.4. Pre-Existing Inventories

If there are any pre-existing inventories for the region or a portion of the area under consideration, it is another tool that is often overlooked. Pre-existing inventories can be especially helpful for organizations whose resources are scarce or diminishing. In these circumstances, it might be easy to update or remove pertinent information from the current emissions inventory. An existing inventory could, at the very least, offer a useful foundation upon which to build a new assessment. Even though there are a lot of unanswered questions regarding the best way to use an earlier inventory, in the end, these choices must be made based on the agency's current knowledge base, the potential goals of a new inventory, and the amount of resources at its disposal.

### 7.2.5. Data Handling

The last component of the emissions inventory is the procedure for handling and retrieving data. The two main methods for handling data effectively are computer-assisted data recovery and manual data recovery. The availability of computer hardware and software, the amount and complexity of data to be processed, the time available, the presence of trained staff, and other factors all play a role in choosing which approach to use. As the database grows, computer data processing becomes noticeably more economical. Additionally, using a computer increases the efficiency of performing repetitive administrative tasks like filing routine status reports, emission summaries, and data storage and sorting. An agency can spend less time processing inventory data and more time collecting and validating it by utilizing an automated data handling system. For large areas with many different emission sources and a complicated inventory, a computerized approach is far superior. Nonetheless, it is advisable to decide to use computer-aided designs early on in the planning phase.

Setting a goal or purpose for the assessment must be done before assembling an extensive emissions inventory. The following action is to make sure that all necessary components are included in the planning process before any source data is gathered. Each component should be continuously examined and updated as efforts to precisely define an accurate source emissions inventory for a region advance in order to ensure the accuracy of the emissions inventory. The resulting emissions inventory will be a helpful tool in the creation of an efficient plan to reduce air pollution if this process is closely monitored.

## 7.3. QUALITY ASSURANCE AND QUALITY CONTROL PROCEDURES FOR EMISSION

Applying the [IPCC] good practice guidelines can help countries choose methods (tiered approaches), emission factors, and activity data in an efficient manner. It is useful for a number of tasks, including choosing suitable techniques and emission factors, assessing and quantifying uncertainty, identifying important categories, recalculating emissions data, and creating plans for quality assurance and quality control.

- **Quality Control:** The inventory development team has put in place a system of regular technical tasks to monitor and manage the inventory's quality as it is being completed. The use of authorized standardized procedures for emission calculations, technical reviews, and accuracy checks are examples of QC procedures.
- **Quality Assurance:** A framework of outside audit and review processes run by staff members unrelated to the inventory process. QA is an impartial, independent review conducted by a third party to evaluate the quality, completeness, accuracy, precision, and representativeness of the inventory as well as the efficacy of the QC program. Nonetheless, the resulting emissions inventory will be a helpful tool in the creation of an efficient air pollution control strategy if this process is closely monitored.

### 7.3.1. Key GHG Quality Reporting Concepts

It's critical to comprehend the fundamental ideas that underpin quality reporting before developing and putting into place controls to enable trustworthy reporting. Relevance, completeness, consistency, transparency, and accuracy are the five main accounting reporting principles that underpin high-quality reporting, according to the World Resource Institute/World Business Council (WRI/WBCSD) for Sustainable Development. The WRI tenets have been embraced by the Climate Registry. These principles have the following definitions:

1. **Relevance**: Make sure the GHG inventory accurately represents the company's GHG emissions and meets the needs of users, both inside and outside the organization, for making decisions.

2. **Completeness**: Every source of greenhouse gas emissions within the designated inventory boundary must be taken into account and reported on. Any specific exclusions should be disclosed and justified.
3. **Consistency**: In order to enable meaningful comparisons of emissions over time, employ consistent methodology. Any modifications to the data, inventory boundary, procedures, or any other pertinent time series elements should be openly documented.
4. **Transparency**: Using a clear audit trail as a foundation, present all pertinent issues in a factual and logical manner. Make appropriate references to the accounting and calculation methodologies and data sources utilized, as well as disclose any pertinent assumptions.
5. **Accuracy**: As far as can be determined, make sure that the quantification of GHG emissions is consistently neither over nor under actual emissions, and that uncertainties are minimized to the greatest extent possible. Reach a level of accuracy that allows users to decide for themselves.

Verifiers evaluate GHG reports according to these five reporting principles, so PSOs should create and put in place controls that produce GHG data that is accurate, transparent, complete, and relevant.

### *7.3.1.1. Customizing Controls to Suit the PSO*

No set procedure or set of "standard" controls exists to guarantee high-quality GHG reporting. The controls must be adjusted to fit the organization's size and reporting structure.

#### *7.3.1.1.1. Reporting Structure*

The distribution of duties and accountability for data and reporting among public sector organizations (PSOs) will be specified by the design structure of the GHG reporting function. It will be up to the organizations to choose between a decentralized and centralized reporting structure.

- In a highly centralized model, a small group of important people at the core of the company will be given reporting responsibilities, such as gathering and documenting GHG consumption data.
- A highly decentralized model will assign reporting roles to multiple people throughout the PSO, and key personnel at the organization's center will then compile and report the data.

The reporting design structure may be based on a decentralized, highly centralized, or hybrid model. The PSO's innate control strengths and weaknesses as well as the kinds of controls required to close control gaps will be influenced by the kind of reporting structure. For instance, because data processing is done by multiple people in a highly decentralized model, there is an inherent risk of inconsistent reporting procedure application. Under these conditions, the PSO should create policies that codify processes throughout the company and set up training to ensure that procedures are followed consistently.

#### *7.3.1.1.2. Size of Organization*

In order to compile GHG reports and assign accountability and responsibilities for data sets to a data owner, PSOs will need to access data sources that offer proof to support emission totals. Creating an efficient and well-managed GHG reporting process requires the data to be transferred from those responsible for data collection to those responsible for data reporting.

The PSO's size and the reporting structure's layout may have an impact on the volume of data handed over.

- A smaller company might only have one data handover, from the person in charge of gathering emissions source data to the one in charge of compiling and reporting all of the PSO's data.
- A larger organization might have multiple data handovers: from the different people in charge of gathering data from emissions sources (e.g., buildings, fleets, paper, and fugitives) to the person in charge of compiling the data to the person in charge of uploading and verifying consumption data for the entire PSO.

The quantity of data handovers will affect the business's intrinsic control strengths and weaknesses as well as the kind of controls required to close control gaps. For instance, there is an inherent risk in a small organization where a key individual controls the majority of the data that there is no separation of duties and that the person in charge of data collection also oversees reviews of the data's accuracy. Under these conditions, the PSO ought to create procedures that allow for formal reviews of data by impartial third parties unaffiliated with data collection, or that record procedures for verifying accuracy twice over.

### *7.3.1.2. GHG Reporting Resources*

- **Infrastructure for Greenhouse Gas Reporting:** The BC government manages SMARTTool, a web-based tool for aggregating, estimating, and reporting greenhouse gas emissions. It offers a standardized method for figuring out and disclosing an organization's emissions. The Data Compiler enters the final, compiled usage data into this tool.

- **GHG Data Sources:** These sources offer proof to back up emission totals. Table 7.1 describes typical data supporting the four emissions.

**Table 7.1.** GHG Reporting Categories and Supporting Data

| Category | Examples of Supporting Data |
|---|---|
| 1. Buildings | • Lists of buildings/office space leased or owned<br>• Electricity and heating invoices or usage reports<br>• Property area totals with % usage by the PSO<br>• Buildings sold or acquired during the reporting period<br>• Leases closed or signed during the reporting period<br>• Subleases signed and to whom (e.g., another PSO, private organization, public organizations not part of the Government Reporting Entity, such as a local government)<br>• At times, Utility Vendor Disclosure Authorization Forms may need to be obtained from landlords to obtain permission to energy usage data for leased buildings |
| 2. Fleet | • Lists of vehicles leased or owned with description of intended use (e.g., operations, business travel); type of vehicle, type of fuel used<br>• Data provided from an outsourced vehicle management company · Fleet purchases made on a vendor specific fuel cards<br>• If applicable, mileage records, gas consumptions totals, odometer readings |
| 3. Fugitives | • Maintenance records<br>• Estimation methodologies |
| 4. Paper | • Invoices from vendors, which contain the type and quantity of paper<br>• Purchasing records from Accounts Payable |

*Source:* https://www2.gov.bc.ca/assets/gov/environment/climate-change/cng/guidance-documents/bc_pso_emission_reporting_quality_control_and_quality_assurance_guidance.pdf.

### 7.3.1.3. Control Types

A PSO can set up a GHG reporting structure that works by implementing a variety of control types. The following is a list of the various control variations along with relevant examples.

#### *7.3.1.3.1. Preventative vs. Detective Controls*

The goal of preventative controls is to stop misstatements before they start. Preventive controls consist of the following examples.

- A mandate that the relevant GHG source data be included with the "unit of measure."
- Training on how to upload data into SMARTTool

The goal of detective controls is to find instances of prior misstatements. Detective Controls do not stop errors from happening; they only show that errors have happened. Detective controls have the following examples.

- **Reconciliations:** On a regular basis, the individual in charge of data capture compares supplier statements to paper purchase invoices that are received.
- Examines consumption data for reasonableness, spots discrepancies, looks into any inaccuracies, fixes them, and records the review process.

A well-managed greenhouse gas reporting process should incorporate both detective and preventative controls, even though the former are typically thought to be more effective.

#### *7.3.1.3.2. Manual vs. Automated Controls*

Manual controls are preset actions that staff members physically start and finish. For example:

- The person in charge of GHG reporting verifies in person that the assets in the fleet inventory match those mentioned in PSO records.
- To make sure they match, year-end consumption data is compared to data owners' totals; management examines and looks into any notable deviations.

Automated controls serve as checkpoints within an information system or application to increase the probability that the control objectives will be met. Employee interaction is not necessary for system controls to operate once they have been set up correctly, but testing and monitoring the controls is a good idea. Spreadsheets can incorporate formulas that are locked down to prevent manipulation and minimize errors in manual calculations and compilation. Because of the possibility of human error, manual controls are typically thought to be less effective than automated controls. Nevertheless, even highly automated reporting processes should include some manual controls to ensure that automated controls are operating as intended.

#### *7.3.1.3.3. Controls, Quality Assurances and Audits*

The activities of quality control are performed by the people who generate the data. Quality Controls are thought to be the most successful because they prioritize accurate reporting from the start, even though they are not infallible. Control actions carried out by people other than those who are primarily in charge of reporting are known as quality assurances. Controls for quality assurance are thought to be helpful in establishing the organization's expectations and tone. Therefore, the lack of strong Quality Assurance controls frequently results in weak Quality Controls since the main people in charge of reporting consider it to be less significant.

An organization's internal audit, or IA, examines the quality and adequacy of the quality assurance and control systems in place as well as the accuracy of the greenhouse gas reports. Individuals outside of the reporting function should conduct internal audits or checks. Because quality assurance and control measures are in place and are working effectively, audits should ideally only find minor errors. Any mistakes discovered during audits ought to raise questions about how well-functioning the quality assurance and control systems are. In addition, some companies might work with an outside auditor before self-certifying and being verified. That being said, it is not necessary. PSOs should strive to enhance quality assurance and controls, and audits should be utilized as a complementary activity to evaluate control efficacy and pinpoint areas in need of improvement. Improving internal confidence in the GHG statement and advancing the integration of GHG reporting into an organization's culture can often be achieved by enlisting the assistance of an internal audit group within the organization to review the quality control and quality assurance procedures as well as the self-certification check-list documentation.

### 7.3.2. GHG Quality Reporting Framework

Accurate GHG reporting requires internal controls and a well-designed process across the entire organization. The GHG Quality Reporting Framework (Framework) is described below. The Framework can be used as a reference to help create and put into place more robust procedures and controls in order to produce accurate and comprehensive GHG reports. Controls are particular actions that guarantee correctness of consumption data, mitigation of risks, and ultimately, compliance of organizations with GHG reporting obligations. The set of steps needed to finish a GHG report is referred to as a GHG reporting process. To produce accurate data, controls are incorporated into the larger GHG reporting process. Creating high-quality GHG reports also depends on the process's design framework. While utilizing the Framework does not guarantee PSOs that regulatory data quality requirements will be satisfied, it does offer guidance that gives reporters flexibility in terms of how PSOs can choose to develop and put into place particular processes and controls.

The Framework (see Figure 7.2) is broken down into parts that are related to entity level controls (shown in green) and important reporting processes (shown in blue). These parts are defined below. Accurate GHG reporting is dependent on entity-level controls as well as reporting process-specific controls. In the following section of the Manual, the main ideas for creating quality controls for GHG reporting are explained, allowing PSOs to create customized controls for every Framework component. Entity-level controls, which are highlighted in green below, include governance establishment and implementation as well as controls that are applicable to the entire organization. They are intended to support PSOs in creating the organizational framework and culture necessary to support high-quality GHG reporting. The following process-specific controls—Design, Capture, Compile, Record, Report and self-certify, and Archive—should be created to support the adherence of consumption data to one or more of the five reporting principles. These controls are indicated in blue below.

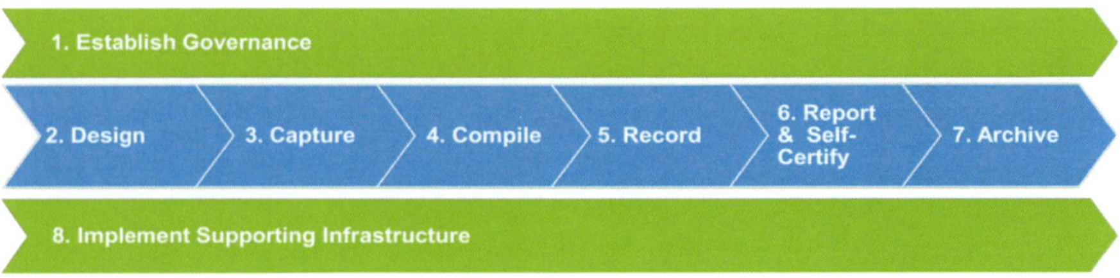


**Figure 7.2.** *GHG reporting framework.*

*Source:* https://www2.gov.bc.ca/assets/gov/environment/climate-change/cng/guidance-documents/bc_pso_emission_reporting_quality_control_and_quality_assurance_guidance.pdf.

### *7.3.2.1. Entity-Level (Organization-Wide) Controls*

The environment of the organization is impacted by governance and supporting infrastructure, which are not always related to a single process or source of emissions. Creating an environment that encourages high-quality reporting throughout the company requires establishing a governance/oversight structure and related systems and tools. These categories of controls could consist of:

- a clear and formalized governance framework for greenhouse gas reporting that is specified in a policy document;
- systems in place that automate the reporting process; and
- automated controls such as lock down functions.

### *7.3.2.2. Process-specific Controls (Key Process Controls)*

The particular sub-processes required for GHG reporting are represented by the Framework components in the diagram's middle band. For each reporting sub-process that is a component of the system, controls are required to address the issues listed in Table 7.2.

**Table 7.2.** *Key Reporting Processes*

| **Process-Specific Controls** | **Related Control Issues** |
|---|---|
| Design | How is the overall reporting structure set up to meet CNGR requirements? |
| Capture | How is all relevant data timely and accurately captured? |
| Compile | How is data compiled and aggregated for reporting? |
| Record | How is data prepared for reporting and recorded into SMARTTool enabling traceability of the data? |
| Report & Self-Certify | How is the GHG report reviewed and validated? How is the submission self –certified? |
| Archive | How is data documented and retained for future verification? |

*Source:* https://www2.gov.bc.ca/assets/gov/environment/climate-change/cng/guidance-documents/bc_pso_emission_reporting_quality_control_and_quality_assurance_guidance.pdf.

## 7.3.3. Establish Governance

By influencing employees' awareness of GHG reporting control, establishing an efficient governance structure for GHG reporting sets the tone for the organization. It serves as the basis for every other component. A governance structure for GHG reporting that is inadequate or ill-defined may result in:

- Roles and responsibilities are not clearly defined, which leads to the repetition or omission of important tasks.
- Reporting tasks being carried out by insufficient or unqualified personnel, which increases errors.
- Inadequate managerial supervision creating a negative atmosphere regarding data quality, which causes reporting errors to go unnoticed.

On the other hand, a well-resourced reporting governance structure promotes accountability, facilitates efficient oversight, and establishes a standard for high-quality GHG reporting throughout the company. It is easier to ensure that reporting responsibilities are fulfilled and not needlessly repeated when roles and responsibilities are clearly defined.

### *7.3.3.1. GHG Reporting Roles*

Depending on the size and complexity of the organization, each PSO may have different roles and responsibilities for those involved in the GHG reporting process. It's common for smaller organizations to be flatter and for one person to play multiple roles. Greater levels of stratification are anticipated in larger organizations, where distinct individuals assume distinct roles within the organization.

- **Designated Representative:** At the highest level of the organization, the Designated Representative is in charge of financial accountability. They should ideally be unrelated to the procedures used to gather and compile the data. Their main duties are to supervise the procedures, policies, and related control activities related to data collection. They also carry out a crucial authorization function by examining and approving the completeness and accuracy of consumption data. In the end, it is the Designated Representative's responsibility to make sure that Data Compilers and Owners have the required knowledge and abilities, that their job descriptions clearly outline their duties, and that they have enough time to finish their data collection and reporting obligations. The Designated Representative is in charge of reporting and attesting to the accuracy of the consumption data needed to finish the GHG report. A person at the Executive level, such as the Chief Financial Officer or someone else typically in charge of public disclosure, must be a Designated Representative.
- **Data Compiler:** In order to guarantee the consistency, accuracy, and completeness of data, the data compiler plays a crucial facilitation role. Either gathering and combining data or supervising data collection from each data owner falls under the purview of the data compiler. They have to make sure the data was uploaded correctly or upload the data into SMARTTool. They are also responsible for verifying that the necessary records and proof are in place for the Designated Representative to self-certify, as well as reviewing the GHG report produced by SMARTTool.
- **Data Owner:** The Data Owner and the Data Compiler should ideally be two different people. The Data Owner's job is to gather and, if necessary, compile usage information for the period's emission category or categories. For instance, gathering energy and heating

bills for their buildings falls under the purview of the building data owner. The Data Owner might also be responsible for uploading data into SMARTTool, depending on how the Data Owner and Data Compiler's roles are divided.

The roles and responsibilities associated with GHG reporting for medium-sized and large PSOs are depicted in Figure 7.3, while those for small PSOs are shown in Figure 7.4.

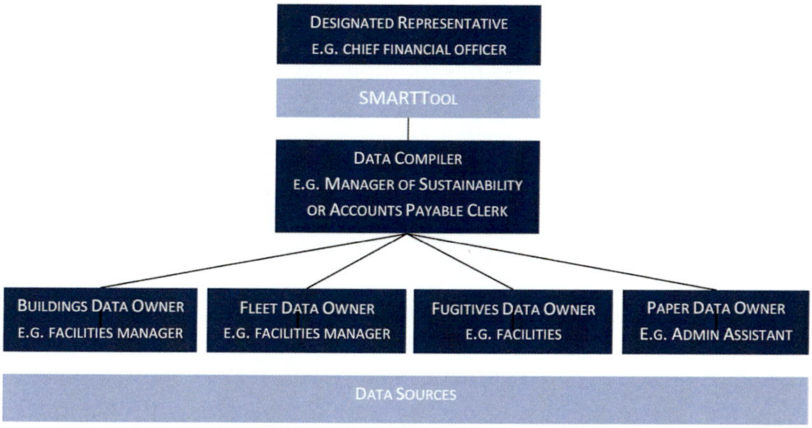


**Figure 7.3.** *Sample roles and responsibilities for GHG reporting in a medium to large PSO.*

*Source:* https://www2.gov.bc.ca/assets/gov/environment/climate-change/cng/guidance-documents/bc_pso_emission_reporting_quality_control_and_quality_assurance_guidance.pdf.

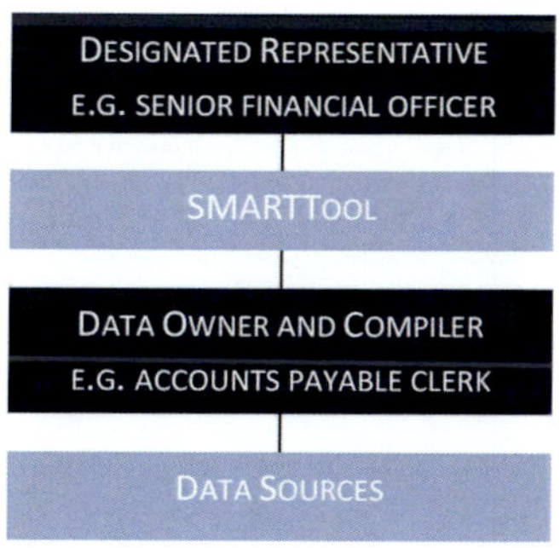


**Figure 7.4.** *Sample roles and responsibilities for GHG reporting in a small PSO.*

*Source:* https://www2.gov.bc.ca/assets/gov/environment/climate-change/cng/guidance-documents/bc_pso_emission_reporting_quality_control_and_quality_assurance_guidance.pdf.

### 7.3.4. Documenting a GHG Reporting Governance Structure and Process

Stakeholders are given guidance by documented reporting procedures on how to carry out their roles and responsibilities with respect to GHG reporting. This may be very significant in the event that:

- Many internal and external parties, including PSO employees and outside data providers like Terasen Gas and B.C. Hydro, are involved in the reporting process.
- The reporting process has multiple data handoffs.
- Data requirements are complex and require detailed direction.

PSOs will want to record procedures so that all parties involved in the process have proper guidance and reporting practices are consistent. Maintaining consistency will reduce the likelihood of mistakes. Additionally, documented procedures reduce "key person risk" by ensuring that staff turnover is consistently reported on.

In order to map data flows from the Data Owner to the Designated Representative and to give guidance on GHG roles and responsibilities throughout the organization, procedures should be sufficiently detailed. Error risk is reduced early in the process by giving Data Owners clear instructions. Table 7.3 describes typical information that a GHG reporting procedures document provides to Data Owners.

**Table 7.3.** Typical Details Provided to Data Owners

| **Procedural Reporting Directions** | **Examples** |
|---|---|
| Name of GHG category | Paper from facility A |
| Type of GHG data required | Invoices from paper purchased |
| GHG data unit of measure | Reams of paper |
| Consumption data source | Invoices pulled from Accounts Payable database |
| Individual charged with data capture of data from data stream | Facilities Manager from facility A: John Smith |
| Reporting period | Invoices of paper purchased within calendar year (January 1st to December 31st) |

| | |
|---|---|
| Reporting timeline | Invoices for previous year to be reported as soon as possible, following calendar year end, and no later than March 31st annually to meet CNGR requirements (or alternatively establishing a quarterly reporting requirement) |
| Data handoff | Invoices to be provided to Manager of EHS |
| Data retention requirements | Original invoices provided by supplier are to be sent with consumption data to Data Compiler for retention; data is to be retained for 7 years (In many cases this work is already being completed by Accounts Payable staff). |
| Data quality control activities | On a quarterly basis invoices pulled from Accounts Payable database will be reconciled against supplier summary statements. (In many cases this work is already being completed by Accounts Payable staff). |

*Source:* https://www2.gov.bc.ca/assets/gov/environment/climate-change/cng/guidance-documents/bc_pso_emission_reporting_quality_control_and_quality_assurance_guidance.pdf.

Every time data is transferred to additional stakeholders involved in the reporting process, a similar level of detail should be recorded. For instance, a procedures document should explain how the Data Compiler aggregates data streams from different Data Owners and what steps are taken to properly categorize data in order to prepare it for reporting into SMARTTool. Furthermore, procedures that give guidance on the steps taken to provide high-quality data should include additional information on activities completed to support the GHG reporting. Typical supporting material for GHG reporting procedures consists of the following:

- Introduction, purpose of document and directions on how to use it;
- The Methodology for Reporting B.C. Public Sector Greenhouse Gas Emissions describes the reporting boundaries, and any interpretations, assumptions, and decisions made around them are described;
- Description of roles and responsibilities for GHG reporting;
- Contact details for the individuals in charge of GHG reporting throughout the company;
- A summary of the governance, oversight, reporting, and change management processes;

- Entity wide QA/QC activities and data retention policy; and
- Infrastructure maintenance processes.

#### *7.3.4.1. Communication and Training*

A procedures document's implementation should be finished in a way that is clearly communicated. Notifying stakeholders ahead of time that they are expected to use the documents and communicating the initiative's value is important. If the document specifies a required procedure, it must also specify the consequences for following the procedure incorrectly. Involving stakeholders in the process of developing procedures will improve their eventual adoption and usage. To ensure that outdated versions are not being used, the Designated Representative should keep procedures in one central location. To ensure that users are aware when they are using an outdated document, version status should be explicitly stated in the procedure. Users also need to have easy access to procedures so they can get information whenever they need it. To facilitate use by individuals lacking technical background in GHG reporting, a glossary ought to be incorporated into the document. Additionally, a support resource can be made available to offer further explanation or direction beyond that which was stated in the procedures document. If it is determined that the stakeholders engaged in GHG reporting lack the necessary skills to carry out their duties, then training ought to be given to them.

### 7.3.5. Develop Quality Control Procedures

The purpose of this section is to offer advice on how to create quality controls that handle hazards during particular stages of the GHG reporting process. Each distinct process step—Design, Capture, Compile, Record, Report and self-certify, and Archive—is defined in this section, along with the associated control areas and topics of discussion.

#### *7.3.5.1. Design*

The process of creating a GHG reporting procedure that will enable the PSO to produce a report that satisfies CNGR regulations is referred to as the "design step." In accordance with legal requirements, this entails defining the PSO's reporting boundary, locating emission sources across all pertinent assets, and recording the protocols for tracking, gathering, reporting, and archiving GHG data. PSOs should create controls in this reporting step that handle risks related to defining the reporting boundary, making sure the reporting design structure is comprehensive, and documenting procedures. Below is a list of relevant control areas.

- **Establish Reporting Boundary:** Reporting boundaries delineate the domains that an entity is accountable for. Reporting requirements for emissions from buildings, fleets, fumigatives, and paper consumption have been imposed under the CNGR. The Scope Summary for B contains details on reporting limits under CNGR. C. Greenhouse gas emissions in the public sector. Applying the reporting requirements in the CNGR to the organization's assets is the main goal of the established reporting boundary step. It is essential to record the interpretations, presumptions, and choices made regarding reporting boundaries under the CNGR in order to create a case history for uniform implementation. For instance, PSOs have to determine the parameters by which buildings fall under the purview of reporting, taking into account the ways in which regulations delineate organizational, operational, and geographic attributes.
- **Completeness of Process Design:** One control area that can increase comfort that data capture, recording, and compilation processes are made to report all necessary source data is the completeness of process design. For instance, what steps does the PSO take to make sure that all pertinent data—such as invoices from corporate outsource fuel suppliers and expense reports—has been gathered for both owned and leased vehicles? How do you know that all possible sources have been explored?
- **Documentation of Procedures:** When carrying out their duties and responsibilities, controls in this area give stakeholders access to pertinent guidelines and system documentation. An impartial third party who was not involved in the documentation's development should assess the draft procedures to ensure that the information is sufficient to support future reporting and external verification even in the absence of the original author. That is, the information should be clear on its own and not require further explanation.

### *7.3.5.2. Capture*

The process of gathering all primary data needed to complete the GHG Report and upload it into SMARTTool is referred to as capture. This could consist of:

- Manually gathering all paper purchase invoices from suppliers.
- Collaborating with a primary supplier to deliver a single, combined invoice for multiple office locations, or.
- Creating an automated procedure to retrieve supplier data from a database that stores purchasing data.

PSOs will create controls in this step to handle risks related to accuracy, timeliness, and completeness. Below is a list of relevant control areas.

- **Completeness:** This part is concerned with making sure the data gathered is comprehensive. Where present, management systems (e.g., Oracle, SAP) that are integrated with primary data capture systems should have manual checks in place to ensure all automatic data transfers occur as designed. As an example, if information systems are used to capture and transfer data, what procedures are in place to gain comfort that this transfer is complete? What protocols are in place to verify that all invoices have been included if the supplier invoice data collection process is done manually? For instance, are there checks made against additional summary statistics? Has a procedure for validation been put in place to reduce human error?
- **Timeliness:** Making sure data is collected in time for the reporting period is referred to as timeliness. This means that when GHG data must begin to be collected in order to satisfy reporting requirements, data owners must be informed in advance. Data Owners ought to be aware of the CNGR regulation beforehand, as it contained comprehensive details regarding the due dates for reporting.
- **Accuracy:** This area deals with making sure the source data is as accurate as possible and bringing any identified inaccuracies down to a manageable level. When creating accuracy controls, keep in mind that people are more likely to put in the time and effort to confirm the accuracy of the data if they are asked to submit primary data documentation or attest to it. To make sure that data is entered accurately and completely, for instance, regular reconciliations are carried out between the original data documentation and the compiled document (such as an aggregated Excel spreadsheet).

### *7.3.5.3. Compile*

The process of gathering all collected data for each emission category is called the Compile step. This could entail compiling information from multiple data sources, categorizing consumption data, and reconciling the data to make sure there are no mistakes. PSOs will create controls in this step to handle risks related to whether or not consumption data has been correctly gathered, categorized in accordance with legal requirements, and reconciled to ensure accuracy. Below is a list of relevant control areas.

- **Accuracy:** This part is concerned with making sure the compilation is done correctly. Including whether the system generates accurate compilations and whether formulas and emissions factors are used in the calculation or processing of data. For instance, it's important to finish routine configuration checks for spreadsheets.
- **Classification:** This area of controls focuses on enforcing consistency and integrity by grouping data into groups. The goal of classification is to guarantee that every source of emissions is appropriately matched to the reporting requirement and applied consistently. For instance, fleet vehicle classification ought to be uniform, and the PSO ought to decide how to handle certain vehicles if there is uncertainty about them. Documented guidelines that staff members regularly adhere to when preparing reports should contain all decisions and interpretations.
- **Reconciliation:** Reconciliation controls check the accuracy of performance by comparing current period records to historical and projected performance. A satisfactory explanation for any anomalies found during this review and analysis should be looked into, and any mistakes found should be fixed. Regular data analysis and reconciliations can, if at all possible, lessen the amount of errors and irregularities that must be corrected as reporting periods come to an end.

### *7.3.5.4. Record*

The process of entering all collected and recorded data into the SMARTTool input sheet and uploading it to the program is referred to as the "Record step." PSOs create controls during the recording step to manage risks related to whether or not the recorded consumption occurred and was validated, as well as whether or not the recorded consumption data matches the CNGR reporting cut-off period. Below is a list of relevant control areas.

- **SMARTTool Entry:** Due to people's frequent pressure to meet reporting deadlines, manual entry errors are common in the recording process. Such errors can be minimized by cross-referencing manually entered data into SMARTTool with source data reports.
- **Cut-off:** The risk that data is not recorded within the appropriate time frame is addressed by cut-off. The mandatory GHG reporting period and the consumption data reporting period are rarely in sync in practice. Data on invoices or usage may cover more than one period; additionally, data may not be received until after the

reporting period or may overlap with data from a prior period. Procedures that remove data when required and estimate data that is not available will need to be developed to handle such scenarios. Correcting cut-off issues may require some degree of judgment; in such cases, either an appropriately experienced individual should be in charge of such actions, or guidance should be given.

- **Occurrence:** The question of whether recorded data accurately depicts actual consumption activities is raised by occurrence. This is particularly important for data that have been estimated or in which there is a significant chance of double counting. Occurrence risks can be adequately covered by automated controls that are able to identify common errors, like double counting. A reasonableness check can be carried out by someone with knowledge of the collected data to ensure that the recorded totals make sense. For instance, since it's improbable that any heating was used, a high natural gas bill in the middle of the summer may not be justified. This could mean that there is a mistake on the invoice and that emissions are shown that were not there.
- **Validation:** Controlling the dependability of data streams is the subject of validation. Data Compilers cannot simply rely on the information they receive from the Data Owner; rather, they, and the Designated Reprehensive in the end, are accountable for comprehensive, accurate, and documented consumption data. Examining the reported data's integrity before entering and uploading it into SMARTTool involves several important steps, including validation.

Review and validation controls address how each data stream can be suitably examined and analyzed before being uploaded into SMARTTool, as well as how this process will be recorded. For instance, assurance testing techniques like interviewing, observing, and corroborating can all be applied to examine and verify the analysis performed on each data stream before uploading it into SMARTTool. The Designated Representative should review the data as it is submitted for recording and uploading. Contextual information about data should be included, such as performance trends in relation to historical or projected data. This enables the Designated Representative to spot any irregularities and raise questions about them. If GHG data is routinely reported to senior management and utilized in decision-making, the Designated Representative will also feel more at ease about the accuracy of the data. Given that the Designated Representative holds a senior management position, it is improbable that they will have the time to thoroughly verify the accuracy of the reported consumption data. Organizations therefore usually depend on proof that controls are in place and that they have been carried out by accountable parties during the reporting process.

### 7.3.5.5. Report and Self-Certify

PSOs will seek to create controls in this step that address risks related to whether or not GHG data has been correctly entered into the SMARTTool, management has examined and verified the report, and the relevant person has completed the self-certification. Below is a list of relevant control areas.

- **Management Review and Self Certification:** The GHG report that is produced must be examined and approved by the PSO for reasonableness, as SMARTTool converts the consumption data into GHG emissions on the PSO's behalf. In the end, the Designated Representative attests that the GHG report has been examined and that the consumption data provided by the PSO are accurate and complete. Once more, the review serves as a reasonableness check to see if the GHG report makes sense, even though the Shared Services B.C. Carbon Measurement and Reporting Branch bears the responsibility for the accuracy of the GHG quantification methodologies and emission factors, which the Designated Representative may not be familiar with. You can contact the Shared Services B.C. Carbon Measurement and Reporting team with any problems you find.

It is also acknowledged that in order to guarantee that procedures and resources are in place to guarantee that data is reported in compliance with requirements, the Designated Representative depends on management, employees, or quality control. Nonetheless, it is the duty of the Designated Representatives to personally review and provide the necessary oversight to guarantee that the PSO's submissions are truthful, accurate, and comprehensive.

### *7.3.5.6. Archive*

The process of recording and keeping track of GHG data in order to furnish adequate proof for upcoming verification is known as the "Archive step." This entails recording the reporting procedure and keeping records for the necessary amount of time in a secure location. In order to comply with verification requirements, PSOs will create controls in this step to guarantee that the necessary documentation has been created and maintained for seven years. Below is a list of relevant control areas.

- **Documentation:** The process of documenting involves creating an audit trail, which starts with the documented design of the greenhouse gas reporting process. This includes creating an evidence trail from source data, such as each fuel delivery invoice, and documenting every control activity that is carried out. The idea behind creating an audit trail is that every reported piece of data should be easily traceable back to its original source and vice versa. The audit trail's documentation ought to be electronic and backed up. In addition to saving time and money during verification, a well-documented audit trail enables organizations to easily transfer reporting roles and responsibilities among employees in the event of a staff turnover. Furthermore, there's a chance that an audit trail for consumption data will intersect with pre-existing financial accountability systems. In such instances, creating and recording the audit trail for consumption data may involve referencing an already-existing financial audit trail.
- **Retention:** Retention controls pertain to the capacity to safely store the GHG report, self-certification document, and any additional pertinent supporting documentation for the seven-year period specified by law. The recorded audit trail should be kept secure and apart from the original documents. Additionally, documents must be kept in a system, and the associated files must be regularly backed up.

## 7.3.6. Implement Supporting Infrastructure

The kind of controls required to ensure the quality and accuracy of data will depend on the kind of information technology systems used to record, aggregate, and report GHG emissions data. Accounts Payable personnel currently use a variety of accounting programs, which can be broadly divided into four categories and used appropriately for this purpose:

1. Excel spreadsheets that are formatted to suit reporting purposes
2. "Custom Build" data systems that have been designed and built to suit the organization
3. "Off the Shelf" emissions management systems purchased as implemented by the supplier
4. Enterprise Resource Planning (ERP) systems that have GHG modules

Of all the system options, using a spreadsheet is thought to be the most manual process, whereas using an ERP system is thought to be the most automated. A lot of systems come with built-in automated controls to improve the accuracy of the data. The quality of the GHG data generally improves with the number of automated controls incorporated into the reporting process. This is due to the fact that automated controls reduce the possibility of human error, which can happen in a process involving manual data manipulation, transcription, and handover. Regardless of the degree of automation achieved in reporting, process controls are necessary to support and enhance automated controls. In actuality, an over-reliance on automated controls may result in the neglect of important process control gaps. The technological options listed above can help with data collection and compilation prior to input into SMARTTool.

#### *7.3.6.1. Building Automated Controls into a Manual Process*

Spreadsheets are usually used by PSOs with a largely manual reporting process to track and report GHG data. Because they are customizable, require human data entry, have limited data backup capabilities, and lack security features, spreadsheets can be error-prone. Therefore, PSOs that depend on spreadsheets ought to try to automate reporting controls by utilizing the features that are already present in spreadsheets. The following are some instances of automated spreadsheet controls.

- Formulas are entered into a spreadsheet to automate any calculations needed for data aggregation or compilation, and formula-containing cells are locked down.
- Labeling every data row and column carefully and securing labeled cells to prevent modification of the report's structure.
- Include units of measure in data row/column labels
- When feasible, include value restrictions for data cells (e.g., fuel supplies are inserted in L).
- Set lock down restrictions on spreadsheet sections that should only be accessible by specific stakeholders (data compiler can only access the spreadsheet's aggregated data, for example).

Manual process controls are still necessary for tasks like validation, reconciliation, and data entry in reporting processes, even if the spreadsheet controls are highly automated.

### 7.3.7. Custom Quality Control Development Process

The high-level, three-step method that PSOs can use to create controls that promote high-quality GHG reporting throughout their company is provided below:

- **Step 1. Identify GHG Reporting Risks:** Perform a process walkthrough of all the data categories that comprise the GHG inventory in order to identify any risks within the current reporting process based on the key control components listed in the GHG Quality Reporting Framework. Interviewing all pertinent parties involved in the GHG reporting process, such as the Data Owner and the Designated Representative, is usually required for this. It also entails tracking the data's journey from capture to recording in SMARTTool and archiving for future reference.
- **Step 2. Define Control Objectives:** The underlying control objective to manage the risk of significant greenhouse gas reporting should be defined once it has been identified. This will give guidance on the controls that are required. The goal of control objectives should be to ensure that the quality of GHG reporting meets at least one of the five essential accounting reporting principles: accuracy, completeness, transparency, consistency, and relevance. Establishing which principles are being addressed makes it easier for PSOs to create the right controls.
- **Step 3. Develop Customized Controls:** Provide specialized controls to address the identified risk and assist in meeting the relevant quality reporting principle(s). When designing a control, keep in mind all the different control options listed in the Key GHG Reporting Quality Concepts section of this chapter (e.g., automated, manual, preventive, detective, etc.). An organization will usually have a number of suitable control options available to it. In such a scenario, PSOs ought to choose controls in accordance with an assessment of the resources that are currently available, the seriousness of the risk that has been identified, and the amount of work necessary to mitigate the risk. It is important to remember that controls and risks are managed collectively when creating controls. Therefore, robust controls in one area of the reporting process could make up for lax controls in another. For instance, a PSO may make up for relatively lax quality controls in data capture and recording by implementing robust quality assurance measures to verify and authenticate the capture and recording of all pertinent data. When creating customized GHG reporting controls, PSOs are encouraged to involve their internal audit team or personnel with experience in controls, as this process of control development only offers high level guidance.

## REFERENCES

1. Alam, T., & Shukla, A. K. (2020). Vehicular emission inventory of Lucknow. In *Lecture Notes in Civil Engineering* (pp. 469–477). Springer Singapore. https://doi.org/10.1007/978-981-15-6463-5_44.
2. Jing, B. Y., Wu, L., Mao, H. J., Gong, S. L., He, J. J., Zou, C., Song, G. H., Li, X. Y., & Wu, Z. (2015). Development of a high temporal-spatial resolution vehicle emission inventory based on NRT traffic data and its impact on air pollution in Beijing – Part 1: Development and evaluation of vehicle emission inventory. *Atmospheric Chemistry and Physics Discussions, 15*(19), 26711–26744. https://doi.org/10.5194/acpd-15-26711-2015.
3. Khummongkol, P., Jermsawatdipong, P., Pungchit, L., Boonpragob, K., Santisirisomboon, J., Decosta, I., & Shrestha, S. (1996). Greenhouse gas inventory in Thailand. In *Greenhouse Gas Emission Inventories* (pp. 201–222). Springer Netherlands. https://doi.org/10.1007/978-94-017-1722-9_12.
4. Kuenen, J. J. P., Visschedijk, A. J. H., Jozwicka, M., & Denier van der Gon, H. A. C. (2014). TNO-MACC_II emission inventory: A multi-year (2003–2009) consistent high-resolution European emission inventory for air quality modelling. *Atmospheric Chemistry and Physics Discussions, 14*(5), 5837–5869. https://doi.org/10.5194/acpd-14-5837-2014.
5. Li, J., Hao, Y., Simayi, M., Shi, Y., Xi, Z., & Xie, S. (2019). Verification of anthropogenic VOC emission inventory through ambient measurements and satellite retrievals. *Atmospheric Chemistry and Physics, 19*(9), 5905–5921. https://doi.org/10.5194/acp-19-5905-2019.
6. Ma, J., & van Aardenne, J. A. (2004). Impact of different emission inventories on simulated tropospheric ozone over China: A regional chemical transport model evaluation. *Atmospheric Chemistry and Physics Discussions, 4*(1), 507–532. https://doi.org/10.5194/acpd-4-507-2004.
7. Obioh, I. B., Momodu, S. A., Siyanbola, W. O., Oketola, F. O., Ibitoye, F. I., Akinbami, J. F. K., & Adegbulugbe, A. O. (1996). Nigeria's greenhouse gases inventory: Estimates for 1990. In *Greenhouse Gas Emission Inventories* (pp. 119–133). Springer Netherlands. https://doi.org/10.1007/978-94-017-1722-9_5.
8. Rada, O. P., Romero, S., & Tejada, F. (1996). National greenhouse gas inventory country study—Bolivia. In *Greenhouse Gas Emission Inventories* (pp. 347–355). Springer Netherlands. https://doi.org/10.1007/978-94-017-1722-9_22.
9. Ruiz, J., Ponce, J., Pizarro, C., Avila, J., Llamas, I., & Culqui, E.

(1996). Peru's greenhouse gas emissions inventory. In *Greenhouse Gas Emission Inventories* (pp. 367–373). Springer Netherlands. https://doi.org/10.1007/978-94-017-1722-9_24.

10. Sari, K. E. (2021). Energy greenhouse gas emission inventory in Batu City. *IOP Conference Series: Earth and Environmental Science, 916*(1), 012003. https://doi.org/10.1088/1755-1315/916/1/012003.
11. Shami, A. A., Al Aawar, E., Baayoun, A., Saliba, N. A., Kushta, J., Christoudias, T., & Lakkis, I. (2022). Updated national emission inventory and comparison with the Emissions Database for Global Atmospheric Research (EDGAR): Case of Lebanon. *Environmental Science and Pollution Research, 29*(20), 30193–30205. https://doi.org/10.1007/s11356-021-17562-8.
12. Touré, S., Antoine, G., Badaye, D., Drissa, D., & Ambroise, K. (1996). Côte d'Ivoire—Preliminary greenhouse gas inventory. In *Greenhouse Gas Emission Inventories* (pp. 99–109). Springer Netherlands. https://doi.org/10.1007/978-94-017-1722-9_3.
13. Yadav, S., Siddiqui, F., & Khanna, A. (2021). Sustainable inventory model with carbon emission dependent demand under different carbon emission policies. In *Soft Computing in Inventory Management* (pp. 163–175). Springer Singapore. https://doi.org/10.1007/978-981-16-2156-7_9.
14. Yang, D., Zhang, Y., Wu, R., Song, Y., & Xu, H. (2022). A review of on-road vehicle emission inventory. *E3S Web of Conferences, 360*, 01027. https://doi.org/10.1051/e3sconf/202236001027.

*Chapter* **8**

# Control of Stationary Source Emissions

## Contents

# INTRODUCTION

The two main types of sources of emissions of air pollutants are usually classified as stationary and mobile sources. Point and area sources are the two main subcategories into which stationary sources are further separated. A single emission source at a facility with a designated location is referred to as a point source. Multiple point sources could be situated on the premises of a facility. Boilers, spray booths, and degreasers are examples of point sources that are typically connected to industrial and manufacturing processes. Area sources include household water heaters, small engines, and consumer goods like hairspray and barbecue lighter fluid. Area sources are small, widely dispersed emission sources that have the potential to produce significant cumulative emissions. The management and reduction of air pollutants released from fixed sources, such as factories, power plants, refineries, and other industrial facilities, is referred to as "controlling stationary source emissions." This is essential for both environmental preservation and observing air quality laws.

## 8.1. CONTROL OF STATIONARY SOURCES, PARTICULATE MATTER

The specific process operations determine the types of air pollutants that stationary sources emit, both organic and inorganic. For instance, in 1993, the combustion of fuel from stationary sources contributed 34.2% of total air pollution, with mobile sources contributing 56.2% (Godish, 1997). Boilers that burn fossil fuels release sulfur dioxide, nitrogen oxides, ash, and, depending on the fuel, vanadium and/or mercury. In addition, "cenospheres," which are light, hollow spheres that can be covered in unburned carbon, can be produced under specific boiler firing conditions. Sulfuric acid ($H_2SO_4$) can be produced by the carbon's adsorption of sulfur trioxide ($SO_3$) and water. A range of metal dusts, such as iron oxides and occasionally fluorides and chlorides, can be released by metallurgical plants. Depending on the product they make, industrial facilities that produce inorganic chemicals will release different waste gases. Petrochemical and organochemical plants may also release odorous organic waste gases. These and other industrial pollutant sources can be divided into a number of groups according to the particular processes they engage in.

- **Process Operations:** Operations in the process that involve partial chemical reactions, such as combustion brought on by unconverted reactants or a reaction whose final yield falls short of the theoretically predicted conversion.
- **Atmospheric Releases:** Atmospheric releases of raw material impurities or secondary components of a process.

- **Auxiliary Losses:** The incidental losses of substances like volatile organic solvents from escaping sources or inorganic substances like hydrogen sulfide and carbon disulfide in rayon manufacturing and fluorine compounds in aluminum production.
- **Waste Emissions:** Emissions from heating, drying, or oxidation processes that have unpleasant smells or contain oxidation compounds.

These categories' releases may come from a range of emission sites and may not be centralized for collection prior to atmospheric entry. To find out how much of the air pollutant is being released and what kind of control measures might be needed, each category and source should be examined separately. There are several types of air release emission points associated with industrial process operations, including vent, fugitive, stack, duct, and area. Release sites that are unconfined in a stack or duct prior to entering the atmosphere are known as fugitive emission points. As a result, the classification of an emission source as a point or fugitive source depends on whether the release was contained or released outside of a stack or duct before entering the atmosphere.

Even though stationary sources contribute significantly to ambient air pollution, controlling these sources is a delicate matter because these sectors produce millions of jobs, thousands of useful consumer goods, and essential goods and services. Pollution from stationary sources must be controlled, but it is impractical to create an economic climate that drives facilities out of business. Therefore, it's important to strike a delicate balance between industry costs and safeguarding the environment and public health. According to the U.S. EPA, common sense measures like installing efficient control technology, altering production procedures, and putting pollution prevention strategies into practice can effectively reduce pollution. Application of suitable stationary source control measures is critical for both successful attainment of air quality standards and compliance with emission standards. Even though the quantity and nature of emissions from stationary sources vary widely, their effects can be lessened by using appropriate control measures. This section's goal is to review stationary source emission control's guiding concepts and procedures.

### 8.1.1. Control Procedures

An effective control strategy can be implemented to reduce emissions from stationary sources. Four steps make up the control strategy needed for an industrial environmental impact: (1) removing the source or operation causing the problem; (2) changing the source's operation; (3) moving the source; and (4) choosing and implementing the best control technology. Carefully designed tall stacks may be one of the methods used to reduce gaseous and particle emissions.

### *8.1.1.1. Exhaust Stacks*

Instead of lowering emissions from a stationary source, exhaust stacks lessen the local effects of pollution by raising the exhaust stream to a level where it can be dispersed more effectively. When costly control technology was not available, high exhaust stacks provided an affordable alternative. Elevated stacks were used for many years keeping the surrounding communities in mind. There was a general consensus that higher stacks lessened the possibility that pollutants would affect nearby populations in any way. Sulfur dioxide ($SO_2$) concentrations at ground level have historically been decreased by utility and smelter operations by using tall stacks (200–400 meters). Ambient air currents and other natural forces diluted and dispersed plumes from higher stacks. Therefore, spreading the pollution over the largest possible area was the goal of tall stacks. Still, this didn't always solve the issue; sometimes it just moved it somewhere else (Godish, 1997). Before worries about the regional and transboundary spread of dangerous toxics emerged, this strategy was employed for a long time. The movement of acid rain, for instance, across political boundaries, is a major source of concern. These are the issues that the United States, Canada, and European nations are currently discussing.

### *8.1.1.2. Plant Operations*

Control technology may be necessary to meet emission standards, but many industrial processes have already reduced emissions through operational modifications. Pretreating process materials, substituting materials or fuel, and altering the manufacturing process are a few of these modifications. Industry has found that a process known as coal washing can result in significant reductions in particulate matter and sulfur emissions. This is an example of how pre-treating raw materials can be a low-cost way to control pollution. This kind of pre-treating raw materials lowers the amount of inorganic sulfur released along with the amount of fly ash released from coal. Using cleaner fuels in place of more conventional ones during the refining process is another method to meet emission regulations. Two fuels that produce less pollution when burned are natural gas and low-sulfur fuel oil. Cleaner fuels, however, may cost more and increase the country's dependency on imported fuels. Paying more attention to plant maintenance can also help reduce emissions from stationary sources. Inadequate maintenance of equipment is often the cause of plants that discharge large amounts of pollutants into the environment. This particularly applies to combustion-based equipment. Reducing the amount of exhaust and fugitive emissions released from vats, valves, and transmission lines requires properly scheduled maintenance. Regular maintenance also lowers the risk of spill-related mishaps by identifying malfunctioning machinery early on.

### *8.1.1.3. Control Technology*

Finally, adding sophisticated control technology to stationary sources can help cut emissions. Control devices have the ability to destroy or recover particulate matter or gaseous compounds for reuse or appropriate disposal. Combustion, adsorption, absorption, and condensation are the pollution control processes that are used to eliminate or seize gases. Thermal incinerators, catalytic incinerators, flares, boilers, process heaters, carbon absorbers, spray towers, and surface condensers are examples of control devices that carry out these processes. The exhaust gas characteristics from emissions tests and the process or site characteristics from a field survey are the most crucial process parameters for choosing air pollution control equipment.

Exhaust gas characteristics:

- Total exhaust gas flow rate
- Exhaust gas temperature
- Required control efficiency
- Particle size distribution
- Particle resistivity
- Composition of emissions
- Corrosiveness of exhaust gas over operating range
- Moisture content
- Stack pressure
- Exhaust gas combustibility and flammability properties

Process or site characteristics:

- Reuse/recycling of collected emissions
- Availability of space
- Availability of additional electrical power
- Availability of water
- Availability of wastewater treatment facilities
- Frequency of startup and shutdowns
- Environmental conditions
- Anticipated changes in control regulations

- Anticipated changes in raw materials
- Plant type – stationary or mobile

The goal of particulate matter control technologies is to eliminate particles from the stream of wastewater. Particulate size and chemical properties are just two of the many variables that affect which particulate control device is best for a given process. The most popular tools for managing particulate matter are cyclone collectors, fabric filters, venturi scrubbers, electrostatic precipitators, and settling chambers. A more thorough explanation of each kind of control technology for both gaseous and particulate emissions can be found in the section that follows.

## 8.1.2. Control Devices for Particulate Emissions

Any liquid or solid that has been finely divided is called particulate matter. Particulate matter is anything that is suspended in ambient air or entrained in effluent gas streams. Examples of particulate matter include dust, smoke, and some types of fine mist. PM10, or particulate matter with a diameter of less than 10 micrometers (μm), is regulated as a criteria pollutant. Controlling smaller, respirable particles that can get past the body's respiratory filters and deeply enter the lungs, where they may pose a health risk to people, is the aim of this kind of regulation. Particles in this size range mostly carry toxic materials such as sulfates, sulfites, nitrates, heavy metals, and polycyclic organic matter. Consequently, modern control devices concentrate on capturing particulate matter with a diameter of less than 10 μm in order to keep particles from entering the surrounding air.

A number of things need to be taken into account when choosing a particulate control device. Particles from an effluent gas stream usually need to be collected, so the right control device will depend on the properties of both the gas stream and the particles. Particle size and resistivity, exhaust flow rate, temperature, moisture content, and different chemical properties of the exhaust stream, such as explosiveness, acidity, alkalinity, and flammability, are among the characteristics that need to be taken into account. For instance, the electrostatic precipitator would be a terrible choice for any exhaust stream that contains high levels of volatile organic compounds (VOCs), even though it has a 99% removal efficiency for the majority of particles smaller than 10μm. Technology should not be used in this dangerous or inappropriate way if a highly explosive, volatile substance is exposed to a control device that is primarily dependent on electrical charge. Using a mix of different devices often results in the highest collection efficiency. Before the exhaust stream enters an electrostatic precipitator, which removes smaller particles, a settling chamber, for instance, can be used to remove larger particles. The most popular control mechanisms for adjusting particulate matter flow into surrounding air are listed below. The most popular tools for reducing particulate emissions are cyclones, fabric filters, electrostatic precipitators, venturi scrubbers, and settling chambers.

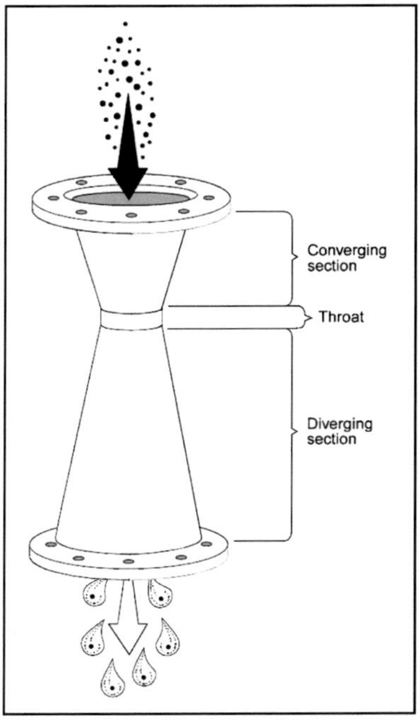


**Figure 8.1.** *Venturi scrubber.*

*Source: Public Domain,* https://commons.wikimedia.org/w/index.php?curid=2741168.

### *8.1.2.1. Venturi Scrubbers*

As seen in Figure 8.1, venturi scrubbers remove solid particles using a liquid stream. In order to atomize the scrubbing liquid and enhance gas-liquid contact, a venturi scrubber speeds up the waste gas stream. A "throat" portion that is integrated into the duct of a venturi scrubber causes the gas stream to accelerate as the duct constricts and then expands. Turbulence and gas velocity both rise as the gas enters the venturi throat. The scrubbing liquid is sprayed into the gas stream before the gas reaches the venturi throat, in the throat, or upwards against the gas flow in the throat, depending on the scrubber design. The turbulence in the throat then atomizes the scrubbing liquid into tiny droplets, increasing the interaction between the droplets and particles. Some designs enhance droplet creation with additional sprays that are atomized by pneumatic or hydraulic means. The drawback of these designs is that, in order to prevent clogging, clean liquid feed is necessary. Following the throat segment, the mixture slows down and more impacts happen, which causes the droplets to group together. Following the particles' absorption by the liquid, an entrainment section—typically comprised of a cyclonic separator and/or a mist eliminator—removes the wetted PM and surplus liquid droplets from the gas stream. Current venturi scrubber designs typically employ the vertical downflow of gas through the venturi throat and include three features: (1) an adjustable throat for the venturi throat to allow for adjustment of the gas velocity and the pressure drop; (2) a "flooded" elbow located below the venturi and ahead of the entrainment separator, to reduce wear by abrasive particles; and (3) a "wet-approach" or "flooded wall" entry section to prevent a dust buildup at a wet-dry junction. In order to prevent dust particles from abrading the venturi throat, a refractory lining is occasionally added. A single-throat venturi scrubber unit's typical gas flow rates are between 0.2 and 28 standard cubic meters per second ($sm^3$/sec) or 500 and 60,000 standard cubic feet per minute (scfm). Higher flows will require the use of multiple throated venturis or multiple venturi scrubbers operating in parallel (Perry, 1984).

Depending on the application, venturi scrubbers have particle collection efficiencies ranging from 70 to more than 99%. In general, PM with aerodynamic diameters of roughly 0.5 to 5μm has higher collection efficiencies. Certain venturi scrubbers have a throat that can be adjusted to regulate the pressure drop and gas stream velocity. Increasing the pressure drop is necessary to increase the efficiency of the venturi scrubber, which raises the energy consumption. Venturi scrubbers have been used to reduce particulate matter emissions from boilers used in commercial, industrial, institutional, and utility settings that burn coal, oil, wood, and liquid waste. They have also been used to control emission sources in the lead, aluminum, iron and steel, mineral products, chemical, wood, pulp and paper, rock products, and asphalt manufacturing industries; they have also been applied to municipal solid waste incinerators. Venturi scrubbers are usually used in situations where high fine particulate matter collection efficiencies are required. As a result, they can be used to manage emission sources that produce large amounts of submicron PM.

Wet scrubbers produce waste as a slurry or wet sludge for PM applications. This makes solid waste disposal and wastewater treatment necessary. The slurry is first treated in order to extract the solid waste from the water. After treatment, the water can be released or used again. But after the water has been removed, what's left over will be a solid or sludge. Solid waste can usually be landfilled if it is nontoxic and inert. The disposal of hazardous waste will be subject to stricter regulations. Solid waste occasionally has value and can be recycled or sold.

- **Advantages of Venturi Scrubbers:**
    - Capable of handling flammable and explosive dusts
    - Can handle mists in process exhausts
    - Relatively low maintenance
    - Simple in design and easy to install
    - Collection efficiency can be varied
    - Provides cooling for hot gases
    - Neutralizes corrosive gases and dusts
- **Disadvantages of Scrubbers:**
    - Effluent liquid can create water pollution problems
    - Waste product collected wet
    - High potential for corrosion problems
    - Requires protection against freezing
    - Final exhaust gas requires reheating to avoid visible plume
    - Collected PM may be contaminated, and not recyclable
    - Disposal of waste sludge may be very expensive

### *8.1.2.2. Electrostatic Precipitators*

An ESP is a PM control device that transfers particles entrained in an exhaust stream onto collection surfaces through the use of electrical forces. When the entrained particles move through a corona—a region where gaseous ions flow—they are charged electrically. The electrical field created by the high-voltage electrodes in the flow lane's center pushes the particles toward the collector plates. The high voltage electrodes consist of rigid "masts" or long wires that run through the axis of each tube and are suspended from a frame located in the upper part of the ESP. Typically, an upper and lower frame support rigid electrodes. In contemporary designs, extra ionization sites are provided by adding sharp points to the electrodes, either at the tube's entrance or all the way along in the shape of stars.

The industrial AC voltage (220 to 480 volts) is converted by the ESP's power supplies to pulsating DC voltage (20,000 to 100,000 volts) as needed. A "corona" is created when there is an electrical breakdown of the gas between the electrodes due to the voltage applied to the electrodes. Because a negative corona can support a higher voltage than a positive corona before sparking happens, the electrodes are typically given a negative polarity. From the electrode to the collecting pipe, the ions produced in the corona travel along electric field lines. Consequently, every electrode-pipe combination creates a charging zone that the particles have to travel through. The electrical forces acting on larger particles (diameters >10μm) are much stronger than those acting on smaller particles (diameters >1μm), because the larger particles absorb many times more ions.

Once the collection plates are completely full, the particulate is extracted using a mechanical process called "rapping" to loosen the particles from the plates. The gathered particulate matter descends into a hopper beneath the apparatus. Recently, sonic horns have been used to clean dry wire-pipe ESPs aurally. Compressed air is typically used to power the horns, which are usually cast metal horn bells. A metal plate that vibrates periodically disrupts the airflow, causing an acoustic vibration. Similar to a rapping system, the hopper receives the collected particulate by sliding downward. As the hopper fills up, it is periodically evacuated. A valve removes dust, which is subsequently transferred into a dust-handling system—like a pneumatic conveyor—and disposed of properly.

In wet ESPs, a liquid spray—typically water—washes the collectors either continuously or sporadically. A drainage system is used in place of the collection hoppers that dry ESPs use. After being collected, the wet effluent is frequently treated on-site. The exhaust gas in a wire-pipe ESP, also known as a tubular ESP, passes vertically through conductive tubes, most of which are running in parallel. The tubes can be arranged to resemble a square, hexagonal, or circular honeycomb. In order to minimize wasted space, square and hexagonal pipes can be stacked closer together than cylindrical pipes. Pipes are generally 7 to 30 cm (3 to 12 inches (in.) in diameter and 1 to 4 m (3 to 12 feet) in length. A source of wash water must be injected or sprayed close to the top of the collector pipes for wet ESPs to function, either continuously or on a timed basis. The rapping mechanism that dry ESPs typically use is replaced by this wash system. Particles are collected by the water and flow into a sump, where the fluid is pumped or drained. To cut down on the total amount of water needed, some of the fluid may be recycled. The remaining material is either pumped into a settling pond or goes through a dewatering phase before the sludge is disposed of.

Wet wire-pipe ESPs, and other ESPs in general, have very low-pressure drops (typically less than 13 millimeters [0.5 in.] water column) because they act only on the particulate to be removed, and only minimally hinder flue gas flow. Consequently, low energy requirements and operating costs are typical. Even with extremely tiny particles, they can achieve very high efficiencies. Additionally, operating costs are not too high. According to AWMA (1992), ESPs can function in vacuum or at high pressures of 1,030 kPa [150 psi]. They can also efficiently handle relatively high gas flow rates. Sticky particles, mists, and explosive or highly resistive dusts can all be collected by wet ESPs. The re-entrainment of particles caused by rapping, which dry ESPs are unable to effectively control, is eliminated by intermittent or continuous washing with a liquid. They can gather high resistivity particles, absorb gases or cause pollutants to condense and cool, and condition the gas stream thanks to the humid atmosphere created by washing in a wet ESP.

In addition to solid particles, liquid or aerosol particles in the gas stream are also collected and offer an additional way to rinse the collection electrodes. Wet wire-pipe ESPs also offer the benefit of lowering "sneakage" because the full gas stream is passed through the collection field. When it comes to hazardous materials, the wet wire-pipe ESP's ability to be tightly sealed to stop material leaks is an especially useful feature. For particles smaller than 10μm, the collection efficiency of an ESP is consistently around 99%. Wire-Plate Type: The design efficiencies of new equipment typically range from 99–99.9%, whereas the actual operating efficiencies of older equipment vary from 90–99.9%. The most significant factor affecting ESP collection efficiency is its size, though there are other factors as well. Treatment duration is based on size. A particle's chances of being collected increase with the length of time it spends in the ESP. Effectively maximizing electric field strength will increase the efficiency of ESP collection. Particle size distribution, dust resistivity, gas temperature, and chemical composition of both the dust and the gas all have an impact on collection efficiency. ESPs of the wire-pipe type typically have expensive initial costs. Wire discharge electrodes (approximately 2.5 mm [0.01 in.] in diameter) are high-maintenance items. The tops of the wires are susceptible to corrosion due to acid condensation and air leakage. Additionally, long weighted wires have a tendency to oscillate, bringing the middle of the wire closer to the pipe, which increases wear and sparking. Rigid electrodes, or "masts," are more common in modern ESP designs, as they virtually do away with the disadvantages of wire electrodes. Because ESPs are extremely sensitive to changes in gas stream conditions (flow rates, temperatures, particulate and gas composition, and particulate loadings), they are generally not appropriate for use in highly variable processes.

**Table 8.1.** *Cumulative PM, PM10, and PM2.5 Collection Efficiencies for Dry ESPs*

| **Application** | **Collection Efficiency (%)** | | |
|---|---|---|---|
| | **Total PM** | **$PM_{10}$** | **$PM_{2.5}$** |
| *Coal-Fired Boilers* | | | |
| Dry bottom bituminous) | 99.2 | 97.7 | 96.0 |
| Spreader stoker (bituminous) | 99.2 | 99.4 | 97.7 |

| | | | |
|---|---|---|---|
| *Primary Copper Production* | | | |
| Multiple hearth roaster | 99.0 | 99.0 | 99.1 |
| Reverbatory smelter | 99.0 | 97.1 | 97.4 |
| *Iron and Steel Production* | | | |
| Open hearth furnace | 99.2 | 99.2 | 99.2 |

*Source:* https://www3.epa.gov/ttncatc1/dir1/fdespwpl.pdf.

Installing ESPs in locations with limited space is challenging as well because these devices need to be relatively large in order to achieve the low gas velocities required for effective PM collection. It takes a relatively sophisticated workforce to perform maintenance, and extra care must be taken to protect workers from the high voltage. During gas ionization, the high voltage also causes the negatively charged electrodes to produce ozone molecules. Wet ESPs increase the complexity of a wash system, necessitating more handling care and treatment in many cases compared to a dry product, particularly if the dust can be recycled or sold.

Wet ESPs are typically required to be built of non-corrosive materials and are restricted to operating at stream temperatures below about 80° to 90°C (170° to 190°F). Wastewater handling needs to be taken into account for wet ESPs. Water containing the particles collected by the ESP may be released from the system to a solids-removing clarifier (part of the plant wastewater treatment system or specifically dedicated to the ESP) and then to final disposal for simple systems with harmless dusts.

More complex systems might need sludge and skimming removal, specialized equipment for clarification, pH adjustment, and/or dissolved solids removal treatment. In order to return the cleaner of the two treated water streams to the ESP, spray water from an ESP preconditioner may be treated independently from the water used to clean the ESP collecting pipes. Recirculating treated water to the ESP could get close to 100%. Additionally, dusts that have resistivities higher than 1,010 ohm cm are not recommended for collection in dry ESPs.

These particles are difficult to charge and, consequently, difficult to gather. On the collecting electrodes, high-resistivity particles also produce ash layers with extremely steep voltage gradients. Positively charged ions are injected into the back corona, or space between the discharge and collecting electrodes, as a result of electrical breakdowns in these ash layers. This lowers the collection efficiency and decreases the charge on the particles in this space. Low-sulfur coal combustion produces fly ash, which is usually difficult to collect due to its high resistivity.

### 8.1.2.3. Fabric Filters

Flue gas is directed through a densely woven or felted cloth in a fabric filter, collecting particulate matter from the gas through a combination of sieving and other mechanisms. Fabric filters come in a variety of shapes and sizes, including sheets, cartridges, and bags. Several fabric filter units are stored in one group. The most popular kind of cloth filter is a bag. The efficiency of collection can be greatly enhanced by the dust cake that the collected PM forms on the filter. Because fabric filters are typically set up in cylindrical bags, they are also known as baghouses. The dimensions of bags can range from 6 to 9 meters (20 to 30 feet) in length and 12 to 30 inches (5 to 12 centimeters) in diameter. To enable bag cleaning or partial bag replacement without completely stopping the fabric filter, groups of bags are arranged in isolable compartments (STAPPA/ALAPCO, 1996). Operating conditions play a significant role in determining the fabric selection. Certain textiles (i.e., polyolefins, nylons, acrylics, and polyesters) are only beneficial between 95° and 150°C (200° and 300°F), which are relatively low temperatures. It is necessary to use more thermally stable fabrics, like fiberglass, Teflon®, or Nomex®, for high-temperature flue gas streams.

Large fabric areas are needed for the practical application of fabric filters in order to prevent an intolerable pressure drop across the fabric. The air-to-cloth ratio, or the volumetric airflow to cloth area, is what determines the baghouse size for a given unit. The choice of air-to-cloth ratio is influenced by the cleaning technique, particulate loading, and characteristics. To prevent an excessive pressure drop caused by an overly heavy dust cake, a higher particulate loading will necessitate the use of a larger baghouse. For instance, a baghouse housing a 250 megawatt (MW) utility boiler might contain five thousand individual bags, totaling close to 46,500 square meters (500,000 square feet) in fabric area. Baghouse performance is determined by the type of fabric selected, the frequency and techniques of cleaning, and the characteristics of the particulates. One can select fabrics that will trap a higher percentage of particulate matter. To achieve this, a membrane with minuscule openings is applied to certain fabrics to improve the removal of submicron particles. But these fabrics are typically more costly. The dust cake can impact a fabric's ability to remove fine particles, so cleaning frequency and intensity are crucial factors in assessing removal efficiency. Excessive frequency or intensity of cleaning procedures can also reduce the fabric filter's removal efficiency. However, too little or ineffective removal will result in an excessively high baghouse pressure drop.

For many years, mechanical shaking of the bags has been a well-liked cleaning technique due to its efficiency and ease of use. Dusty gas enters the shaker through an inlet pipe during normal operation. When very large particles hit the baffle plate in the inlet duct and land in the hopper, they are taken out of the stream. The gas that is full of particulates is pulled into the filter bags from under a floor-mounted cell plate. The gas travels through the outlet pipe and leaves the bags after entering. A filter cake forms when the particles gather on the inside surface of the bags. The tops of the bags are fastened to a shaker bar in mechanical shaking units, which is rapidly moved (often horizontally) to clean the bags. The shaker bars can be manually operated or driven by mechanical motors in situations where frequent cleaning is not necessary (Figure 8.2).

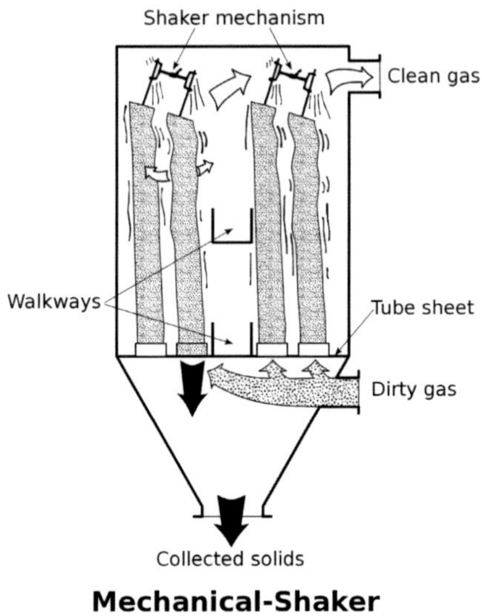


**Figure 8.2.** *Mechanical shaker baghouse.*

*Source: By Goran tek-en, CC BY-SA 4.0,* https://commons.wikimedia.org/w/index.php?curid=30458371.

Another well-liked technique for cleaning fabric filters is reverse-air cleaning, which has been refined and used extensively over time. It is a more delicate, though occasionally ineffective, cleaning method than mechanical shaking. Shaker-cleaned fabric filters and reverse-air fabric filters function similarly in most cases. The bags are normally closed on top and open on the bottom, allowing the gas to flow from inside to the outside while collecting dust inside. On the other hand, the exterior of some reverse-air designs tends to gather dust. Reverse-air cleaning is accomplished in both designs by pushing clean air through the filters in the opposite direction of the dusty gas flow. The bag flexes and the filter cake cracks as a result of the gas flow changing direction. During reverse-air cleaning, the bags used for internal cake collection are permitted to partially collapse. Typically, the bags are stitched with rings or another type of support to keep them from completely collapsing. The dust cake can drop off the bags and into the hopper thanks to the support. Because felted fabrics retain dust more than woven fabrics, the reverse flow of the gas also helps with cake release. They are therefore more challenging to clean. Reverse-air systems typically do not use felts because of this (Figure 8.3).

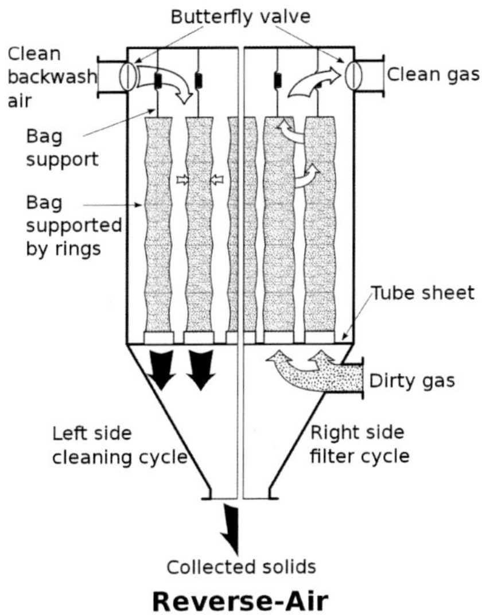


**Figure 8.3.** *Reverse air baghouse.*

*Source: By Goran tek-en, CC BY-SA 4.0,* https://commons.wikimedia.org/w/index.php?curid=30458372.

The flow through the filters can be reversed using a few different techniques. The most popular method, similar to mechanical shaker-cleaned fabric filters, is to designate distinct sections within the fabric filter. This allows each compartment to be cleaned and isolated independently while the remaining compartments continue to treat the dusty gas. One way to supply the reverse flow air is by using cleaned gas from the other compartments or a secondary fan. Only in situations where the dust readily separates from the fabric can reverse-air cleaning be employed alone. Reverse-air is frequently used in combination with sonic horns, pulsing, or shaking. Sonic horns are being used more frequently to improve mechanical shaker and reverse-air fabric filter collection efficiency. Compressed air is used in sonic horns to cause a metal diaphragm to vibrate, emitting a low frequency sound wave from the horn bell. The fabric area and the number of baghouse compartments determine how many horns are needed.

Generally, each compartment needs one to four horns running at 150 to 200 hertz. At 275 to 620 kiloPascals (kPa) (40 to 90 pounds per square inch gage [psig]), compressed air is supplied to power the horns. For approximately 10 to 30 seconds every cleaning cycle, sonic horns sound. The pressure drop across the filter fabric is reduced by 20–60% when the residual dust load on the bags is greatly reduced through sonic horn cleaning. Additionally, it reduces the mechanical strain on the bags, extending their useful life. As previously mentioned, this can reduce the OandM cost by 1% to 3% per year.

At coal-burning utilities, baghouse compartments are commonly equipped with fabric filters and can be easily modified with sonic horns. Another cleaning technique that was created in the 1950s to improve the removal of leftover dust is reverse-jet. Using this technique, a ring with a small opening around the bag receives reverse air via piping. When air passes through the slot, a fast-moving air stream is created, which causes the bag to flex at that location. A motor and cable system that travels up and down the bag powers a carriage that holds the ring. This technique offers a great way to remove any remaining dust. However, because of its complexity, there is a lot of maintenance involved. Additionally, the bags experience more wear due to air impingement. As a result, reverse-jet cleaning techniques have become less common in recent years.

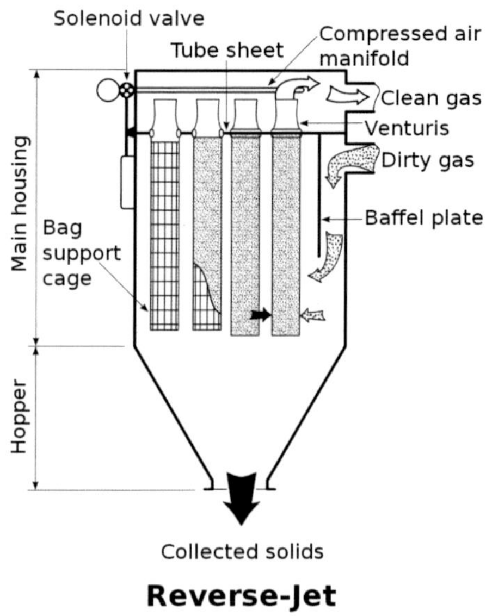


**Figure 8.4.** *Pulse jet baghouse.*

*Source: By Goran tek-en, CC BY-SA 4.0,* https://commons.wikimedia.org/w/index.php?curid=30458370.

Compared to other types of fabric filters, pulse-jet cleaning is a relatively new concept, having only been used for the last 30 years. But because it can handle heavy dust loadings, run at a constant pressure drop, and take up less space than other kinds of fabric filters, this cleaning mechanism has been more and more popular. Only external cake collection devices can be used with fabric filters that have been cleaned by pulse jet. The bags are held up by internal retainers known as cages and are closed at the bottom and open at the top. Diffusers are often used to stop oversized particles from damaging the bags as part of the process of particulate-laden gas flowing into the bag. The gas moves through the bags, entering from the outside and exiting through the gas exhaust. After that, the particles are gathered on the outside of the bags and fall into a hopper that is situated beneath the fabric filter (Figure 8.4).

A brief burst of high pressure air, lasting between 0 and 1 seconds and 415 to 830 kiloPascals (kPa) (60 to 120 pounds per square inch gage [psig]), is injected into the bags during pulse-jet cleaning. This cleaning cycle air is controlled by a remote timer that is linked to a solenoid. After that, the pulse is forced through a venturi nozzle at the top of the bags, creating a shock wave that extends to the bag's bottom. The dust cake is dislodged by the wave as it flexes the fabric and pushes it away from the cage before snapping it back. The solenoid valve regulates the air burst, which is then released into blow pipes with nozzles above the bags. Usually, the bags are cleaned one row at a time.

Pulse-jet cleaning has a number of special qualities. The cleaning process does not require stopping the flow of dusty gas because the cleaning pulse is so short. As other bags are cleaned, the remaining bags carry out additional filtering duties. Pulse-jet cleaning generally has no effect on the pressure drop or performance of the fabric filter. This allows solenoid valves to be the only major moving parts, allowing the pulse-jet fabric filters to run continuously. The intensity and frequency of pulse-jet cleaning are higher than those of the other techniques for cleaning fabric filters. Between each pulse of the bag and this thorough cleaning, almost all of the dust cake is removed. As a result, the filtration function of pulse-jet filters is independent of a dust cake. Pulse-jet fabric filters use felted (non-woven) fabrics because they can achieve high collection efficiencies without the use of a dust cake. Nevertheless, it has been discovered that after being cleaned, woven fabrics with pulse-jet fabric filters leak a lot of dust.

Pulse-jet cleaned fabric filters don't require additional compartments to maintain proper filtration while being cleaned because the bags are cleaned without having to be isolated. They can also treat higher gas flow rates with higher dust loadings due to the thorough and frequent cleaning. As a result, higher gas-to-cloth ratios can be achieved because fabric filters cleaned using the pulse-jet method can be smaller than other types of fabric filters when treating the same amount of gas and dust. High collection efficiencies are generally offered by fabric filters for both coarse and fine (submicron) particulates. The typical efficiency range for new equipment design is 99–99.9%. The actual operating efficiencies of older, still-in-use equipment range from 95–99.9%. The efficiency of fabric filter collection is determined by multiple factors, such as the cleaning mechanism, fabric characteristics, particle characteristics, and gas filtration velocity. Particle size and filtration velocity are generally correlated with increased collection efficiency. The effluent particle concentration from a fabric filter is almost constant for a given combination of dust and filter design, but the overall efficiency is more likely to change with particulate loading. Fabric filters are therefore more appropriately categorized as constant outlet devices than constant efficiency devices. Because part of the fabric filter is always being cleaned, constant effluent concentration is possible. The cleaning mechanisms in cloth filters cause the collection efficiency to fluctuate over time. Every cleaning cycle loosens the particles that are still on the filter and removes a minimum of some of the filter cake. The lost filter cake and loose particles are forced through the filter by the gas flow, which reduces the filtering capacity when filtration resumes. Efficiency rises as more particles are captured and held until the next cleaning cycle. Fabric filter average collection efficiencies are typically ascertained through tests that encompass multiple cleaning cycles with a consistent inlet loading.

Fabric filters are not very sensitive to changes in the conditions of the gas stream. For filters that are cleaned continuously, significant variations in the inlet dust loadings have little effect on efficiency or pressure drop. Because filter outlet air is so clean, it can frequently be recirculated inside the plant to save energy. The material that has been collected is gathered in a dry form for disposal or further processing. Component rusting and corrosion are typically not issues. Using a baghouse is a fairly easy process. Contrary to electrostatic precipitators, fabric filter systems don't need high voltage, which makes maintenance easier and allows for the careful collection of flammable dust. Submicron smokes and gaseous pollutants can be collected with high efficiency by using specific fibrous or granular filter aids (precoating). Because filter collectors come in so many different configurations, there is a wide range of dimensions and possible locations for the inlet and outlet flanges to meet installation needs. Particles with resistivities that are either too high or too low to be collected using electrostatic precipitators are collected using fabric filters. Therefore, fabric filters might be a good option for gathering fly ash from low-sulfur coals or fly ash with high levels of unburned carbon. These types of coals have high and low resistivities, respectively, and are therefore more challenging to gather using electrostatic precipitators.

### *8.1.2.4. Settling Chambers*

This kind of technology belongs to the category of air pollution controls known as "precleaners." Precleaners get their name from the fact that they are frequently used to remove larger, abrasive particles from the inlet loading of particulate matter (PM) to downstream collection devices. Other names for settling chambers include outfall chambers, expansion chambers, gravity collectors, and settling chambers. This is due to the fact that settling chambers can often be used in conjunction with other control devices because they are highly effective in removing only large particles. The earliest and most basic type of mechanical collectors are settling chambers, which use gravitational settling as a collection mechanism. Typically, settling chambers are constructed as long, rectangular, horizontal spaces with an entrance at one end and an exit on the side or top of the other. The chamber's flow needs to be consistent and free of macroscopic mixing. At the chamber's inlet, flow straighteners can help to improve uniform flow. The settled-out material is collected using hoppers, though screw conveyers and drag scrapers have also been used. In order to reduce turbulence, cause dust re-entrainment, and hinder the device's ability to discharge dust effectively, the dust removal system needs to be sealed to keep air from leaking into the chamber.

The expansion chamber and the multiple-tray chamber are the two main varieties of settling chambers. The gas expands in a sizable chamber inside the expansion chamber, greatly reducing the gas stream's velocity. Larger particles are able to separate from the gas stream due to the decrease in velocity. An expansion chamber with several thin trays placed closely together inside the chamber, allowing gas to move horizontally between them, is known as a multiple-tray settling chamber. When comparing a multiple-tray chamber to a simple expansion chamber, the collection efficiency usually improves because the particles have to fall a shorter distance before being collected, even though the gas velocity is slightly increased. For the collection of small particles (<15μm), multiple-tray settling chambers require less volume than expansion-type settling chambers.

The longer the waste gas stays in the settling chamber, the more efficient the chamber becomes. Settling chambers are therefore frequently run at the lowest feasible gas velocities. In actuality, the gas velocity needs to be sufficiently low to keep dust from reentering the chamber without going beyond reasonable limits. The ideal gas velocity within the unit, which should be less than 3 meters per second (m/s) (10 feet per second (ft/sec), and ideally less than 0.3 m/s (1 ft/sec), is what typically determines the unit size. The most frequent cause of settling chamber failure is clogged chambers from accumulated dust. Hopper bridging or a broken discharge seal can cause plugging in expansion settling chambers. The individual gas passages in multiple-tray settling chambers may become clogged. The use of hopper level indicators or ongoing dust discharge monitoring can stop or reduce these kinds of failures. A planned internal inspection can identify areas where condensation and air leakage may occur, both of which can lead to hopper bridging. Typically, a settling chamber's instrumentation consists solely of a differential static pressure indicator. Plugging may be indicated by an increase in the static pressure drop.

Particle size and settling chamber design both affect how well particles are collected in settling chambers. The best use cases for settling chambers are large and/or dense particles. When the settling velocity is higher than roughly 13 centimeters per second (cm/s) (25 feet per minute [ft/min]), particles may be removed by gravitational force. This generally holds true for particles larger than 50 μm in the case of low particle density and smaller than 10 μm in the case of a reasonably high material density. Smaller particles would need longer horizontal flow distances, which would result in larger chamber volumes. For PM with an aerodynamic diameter of less than or equal to 10μm (PM10), the collection efficiency is usually less than 10%. Lower volume requirements apply to multiple-tray chambers when collecting particles as small as 15μm.

Advantages of Settling Chambers:

- Low capital cost
- Very low energy cost
- No moving parts
- Few maintenance requirements
- Low operating costs
- Excellent reliability
- Low pressure drop through device
- Device not subject to abrasion due to low gas velocity
- Provide incidental cooling of gas stream
- Dry collection and disposal

Disadvantages of Settling Chambers:

- Relatively low PM collection efficiencies
- Unable to handle sticky or tacky materials
- Large physical size
- Trays in multiple-tray settling chamber may warp

### *8.1.2.5. Cyclones*

The term “precleaner” is also used to describe this kind of air pollution control because it also works by removing larger, abrasive particles to lower the amount of PM that enters downstream collection devices. In addition to cyclones, there are centrifugal separators, inertial separators, cyclone collectors, and cyclone separator separators. Multiple tube cyclones, multicyclones, or multiclones are terms used to describe systems in which numerous small cyclones are functioning in tandem.

Particles are extracted from the gas stream by cyclones using inertia. Centrifugal force is applied to the gas stream by the cyclone, which is often housed in a conical-shaped chamber. Inside the cyclone body, a double vortex is created to power the storm. Particulates are driven to the inner surface of the cyclone walls by the downward, circular motion of the incoming gas. The gas spirals up through the cyclone’s center and out of the top outlet after turning at the bottom of the storm. The fluid drag force of the gas passing through and out of the cyclone opposes the centrifugal force of the spinning gas, which pushes particles in the gas stream toward the walls of the cyclone. When larger particles approach the cyclone walls and gather, their inertial momentum overcomes the fluid drag force. Smaller particles exit the cyclone along with the departing gas when the fluid drag force exceeds their inertial momentum. The bigger particles that make it to the cyclone walls also descend into a bottom hopper due to gravity. Cyclones have a more intricate gas flow pattern than momentum separators, but they still use the same separation mechanism. This makes them more efficient.

Generally speaking, there are four different types of cyclones based on how the gas stream is introduced into the machine and how the dust collected is released. Tangential inlet, axial discharge, axial inlet, tangential inlet, peripheral discharge, and axial inlet, peripheral discharge are the four types. The most prevalent kinds are the first two.

Because pressure drop directly affects operating costs and control efficiency, it is a crucial parameter. Increasing inlet velocities for a given cyclone can result in higher control efficiencies, but doing so also causes a pressure drop. Generally speaking, the ideal operating velocity is 18.3m/s, or 60 feet per second. Cyclones typically experience pressure drops between 0.5 and 1 kilopascals (kPa) (2 and 4 in. $H_2O$) for high throughput, low efficiency units, 1 to 1.5 kPa (4 to 6 in. $H_2O$) for conventional medium-efficiency units and 2 to 2.5 kPa (8 to 10 in. $H_2O$) for devices with high efficiency. Many cyclones with many tubes sharing a common gas inlet and outlet in the chamber can be used when both high efficiency (requiring a small cyclone diameter) and large throughput are required. Axial inlet vanes allow the gas to enter the tubes and give it a circular motion. The wet cyclonic separator is another high-efficiency device that boosts control efficiency by combining centrifugal force and water spray.

PM larger than 10 micrometers (μm) in aerodynamic diameter is mainly controlled by cyclones. Nonetheless, there exist high-efficiency cyclones that are intended to function well for particulate matter with an aerodynamic diameter of less than or equal to 10 μm and less than or equal to 2.5μm ($PM_{10}$ and $PM_{2.5}$). While cyclones can be employed to gather particles larger than 200μm, less abrasive and generally more satisfactory options are gravity settling chambers or basic momentum separators. Particle size and cyclone design have an impact on cyclone collection efficiency. Particle size and/or density, inlet duct velocity, cyclone body length, number of gas revolutions within the cyclone, ratio of cyclone body diameter to gas exit diameter, dust loading, and smoothness of the cyclone inner wall are all known to increase cyclone efficiency. Increases in (1) gas viscosity, (2) body diameter, (3) gas exit diameter, (4) gas inlet duct area, and (5) gas density will result in a decrease in cyclone efficiency. Air leakage into the dust outlet is a common factor contributing to reduced control efficiencies in cyclones. Three categories of cyclones are commonly used to determine control efficiency ranges for individual cyclones (i.e., high-throughput, conventional, and highly efficient). For traditional single cyclones, the control efficiency range is estimated to be 70–90% for PM, 30–90% for PM10, and 0–40% for $PM_{2.5}$.

Compared to traditional cyclones, high efficiency single cyclones are intended to achieve greater control over smaller particles. High efficiency single cyclones can remove particles as small as 5 μm with up to 90% efficiency; larger particles can achieve even higher efficiencies. For high efficiency single cyclones, the control efficiency ranges are 20–70% for PM2.5, 60–95% for PM10, and 80–99% for PM. Increased pressure drops in more efficient cyclones mean that more energy is needed to force the waste gas through the cyclone. Rather than trying to meet a predetermined control efficiency, cyclone design is typically dictated by a pressure-drop limitation. While some collection of smaller particles does occur, high throughput cyclones are only guaranteed to remove particles larger than 20μm. For high-throughput cyclones, the control efficiency ranges are 0 to 10% for $PM_{2.5}$, 10 to 40% for PM10, and 80–99% for PM. It has been reported that multicyclones can collect particles as small as 5μm with an efficiency of 80–95%.

There are numerous uses for cyclones. Although cyclones by themselves typically fall short of meeting strict air pollution regulations, they do play a crucial role in precleaning more costly final control devices like fabric filters or electrostatic precipitators (ESPs). Cyclones are used in many process applications to recover and recycle food products and process materials like catalysts, in addition to their use in pollution control. Cyclones are widely used to collect salable or useful material, after crushing, grinding, and calcining operations in the mineral and chemical industries, and after spray drying operations in the food and chemical industries. Cyclones are frequently used as an initial step in the management of particulate matter emissions from sinter plants, roasters, kilns, and furnaces in the ferrous and nonferrous metallurgical industries. Cyclones remove particulate matter (PM) from the fluid-cracking process to enable catalyst recycling. Multiple cyclones, typically used upstream of a wet scrubber, ESP, or fabric filter, are commonly used in fossil fuel and wood waste fired industrial and commercial fuel combustion units. These cyclones are more efficient at collecting fine particulate matter (<2.5 μm) than a single cyclone. To increase the effectiveness of PM control, collected fly ash may occasionally be reinjected into the combustion unit.

Advantages of Cyclones:

- Low capital cost
- No moving part
- Few maintenance requirements
- Low operating costs
- Relatively low-pressure drop
- Dry collection and disposal
- Relatively small space requirements

Disadvantages of Cyclones:

- Relatively low PM collection efficiencies
- Unable to handle sticky or tacky materials
- High efficiency units may experience high-pressure drops

### *8.1.2.6. Momentum Separators*

Momentum separators are an improvement over settling chambers because they work by using strategically positioned baffles to force exhaust gases to abruptly change direction inside a gravity settling chamber. Because it is frequently used to remove larger, abrasive particles from the air, it is also referred to as a "precleaner," reducing the amount of particulate matter (PM) that enters downstream collection devices. Other names for momentum separators include knock-out chambers, baffle chambers, and impingement separators.

Momentum separators work by using strategically positioned baffles to force waste gas to abruptly change direction inside a gravity settling chamber.

Usually, the baffles force the gas to flow abruptly upward after it has initially flowed downward. Larger particles cross the gas flow lines and gather in a hopper at the bottom of the chamber as a result of the downward action of inertial momentum and gravity on the particles. Momentum separator designs must include enough volume to permit the settling of materials isolated from the high-velocity gas stream and materials durable enough to withstand severe abrasion. Similar to other mechanical collectors, the design needs to incorporate techniques for stopping the discharge of dust from hoppers in order to stop air leaks. The techniques could involve using flapper valves, rotary air locks, or other positive sealing devices. Leakage of air through the shell or hopper affects dust discharge, alters gas distribution, and can lead to corrosion or condensation. Construction materials need to be highly abrasion resistant due to the high velocities required to separate the particles from the gas stream and the impaction of these particles on surfaces that direct the gas flow. Particle size and momentum separator design both affect how well a momentum separator collects particles. Increases in particle size and/or density, gas stream velocity, and the quantity of turns, baffles, or other abrupt changes in gas flow direction all generally result in higher momentum separator efficiency. For the purpose of controlling fly ash, the EPA provides fractional collection efficiency curves for two louvered momentum separators. The fractional collection efficiencies are as follows: ≤5% for particles up to 5 μm in size, 10–20% for particles up to 10 μm in size, and up to 99% for particles up to 90 μm in size. Even though particulate control devices are made to collect and hold pollutants, the pollution must still be properly disposed of. Scrubber-generated wastewater is typically sent to treatment facilities, while collected particulate matter is typically disposed of in a landfill. To create more efficient methods of recycling and reusing this collected particulate matter, research is currently underway.

## 8.2. CONTROL OF STATIONARY SOURCES, GASEOUS EMISSIONS

Dangerous air pollutants (HAPs) and other criteria are primarily produced by stationary sources, which include power plants, chemical production facilities, oil refineries, manufacturing plants, printers, dry cleaners, and even home wood stoves. Particulate matter (PM) or gases such as sulfur oxides, nitrogen oxides, volatile organic compounds, or other HAPs listed gases can be released as pollutants.

### 8.2.1. Gaseous Emissions

Using add-on control devices to either recover or destroy the gaseous pollutant is the preferred method of controlling it. Add-on equipment uses a variety of control methods, such as condensation, adsorption, combustion, and absorption. Presently in use are flares, boilers, process heaters, thermal or catalytic incinerators, and flares. A brief overview of gaseous control devices and the methods for removing CO, NOx, and other dangerous gases is given in the section that follows.

### *8.2.1.1. Thermal Incinerators*

It is common practice to eliminate volatile organic compounds (VOCs) using thermal incinerators. Generally speaking, incineration entails destroying waste—liquid, solid, or gaseous—by carefully controlling a high-temperature fire. With 99.9% efficiency possible, incinerators are one of the most effective and tried-and-true ways to destroy volatile organic compounds. When high efficiencies are required and the waste gas content exceeds 20% of the LEL, thermal incinerators are frequently the best option (Figure 8.5).

**Figure 8.5.** *Preassembled process unit for air pollution control, i.e., a thermal oxidizer, being installed at a work site.*

*Source: By Combustion2016 – Own work, CC BY-SA 4.0,* https://commons.wikimedia.org/w/index.php?curid=55395659.

The ability to recover the energy and heat generated during the process and utilize it to power other operations within the facility is one benefit of thermal incineration. The process of oxidizing combustible materials by raising their temperature above their auto-ignition point in the presence of oxygen and keeping it there long enough to allow the material to completely burn into carbon dioxide and water is known as incineration, also known as thermal oxidation. The temperature, time, turbulence (for mixing), and oxygen availability all have an impact on how quickly and efficiently fuel burns. The fundamental design parameters for VOC oxidation systems are provided by these factors. The only component of a straight thermal incinerator is a combustion chamber; no heat exchanger is used to recover heat from exhaust air; this kind of incinerator is also known as a recuperative incinerator. The core of the thermal incinerator is a nozzle-stabilized flame that is fed additional air when needed, waste gas compounds, and auxiliary fuel. The waste gas is heated from its preheated inlet temperature to its ignition temperature as it passes through the flame. The ignition temperature is typically ascertained empirically and varies for various compounds. The temperature at which the rate of combustion reaction surpasses the rate of heat loss, leading to an increase in the temperature of the gases, is known as the ignition temperature.

Therefore, if the temperature of any organic/air mixture is raised to a high enough level, it will ignite. Reactor temperature is determined by the necessary degree of volatile organic compound (VOC) control that must be attained during the waste gas's stay in the thermal combustion chamber. Reactor temperature must be higher the shorter the residence time. The combustion chamber volume divided by the volumetric flow rate of the gas determines the nominal residence time of the reacting waste gas in the chamber. The majority of thermal units are made to allow waste gases to reside for no longer than one second at temperatures between 650° and 1,100°C (1,200° and 2,000°F). The residence time cannot be readily altered once the unit is designed and constructed, so the specific gaseous species and desired level of control determine the necessary reaction temperature. Research grounded in empirical field test data indicates that, in order to guarantee 98% destruction of non-halogenated organics, commercial incinerators should typically operate at 870°C (1,600°F) for a nominal residence time of 0.775 seconds.

Efficiency of VOC destruction is determined by design parameters (i.e., compound type, degree of mixing, inlet VOC concentration, chamber temperature, and residence duration). The typical design efficiencies of thermal incinerators vary from 98 to 99.99% and higher, contingent upon the specifications of the system and the features of the polluted stream. In order to achieve >98% control or 20 parts per million by volume (PPMV), specific design conditions must be met. At the exit, it is advised to maintain a combustion temperature of 870°C (1,600°F), a residence time of 0.75 seconds, and adequate mixing. The recommended parameters for halogenated volatile organic compounds (VOCs) are a combustion temperature of 1,100°C (2,000°F), a residence time of 1,100 seconds, and the use of an acid gas scrubber at the outlet. Reaction rates drop, maximum VOC destruction efficiency drops, and an incinerator outlet VOC concentration of 20 PPMV or less may be attained for vent streams with VOC concentrations below about 2,000 PPMV.

The literature typically lacks controlled emissions and/or efficiency test data for PM in incinerators. On the other hand, there are emission factors for PM in phthalic anhydride processes using incinerators. It was discovered that these processes' PM control efficiencies ranged from 79–96%. Incinerators used as PM control devices were reported to achieve 25–99% control efficiency of particulate matter 10 microns or less in aerodynamic diameter (PM10) at point source facilities in the EPA's 1990 National Inventory. These devices had reported VOC control efficiencies ranging from 0–99.9%.

**Table 8.2.** *Thermal Incinerator PM10 Destruction Efficiencies by Industry*

| **Industry/Types of Sources** | **$PM_{10}$ Control Efficiency (%)** |
|---|---|
| *Petroleum and Coal Products*<br>Asphalt roofing processes (blowing, felt, saturation); mineral calcining; petroleum refinery processes (asphalt blowing, catalytic cracking, coke calcining, sludge converter); sulfur mfg. | 25–99.9% |
| *Chemical and Allied Products*<br>Carbon black manufacturing; charcoal manufacturing; liquid waste disposal; miscellaneous chemical manufacturing processes; pesticide manufacturing; phthalic anhydride manufacturing (xylene oxidation); plastics/synthetic organic fiber manufacturing; solid waste incineration (industrial) | 50–99.9% |
| *Primary Metals Industries*<br>By-product coke processes (coal unloading, oven charging and pushing, quenching); gray iron cupola and other miscellaneous processes; secondary aluminum processes (burning/drying, smelting furnace); secondary copper processes (scrap drying, scrap cupola and miscellaneous processes); steel foundry miscellaneous processes; surface coating oven | 70–99.9% |

| | |
|---|---|
| *Electronic and Other Electric Equipment*<br>Chemical mfg miscellaneous processes; electrical equipment bake furnace; fixed roof tank; mineral production miscellaneous processes; secondary aluminum roll/draw extruding; solid waste incineration (industrial) | 70–99.9% |
| *Electric, Gas, and Sanitary Services*<br>Internal combustion engines; solid waste incineration (industrial, commercial/ institutional) | 90–98% |
| *Stone, Clay, and Glass Products*<br>Barium processing kiln; coal cleaning thermal dryer; fabricated plastics machinery; wool mfg. | 50–95% |

*Source:* https://www3.epa.gov/ttncatc1/dir1/frecup.pdf.

Moreover, low-concentration, high-flow organic vapor streams are typically not economically viable for thermal incinerators. Because thermal incinerators do not recover waste heat energy from the exhaust gases, they are typically less cost-effective annually than recuperative or regenerative incinerators. The amount of extra fuel needed can be decreased by using this heat to pre-heat incoming air. If more heat energy is available, it can be applied to other process heating requirements.

### *8.2.1.2. Catalytic Incinerators*

The main distinction between thermal and recuperative incinerators and catalytic incinerators is that the gas flows through a catalyst bed following its passage through the flame area. By speeding up the rate of oxidation reaction, the catalyst allows conversion to occur at lower reaction temperatures than in thermal incinerator units. Catalysts therefore enable smaller incinerator sizes. The two most common catalysts for burning VOCs are palladium and platinum. For gas streams containing chlorinated compounds, alternative formulations include metal oxides. In a catalytic incinerator, the gas stream is introduced into a mixing chamber where it is additionally heated. The waste gas is typically preheated by post combustion gas in a recuperative heat exchanger. The catalyst bed is subsequently traversed by the heated gas. Through gas diffusion, oxygen and volatile organic compounds (VOCs) are drawn to the catalyst surface and adsorbed onto the catalyst's active sites, where oxidation takes place. The gas then diffuses the products of the oxidation reaction back into the gas stream after desorbing them from the active sites.

Particulate matter has the ability to quickly coat the catalyst, preventing the catalyst's active sites from facilitating the oxidation of gas stream pollutants. Over time, the blinding effect of PM on the catalyst will cause the catalyst to deactivate. The catalyst does not require a large PM to blind it because nearly all of its active surface is contained in relatively small pores. Because catalysts differ greatly in terms of pore size and volume, there are no set standards for the PM concentration and size that they can withstand. It's likely that the catalyst manufacturers have this information.

Catalytic incineration systems are distinguished by the way the catalyst and the VOC-containing stream are brought into contact. Systems with fluid beds and fixed beds are both in use. A packed-bed catalyst or a monolithic catalyst can be used in fixed-bed catalytic incinerators. The catalyst monolith is the most widely used technique for coming into contact with the VOC-filled stream. The catalyst in this scheme is a porous solid block with parallel channels that are not intersecting and are oriented in the direction of the gas flow. The benefits of monoliths include low overall pressure drop and little attrition from thermal expansion and contraction during startup and shutdown. A straightforward packed-bed method serves as a second contacting scheme, wherein catalyst particles are supported in a tube or in shallow trays that allow gases to pass through. Due to the catalyst particles breaking because of thermal expansion and the scheme's intrinsically high pressure drop when compared to a monolith, this method is not widely used. However, the tray type arrangement of a packed-bed scheme, where the catalyst is pelletized, is used by several industries (i.e., heat-set web-offset printing). When there are significant concentrations of contaminants like silicon compounds or phosphorous compounds, a pelletized catalyst is beneficial.

High mass transfer rates are a benefit of fluid-bed catalytic incinerators, despite the fact that there is a little bit more pressure drop overall than with a monolith. A high bedside heat transfer in comparison to a typical gas heat transfer coefficient is another benefit of fluid beds. Higher heat release rates per unit volume of gas processed are made possible by the increased heat transfer rate to heat transfer tubes submerged in the catalyst bed. As a result, waste gas with higher heating values may be processed without reaching the catalyst bed's maximum allowable temperature. The amount of heat transfer through embedded heat transfer surfaces determines how much the gas phase temperature rises from the gas inlet to the gas outlet in these reactors. The rate of reaction at the catalyst surface and the rate of heat exchange between the catalyst and embedded heat transfer surfaces determine the catalyst temperatures.

Generally speaking, fixed-bed or monolithic catalysts are less tolerant of PM in the gas stream than fluid-bed catalysts. This is because the fluidized catalyst pellets are constantly abrading, which continuously removes PM from the catalysts' exterior. The slow attrition of catalyst is one drawback of a fluid bed. To address this issue, scientists have created attrition-resistant catalysts in recent years. Many stationary sources can have their emissions reduced with the use of catalytic incinerators. Catalytic incineration is widely used by many industries in this category, and solvent evaporation processes related to surface coating and printing operations are a major source of VOC emissions. Additionally, the following sources of emissions are managed by catalytic incinerators:

- Varnish cookers
- Foundry core ovens
- Filter paper processing ovens
- Plywood veneer dryers
- Gasoline bulk loading stations
- Process vents in the synthetic organic chemical manufacturing industry (SOCMI)
- Rubber products and polymer manufacturing
- Polyethylene, polystyrene, and polyester resin manufacturing

The best conditions for catalytic oxidation are lower exhaust volumes, little variation in the kind and amount of volatile organic compounds (VOCs), and the absence of catalyst poisons and other fouling contaminants like silicone, sulfur, heavy hydrocarbons, and particulates. The composition and concentration of VOCs, the operating temperature, the oxygen content, the properties of the catalyst, and the space velocity all affect how efficiently VOCs are destroyed. The volumetric flow of gas entering the catalyst bed chamber divided by the catalyst bed's volume is the standard definition of space velocity. The temperature of the catalyst during operation has a significant impact on the relationship between space velocity and VOC destruction efficiency.

The destruction efficiency of volatile organic compounds (VOCs) decreases with increasing space velocity and increases with increasing temperature. As an example, a catalytic unit operating at about 450°C (840°F) with a catalyst bed volume of 0.014 to 0.057 cubic meter ($m^3$) (0.5 to 2 cubic feet ($ft^3$) per 0.47 standard cubic meters per second ($SM^3$ /sec) (1,000 standard cubic feet per minute (scfm) of off gas passing through the device can achieve 95% VOC destruction efficiency. Higher destruction efficiencies of 98–99% are feasible; however, they usually need to be designed for a specific site and call for larger catalyst volumes and/or higher temperatures. Incinerators as a class, including catalytic incinerators, were reportedly used as PM control devices in the EPA's 1990 National Inventory. It was reported that these devices were reportedly achieving a 25–99% control efficiency of PM10 at point source facilities.

The devices' reported VOC control efficiency varied from 0 to 99.9%; however, it is presumed that reports indicating higher efficiencies (above 99%)

are related to thermal incinerators. These wide ranges of control efficiencies are due to the fact that they cover both facilities that have low PM emissions and are primarily focused on controlling VOC emissions as well as those that have no VOC emissions and only control PM.

Advantages of Catalytic Incinerators:

- Lower fuel requirements
- Lower operating temperatures
- Little or no insulation requirements
- Reduced fire hazards
- Reduced flashback problems
- Less volume/size required

Disadvantages of Catalytic Incinerators:

- High initial cost
- Catalyst poisoning is possible
- Particulate often must first be removed
- Disposal problems for spent catalyst

When used properly, catalytic incinerators provide a number of benefits. However, the sensitivity of catalytic incinerators to VOC inlet stream flow conditions and catalyst deactivation limits their applicability for many industrial processes; therefore, choosing a catalytic incinerator needs careful consideration.

#### *8.2.1.3. Flares*

A specially designed burner tip, auxiliary fuel, and steam or air to promote mixing are used in the flaring process, a VOC combustion control method, to burn VOCs in an open flame in the open air for nearly complete (>98%) VOC destruction. The factors that control the degree of combustion in a flare include the temperature of the flame, the duration of time spent in the combustion zone, the turbulence that mixes the components of the gas stream to finish the oxidation reaction, and the availability of oxygen to form free radicals. If every volatile organic compound is transformed into carbon dioxide and water, combustion is finished. Certain volatile organic compounds (VOCs) remain unaltered or transform into other organic compounds like acids or aldehydes as a result of incomplete combustion.

Flares are commonly classified into two categories: (1) flare tip height (either elevated or on the ground); and (2) flare tip enhancing mixing method (either steam-assisted, air-assisted, pressure-assisted, or non-assisted). When an open flame, or ignition source, is situated close to a process unit, elevating the flare can avoid potentially hazardous circumstances below ground level. In order to lessen the effects of heat, noise, smoke, and unpleasant odors, elevating the flare also enables the combustion products to be dispersed above working areas. Diffusion flames are used in most flares to facilitate combustion. When air diffuses past the fuel/combustion product stream's edge and toward the fuel flow's center, a diffusion flame forms, encircling a core of fuel gas with an envelope of combustible gas mixture.

A steady flame zone is created around the gas core above the burner tip by this mixture when it ignites. Warm combustion products from the flame zone diffuse into this inner gas core, heating it. When tiny, hot carbon particles form—which are responsible for the flame's distinctive brightness—cracking may take place. Smoking happens when there is a lack of oxygen in the air and when the carbon particles cool down below the point of ignition. Combustion product vortices have the ability to form around burning gas particles in large diffusion flames, effectively stopping the flow of oxygen. Frothing is brought on by this localized instability, and soot formation may follow. Good mixing and a sufficient air supply are necessary for complete combustion and the least amount of smoke, as in all combustion processes. The ability of each flare design to combine air and combustibles is what essentially sets them apart. Single burner tips that are raised above ground for safety purposes and burn the released gas in a diffusion flame are known as steam-assisted flares. In refineries and chemical plants, they are the most common type of flare and are said to make up the majority of flares that are installed. This kind of flare system injects steam into the combustion zone to create turbulence for mixing and to inject air into the flame, ensuring an adequate air supply and good mixing. To achieve smokeless operation, certain flares rely on forced air to supply the necessary combustion air and mixing. Inside, but close to the top of a steel cylinder measuring at least 0.6 meters (24 inches) in diameter, is a spider-shaped burner with numerous tiny gas apertures that powers these flares.

The cylinder's bottom fan provides combustion air. The amount of combustion air can be adjusted by changing the fan speed. The main benefit of using air-assisted flares is that they can be used in situations without steam. Large flares don't usually require air assistance because it isn't cost-effective to use air assistance when the gas volume is high. However, the quantity of large flares that do require air assistance is growing. The flare tip of an unassisted flare is devoid of any additional features that would improve the air-to-flame mixing. Its application is restricted to gas streams that burn easily without emitting smoke and have low heat content and carbon/hydrogen ratios. These streams are more resistant to cracking, have lower combustion temperatures that reduce cracking reactions, and require less air for complete combustion. Pressure-assisted flares encourage mixing at the burner tip by utilizing the vent stream pressure.

These days, a number of suppliers sell burner tips with high pressure drops that are proprietary. These flares can be used in streams that previously needed steam or air assistance for smokeless operation, provided there is enough vent stream pressure available. The burner arrangement for pressure-assisted flares is typically (though not always) at ground level, so they must be placed in a remote area of the plant with plenty of room. They have several burner heads that are set up to run in different sequences according to the amount of gas released. The features of the vent gas usually determine the burner heads' size, shape, quantity, and grouping. The burner heads of an enclosed flare are housed inside an internally insulated shell. In addition to providing wind protection, the shell lowers heat radiation, brightness, and noise. For esthetic or safety reasons, enclosed, or ground-based, flares are typically used instead of elevated ones. When air or steam assistance is not needed, a high nozzle pressure drop is typically sufficient to provide the mixing required for smokeless operation. Within this framework, enclosed flares can be regarded as a unique class of flares, either assisted or not assisted by pressure. In order to create a sufficient draft and distribute the thermal plume and provide enough air for smokeless combustion, the height must be appropriate. These flares are always pointed at the earth.

While dependable and effective operation can be achieved over a broad range of design capacity, enclosed flares are typically used to combust continuous, constant flow vent streams and have a lower capacity than open flares. Because they are shielded from wind effects, stable combustion can be achieved with lower heat content vent gases than with open flare designs (1.9 to 2.2 $MJ/SM^3$ [50 to 60 Btu/SCFJ). Landfills usually use enclosed flares to burn off waste gas. Nearly any VOC stream can be managed with flares, which are also generally capable of withstanding significant variations in VOC concentration, flow rate, heating value, and inert species content. Although flaring can be used in batch, continuous, and variable flow vent stream applications, its main function is as a safety measure to regulate a significant amount of pollution that arises from unfavorable conditions. The petrochemical and petroleum industries are the main users of flares.

Most chemical plants and refineries already have flare systems in place that are meant to release large amounts of gas in the event of an emergency process upset. In addition to being intended for emergency releases, these large diameter flares can also be used to regulate vent streams resulting from different process operations. The majority of the low molecular weight volatile organic compounds (VOCs) and high heating values found in flared gases from refineries, the petroleum industry, and the chemical sector. Low-heat carbon monoxide and inert species make up the flares used to control waste gases from blast furnaces. The composition of gases flared from coke ovens is intermediate compared to the other two groups, and their heating value is moderate.

Depending on the kind of flare arrangement (i.e., flares to treat waste gases can vary up to approximately 50,000 kilograms per hour (kg/hr) (100,000 pounds per hour (lb/hr) of hydrocarbon gases for ground flares and approximately 1 million kg/hr (2 million lb/hr) or more for elevated flares, depending on the source of the waste stream and the type of flare. Incinerators are not concerned about flares' safety because the waste gas they produce has a low level of organic material. This is so because flaring doesn't have an enclosed combustion chamber that could produce an explosive environment; instead, it is an open combustion process. However, because incinerators have an enclosed combustion chamber, the concentration of the waste gas must be significantly below the lower flammable level (also known as the lower explosive limit, or LEL) of the particular compound being controlled in order to prevent the possibility of an explosion. Typically, this means a safety factor of four (i.e., 25% of the LEL) is put to use. More than 11 MJ/scm (300 Btu/SCF) of heating value must be present in the waste gas stream. Should the waste gas fall short of this threshold, auxiliary fuel needs to be added in a big enough amount to cover the shortfall.

A knockout drum removes liquids that might be in the vent stream gas or that might condense out in the collection header and transfer lines. Usually found in a vertical vessel inside the flare stack base, or in a horizontal or vertical vessel near the flare base, is the knockout or disentrainment drum. Liquid in the vent stream has the potential to put out the flames or to produce smoking and uneven combustion. Furthermore, the use of flaring liquids may produce a spray of burning chemicals that could fall to the ground and endanger public safety.

Advantages of Flares:

- Economical way to dispose of gas
- Does not require auxiliary fuel to support combustion
- Used to regulate waste streams that are sporadic or variable.

Disadvantages of Flares:

- Produces undesirable noise, smoke, heat radiation, and light
- Source of SOx, NOx, and CO
- Cannot treat waste streams with halogenated compounds
- Released heat from combustion is lost

### 8.2.1.4. Boilers and Process Heaters

Production facilities frequently use boilers and process heaters to produce power and heat. By recycling the pollutant for fuel, they can be used as an efficient pollution control device in addition to their primary function of supporting plant operations. But only contaminants that don't interfere with the burner unit's operation can be utilized as fuel. An exhaust stream, for instance, may be utilized as additional fuel, but only insofar as its fuel value keeps the combustion process going. The heating values of all volatile organic compounds (VOCs) vary. The exhaust may serve as the plant's main fuel source if the heating value is high and the pollutant stream is large. Low heating value gas streams, however, can only be utilized with a boiler or heater that is small enough to hold the lower fuel value.

Boilers and process heaters often achieve removal efficiencies of 98% or higher when used as pollution control devices. Since these devices are already necessary for the plant to operate, using them for pollution control has the advantage of little additional capital cost. Additionally, by recycling and reusing products from the refining process, exhaust streams lower fuel costs.

### 8.2.1.5. Adsorbers

The main purpose of adsorbers has been to reduce VOC emissions. When VOC emissions are controlled, concentrations usually drop from 400–2,000 parts per million (ppm) to less than 50 ppm. The range of VOC concentrations can now be expanded by adsorption technology from 20 ppm to one-fourth of the Lower Explosive Limit (LEL). Small concentrations like these at the lower end of this range might be hard or expensive to regulate with another technology, or even with all adsorbents combined. For absorbers at the upper end of the range, it may be more cost-effective to use incinerators, membrane separators, and condensers.

The adsorber not only controls emissions but also enables VOC recovery. The cost of emission control can be greatly reduced by recovering a marketable or reusable volatile organic compound. Adsorbers can also raise the VOC concentration to a point where it is economically feasible to destroy the material by burning it or to recover it using a membrane or condenser. The pollutant is adsorbed on the surface (usually the internal surface) of an adsorbent material granule, bead, or crystal during the adsorption process. It is not taken up by a chemical process. This distinction is significant. The material that has been adsorbed is held somewhat loosely physically and is easily released or desorbed when heated or suctioned. An absorber, on the other hand, combines chemically with the material being absorbed to hold the absorbed material more firmly and necessitates more energy to release. Sadly, "absorbent carbon" is the name of one of the adsorbents. This enduring misnomer dates back to the period before the 1920s, when adsorption was first understood. "Activated carbon" is a better term. The pyrolysis of coal, wood, bark, coconut husks, etc., releases carbon to eliminate every gas or vapor of volatile material, leaving only carbon. The pores on this carbon could then enlarge if it is partially oxidized.

Systems for adsorption of carbon can be non-regenerative or regenerative. In a regenerative system, multiple carbon beds are present. Another bed is cleaned and ready for use later while the first bed is actively used to remove pollutants. The accumulated pollution can be removed from the bed using steam, after which it is recycled or destroyed. The same activated carbon material can be reused repeatedly by "regenerating" the carbon. When the concentration of the pollutant in the gas stream is relatively high, regenerative systems work best. Every six months to five years, or whenever there is a bed fire, carbon needs to be replaced. The type of carbon, the frequency of regeneration, and the operating temperature all affect the frequency. Conversely, nonregenerative systems typically use thinner activated carbon beds, which are disposed of once they are saturated with the pollutant. Non-regenerative carbon adsorbers are typically only used when the pollutant concentration is very low because the carbon coated pollutant residue causes an issue with solid waste (Figure 8.6).

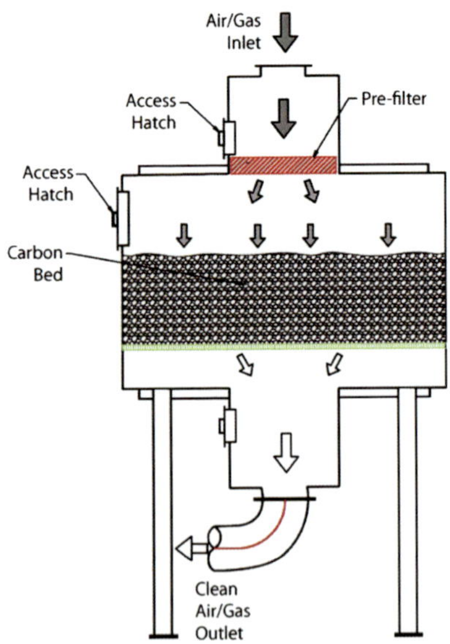


**Figure 8.6.** *Deep bed carbon adsorber flow diagram.*

*Source:* https://www.monroeenvironmental.com/air-pollution-control/carbon-adsorbers/#.

The term "adsorption isotherm" refers to the adsorption capacity of each adsorbent. In the past, the isotherm was used to calculate how many pounds of pollution could be adsorbed at a particular temperature for every pound of adsorbent when measuring carbon. Nevertheless, the adsorption isotherm also depends on the total ambient pressure surrounding the adsorbent and the pollutant concentration as determined by the partial pressure of the vapor. Pressure affects how many molecules can be held in the "well," with higher pressure being correlated with molecules that are adsorbed closer together. The amount of material that can be adsorbed per pound of adsorbent is still the definition of the adsorption isotherm.

Pressure swing adsorption, vacuum regeneration, and thermal swing regeneration are the three types of regeneration techniques. The temperature isotherm is used by thermal regeneration systems, the total and partial pressure isotherms by vacuum regeneration systems, and the total ambient pressure isotherm by pressure swing regeneration systems. In reality, each of these variables functions to create the "isotherm." As a result, rather than being a constant or a two-dimensional plot, the "isotherm" is actually a complex function that forms at least a three-dimensional surface. A linear or highly non-linear function, contingent on the adsorbent, pressure, and material being adsorbed, can be found in the isotherm. The amount of adsorption area that the VOC molecules can reach determines the adsorption isotherm as well. Each of these adsorbents has pores on the surface that provide access to the interior space. The majority of the adsorption area is found inside. There are pores in carbon that lead to smaller pores and so on. These micropores contain a large portion of the internal surface area, and this process appears to go on indefinitely in carbon.

For certain kinds of substances, the adsorbent's affinity can be substantially higher than its affinity for other types. Because of this, molecules with lower affinity either get held with a lower adsorption capacity or are released in favor of molecules with higher affinity when there is a stronger affinity for some of the molecules. With an affinity for both polar and non-polar molecules, carbon maintains some adsorption area despite not being completely hydrophobic or hydrophilic. For this reason, humidity has a discernible impact on carbon. A study has even revealed that when steam condenses and the water replaces the VOC is desorbing from carbon.

Every adsorbent has an MTZ, or mass transfer zone. Within the MTZ, the airborne VOC concentration decreases to the lowest possible vapor pressure in balance with the desorbed adsorbent, from 100% of the inlet vapor concentration. The MTZ is typically much shorter than the depth of the bed, which is defined as the dimension of the adsorbent in the direction of airflow. As the bed fills up, the MTZ passes through it and the absorption isotherm fills up as well. The bed is at capacity upstream of the MTZ, but it is essentially still desorbed downstream. Breakthrough happens when the MTZ leaves the adsorption bed and is detected as a rise in the VOC vapor concentration at the downstream end of the bed. This can happen at varying times depending on the VOC.

All types of adsorbents have something in common with gas chromatography columns: they employ adsorption. The material that has been adsorbed will naturally desorb and move downstream with each system's gas flow. This is due to the pressure gradient, which is purposefully very low in gas chromatography adsorption columns and rather high when adsorbers are used as emission control technologies. Since the migration of pollutants is usually so slow, it can be disregarded when emission levels are being regulated. This property, however, establishes a strict maximum amount of time for adsorption prior to bed regeneration. Since the equilibrium between the vapor pressure and the adsorbent is constantly changing, breakthrough is actually a gradual process. Therefore, any discernible increase in the effluent concentration can be considered a breakthrough.

In addition, the MTZ will move to breakthrough without relying on the VOC loading. A VOC input spike and the MTZ migration will facilitate the VOC's spread and lower the concentration that will be desorbed upon breakthrough. An illustration of how an adsorber could effectively utilize the MTZ's migration characteristic is the airflow from a hood in a mercaptan handling plant. Since natural gas has no smell, mercaptan is added to it to create an olfactory cue, or odor, that indicates a leak. People would notice the mercaptan spill and become concerned if emissions were detected in quantities greater than one part per billion, as they might believe that a natural gas leak was occurring. Consequently, hoods that handle mercaptan are outfitted with an appropriately sized adsorber. Any spill vapors are absorbed by this adsorber, and the mercaptan is released at a reduced concentration when the MTZ migrates to breakthrough. This is as a result of the mercaptan's intermittent emission. Breakthrough occurred at a lower concentration, below the smell threshold, because the adsorber did not reach its full capacity. Desorbing or regeneration is therefore not required unless spills occur too frequently.

The term "conditioning cycles" refers to the number of cycles—typically five or so—needed to attain a steady state of adsorption and regeneration. Following the conditioning cycles, carbon can absorb roughly 50% of the quantity that the virgin material was able to absorb. About 90% of the amount adsorbed by the virgin material can be absorbed by zeolite. Polymers can range from 50–90%. The "working capacity" of the adsorbent is also determined by this "conditioned isotherm." The continuous recapture of molecules (i.e., depressed vapor pressure) in the micro-pores is thought to be the reason why carbon and certain polymers will usually desorb only half as much as the virgin material. Only pyrolysis, which would destroy the VOC and either destroy the polymer or change the properties of the carbon, could remove VOC from these micropores. The difference between the adsorbent's full adsorption isotherm and desorbed conditioned level is what defines its "working capacity." As such, when sizing the adsorbent bed, the conditioning effect needs to be taken into account. The amount of adsorbent in the bed may need to be doubled for the conditioning effect. The capacity of the bed(s) can be determined by the concentration of volatile organic compounds (VOCs), the air flow rate, the weight of the adsorbent in the bed(s), the type of adsorbent, and the "working capacity" of the adsorbent, since regeneration should only happen roughly every 8 or 16 operating hours. In this sense, the price of installing, running, and maintaining an adsorber can be greatly impacted by its "working capacity."

### *8.2.1.6. Absorbers*

Absorption is the process of dissolving a gaseous pollutant in a liquid in order to remove it. The most used absorbent fluid is water. Similar to how sugar is absorbed in a glass of water when stirred, the gas stream is mixed into the solution as it passes through the liquid. Recovering products or purifying gas streams with high concentrations of organic compounds are frequent uses for absorption. However, the quantity of wastewater produced during the absorption process is a significant drawback. Column absorbers and scrubbers are criticized for producing wastewater, which turns an air pollution issue into a water pollution issue. The goal of absorption equipment is to combine as much liquid and gas as possible. Scrubbers, another name for absorbers, come in a variety of commercial configurations. The most widely used absorption apparatuses are venturi scrubbers, spray chambers, packed columns, and spray towers.

A column of inert (non-reactive) material, like plastic or ceramic, fills the packed-column absorber, increasing the liquid surface area available for mixing the absorbing liquid and the pollutant gas. The inert substance enhances the column's capacity for absorption. A counter-current flow is created as the gas and liquid are introduced at opposite ends of the column, effectively mixing the two materials. Absorbers can typically achieve removal efficiencies of over 95%. Packing material in the shape of Raschig rings, spiral rings, or Berl saddles, which offer a lot of surface area for liquid-particle contact, are layered inside a chamber to form packed-bed scrubbers. The packing is supported by a plate close to the scrubber's bottom and kept in place by wire mesh retainers. Equal amounts of cleaning liquid are added above the packing and run through the bed. It should be highlighted that the pollutant to be absorbed needs to be soluble in the fluid in order for the liquid to coat the packing and create a thin film. The gas stream flows up the chamber in vertical designs (packed towers), countercurrent to the liquid. Certain packed beds are engineered to allow for cross-current, or horizontal gas flow across the packing.

Physical absorption is dependent on the specifics of the pollutant(s) in the gas and liquid stream as well as the density and viscosity of the gas stream and liquid solvent (i.e., diffusivity, solubility in equilibrium). Since these characteristics are temperature-dependent, lower temperatures typically encourage the solvent to absorb gases. Increased liquid-gas ratios, higher contacting surfaces, and higher concentrations in the gas stream all improve absorption. Although the rate-limiting step is usually the physical absorption rate rather than the rate of the chemical reaction, chemical absorption may still be restricted by the rate of reaction.

### *8.2.1.6.1. VOC Control*

One process that is frequently used in chemical processing is absorption. It is used as a raw material and/or a product recovery technique in separating and purifying gaseous streams containing high concentrations of organics (i.e., in natural gas purification and coke by-product recovery operations). The organics in the gas stream dissolve in a liquid solvent during absorption. Scrubbers, packed or plate columns, or countercurrent spray towers facilitate the interaction between the absorbing liquid and the vent gas. There are a few restrictions on the use of absorption as the main method of controlling organic vapors. The availability of an appropriate solvent is one of the factors. Only soluble volatile organic compounds (VOCs) can be eliminated from any given absorbent liquid; the VOC must be soluble in the liquid to be removed. Water, mineral oils, and other non-volatile petroleum oils are a few typical solvents that can be used to capture volatile organic materials. The availability of vapor/liquid equilibrium data for the particular organic/solvent system in question is another factor that influences the suitability of absorption for controlling organic emissions. These kinds of data, which are required for the design of absorber systems, are not easily accessible for unusual organic compounds.

The solvent that is selected to eliminate the pollutant(s) should be reasonably priced, have low vapor pressure, low viscosity, and high solubility for the vapor or gas. VOCs with relatively high water solubilities are absorbed by water. Hydrophobic volatile organic compounds can dissolve in water when amphiphilic block copolymers are added. For volatile organic compounds (VOCs) with low water solubilities, other solvents, such as hydrocarbon oils, are also utilized; however, these oils are only used in large quantities in certain industries (petrochemical plants and petroleum refineries). The way the material extracted from the absorber is treated or disposed of is another factor to take into account when using absorption as a control technique. Most of the time, the VOC-containing scrubbing liquid is recovered through a process called stripping, which involves the desorbed VOC from the absorbent liquid, usually at high temperatures and/or under vacuum. After that, a condenser recovers the VOC as a liquid.

Most of the PM is impacted by the gas stream in packed-bed scrubbers because it is compelled to travel around the packing material. The PM is gathered by the liquid on the packing material, which then travels through the chamber and down the tower to the drain. It is customary to place a mist eliminator, also known as a "de-mister," above or after the packing and scrubbing liquid supply. The mist eliminator will extract any scrubbing liquid and wetted PM entrained in the departing gas stream, returning them to drain through the packed bed. High PM concentrations in a packed-bed scrubber can clog the bed, which is why these devices are only suitable for streams with relatively low dust loadings. Because the packing on packed-bed scrubbers is harder to reach and maintain than on other scrubber designs, plugging is another major issue. There are mobile bed scrubbers that are filled with low-density plastic spheres that can move around inside the packed bed. The increased movement of the packing material in these scrubbers makes them less prone to plugging. Packing-bed scrubbers are generally a better fit for gas scrubbing than PM scrubbing due to the high maintenance needed to control PM.

The following variables typically determine whether gas absorption is a suitable method for controlling pollution.

1. Availability of suitable solvent;
2. Required removal efficiency;
3. Pollutant concentration in the inlet vapor;
4. Capacity required for handling waste gas; and
5. Recovery value of the pollutant(s) or the disposal cost of the unrecoverable solvent.

The chemical, aluminum, coke and ferroalloy, food and agriculture, and chromium electroplating industries are the main industries that use packed-bed scrubbers. In flue gas desulfurization (FGD) systems, these scrubbers have seen limited use; however, to prevent flooding, the flow rate of the scrubbing solution needs to be carefully regulated. Packed towers typically offer more cost-effectiveness than impingement plate towers when it comes to absorption-based pollution control. However, when internal cooling is required or low liquid flow rates would not sufficiently wet the packing, the impingement plate design is sometimes preferred over packed-tower columns. The main purpose of absorbers is to regulate inorganic gases, vapors, and fumes (i.e., hazardous air pollutants (HAPs) in particulate form (PMHAP), particulate matter (PM), including PM less than or equal to 10 micrometers (µm) in aerodynamic diameter (PM10), PM less than or equal to 2.5µm in aerodynamic diameter ($PM_{2.5}$), and chromic acid, hydrogen sulfide, ammonia, chlorides, fluorides, and $SO_2$), volatile organic compounds (VOCs). In order to separate and purify gaseous waste streams with high concentrations of volatile organic compounds (VOCs), absorption is also frequently utilized as a raw material and/or product recovery technique. This is particularly true for water-soluble compounds like methanol, ethanol, isopropanol, butanol, acetone, and formaldehyde. An amphiphilic block copolymer dissolved in water can be used to absorb hydrophobic volatile organic compounds. It is, however, much more than that as an emission control method.

#### *8.2.1.7. Condensers*

The process of turning a gas or vapor into a liquid is called condensation. Any gas can be made into a liquid by applying more pressure or lowering its temperature. Therefore, by lowering the temperature of the gas to a point where it condenses and gathers as a liquid, a condenser eliminates gaseous pollution from the air. When water droplets form on the exterior of a cold glass of water, it serves as a basic illustration of the condensation process. Water vapor from the surrounding air condenses onto the glass's surface and becomes liquid due to the cold temperature of the material. Heat extraction is the most widely used method, but both raising pressure and removing heat from the system can result in condensation. Valuable products can also be extracted from a waste stream using condensers. Condensers generate wastewater that needs to be disposed of, just like scrubbers do. Condensers are therefore typically utilized in conjunction with other control mechanisms.

Surface and contact condensers are the two main types of condensers. The coolant and vapor stream are physically combined in a contact condenser. They emerge as a single exhaust stream from the condenser. In a surface condenser, tubular heat transfer surfaces keep the coolant apart from the vapors, and distinct exits allow it to exit the unit. Shell and tube heat exchangers are a common term for surface condensers. The devices also function as heaters because they raise the coolant's temperature. These kind of condensers typically produce removal efficiencies ranging from 50–95%, contingent upon application and design.

### *8.2.1.8. Inorganic Gas Control*

The most popular solvent for eliminating inorganic pollutants is water. Enhancing the absorbing solution's chemistry to make it react with the pollutant can help remove pollutants more effectively. Sodium hydroxide, or NaOH, is the most widely used scrubbing liquid for controlling acid-gas (i.e., HCl, $SO_2$, or both), though calcium hydroxide (slaked lime, $Ca[OH]_2$) and sodium carbonate ($Na_2CO_3$) are also employed. Neutral salts are created when the acid gases react with the alkaline compounds in the scrubbing solution. The solubility of the acid gases in the scrubbing liquid determines how quickly they are absorbed.

## 8.2.2. Sulfur Dioxide Emissions Controls

There are several technologies available that are intended to extract $SO_2$ from the flue gas that power plants produce. The level of commercial readiness exhibited by these technologies varies. Some have only recently been demonstrated at commercial plants, while others can boast operational experience spanning tens of thousands of hours. Historically, wet scrubbers that eliminate sulfur dioxide ($SO_2$) emissions from sizable electric utility boilers—primarily coal combustion—have been referred to as flue gas desulfurization. However, the terms FGD, acid gas, and acid rain control are used interchangeably to classify a wide range of control system designs due to the necessity of controlling acid emissions from industrial boilers and incinerators as well as the evolution of different types of acid control systems. Reducing $SO_2$ emissions also involves the use of FGD systems.

From processing facilities like pulp and paper mills, smelters, acid plants, and refineries. Depending on how the sorbent is handled after it has absorbed $SO_2$, commercially available FGD technologies can "conventionally" be categorized as once-through and regenerable (Scrubber Myths and Realities). In once-through technologies, the sorbent—which, in the case of gypsum, must be disposed of as waste or used as a by-product—permanently binds the $SO_2$. In regenerable technologies, sulfuric acid, elemental sulfur, or liquid sulfur can be produced by processing the $SO_2$ that is released from the sorbent during the regeneration step. In the step where $SO_2$ is scrubbed, the regenerate sorbent is recycled. Wet and dry technologies are additional categories for both regenerable and once-through technologies. Wet processes result in the production of wet slurry waste or by-products, and the flue gas that exits the absorber is saturated with moisture. Dry processes result in dry waste material production and moisture-free flue gas exiting the absorber.

Once-through wet FGD processes can result in the production of salable by-products or slurry waste, depending on the process configuration and local market conditions at the plant site. Before being disposed of or sold, this waste or byproduct needs to be dewatered in some way (if it's a salable by-product) (Figure 8.7).

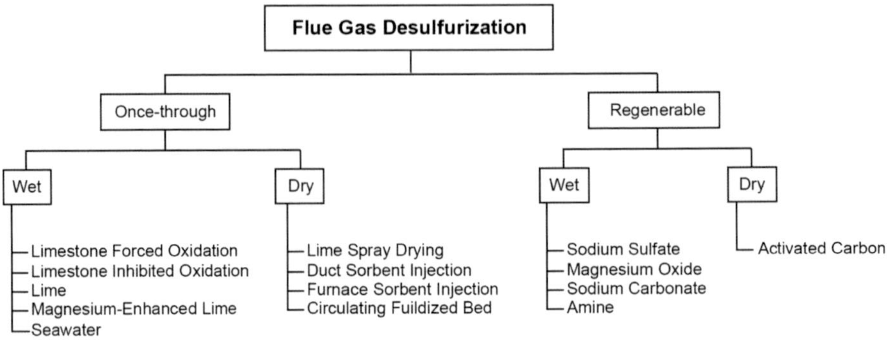


**Figure 8.7.** *Flue gas desulfurization technology tree.*

*Source:* https://www.4cleanair.org/wp-content/uploads/APTI/452combined.pdf.

### *8.2.2.1. Wet FGD Technologies*

Flue gas interacts with an alkaline slurry in an absorber during wet FGD processes. The absorber could be a tray tower or a spray tower, depending on the manufacturer and the intended process setup. Nonetheless, the counterflow vertically oriented spray tower is the absorber application that is utilized the most frequently. A wide range of wet FGD procedures have developed to capitalize on unique qualities of different sorbents and/or byproducts. Every wet FGD procedure covered in this article is one-time only (i.e., non-regenerable). In an absorber, limestone slurry comes into contact with flue gas containing $SO_2$. There are two steps involved in preparing limestone slurry. Limestone is first ground into a fine powder with the appropriate distribution of particle sizes. It happens in a crushing station, i.e., ball mill (fine crushing increases a limestone's rate of dissolution).

Subsequently, water is combined with this fine powder in a slurry preparation tank. The absorber reaction tank receives a pumping of the sorbent slurry from this tank. As previously indicated, the absorber is typically a counterflow tower, with an array of spray nozzles spraying limestone slurry downward and flue gas flowing upward. Both sorption and reaction with the slurry remove $SO_2$ from the absorber. In a reaction tank, which offers retention time for finely ground limestone particles to dissolve and react with the dissolved $SO_2$, reactions started in the absorber are finished. From the reaction tank, the slurry is pumped back into the absorber. The waste is concentrated by the waste handling equipment after reaction products from the reaction tank are pumped there. The concentrated waste is sent from the waste handling equipment to be disposed of (ponded or stacked) or processed to create a marketable by-product known as gypsum (calcium sulfate dihydrate).

- **Limestone Forced Oxidation:** Limestone forced oxidation (LSFO) has emerged as the go-to method for wet FGD technology globally over time because it reduces scaling issues in the absorber. When the slurry oxidation level, or the fraction of calcium sulfate in the slurry, is higher than 15%, gypsum scale usually forms naturally. By forcing the oxidation of calcium sulfite to calcium sulfate (in-situ oxidation) or ex-situ oxidation (blow air into the reaction tank), LSFO prevents scaling. As a result, the concentration of gypsum in the slurry recycled to the absorber decreases. The gypsum thus formed is removed as usual. Nearly total oxidation (more than 99%) is attained in LSFO systems used to create marketable gypsum.
- **Limestone-Inhibited Oxidation:** Limestone-inhibited oxidation (LSIO) is another wet limestone process designed to suppress oxidation in the absorber. In order to keep the slurry oxidation level in the absorber from rising above 15%, emulsified sodium thiosulfate ($Na_2S_2O_3$) is added to the limestone slurry feed in LSIO. Because LSIO has a higher sulfite content than LSFO, solids dewatering is generally more challenging there. Applications involving high sulfur coals benefit greatly from the efficiency of the LSIO chemistry.
- **Lime and Magnesium-Lime:** In a counterflow spray tower, calcitic lime slurry is used in the lime process. Although it costs more, this slurry is more reactive than limestone slurry. A unique kind of lime is used in the magnesium-enhanced lime (MEL) process, which is a variation on the lime process. Compared to calcitic lime, MEL can remove $SO_2$ with high efficiency in much smaller absorber towers. Furthermore, MEL requires less slurry to remove $SO_2$ to the same extent as LSFO.

### *8.2.2.2. Dry FGD Technologies*

For sources that burn coal with a sulfur content of low to medium, LSD is used to control $SO_2$ emissions; higher sulfur coals are sometimes used as well. A few of the problems that prevent spray dryers with high sulfur coals from being used are as follows: in these technologies, flue gas that contains $SO_2$ interacts with an alkaline (usually lime) sorbent. This leads to the production of dry waste, which is typically simpler to dispose of than waste from wet FGD processes. The sorbent can be introduced into the flue gas as a dry powder via duct sorbent injection (DSI), furnace sorbent injection (FSI), and circulating fluidized bed (CFB) or as an aqueous slurry through lime spray drying (LSD). For the sorbent to react with $SO_2$, LSD and CFB require special absorber vessels, whereas sorbent delivery equipment is the only new hardware needed for DSI and FSI. Sorbent recirculation can be employed in dry processes to maximize its utilization. The possible influence of the chloride present in the coal on the performance of the spray dryer, as well as the capacity of the current particulate control device to manage the elevated loading and meet the necessary efficiency standards.

In a spray dryer vessel, a finely atomized fresh lime slurry is misted with hot flue gas. In a slaker, which is typically a ball mill, fresh lime slurry is made at a nominal solids concentration. Finely dispersing lime slurry into flue gas is accomplished using rotary atomizers or two-fluid nozzles. For low sulfur coals, spray dryers are usually run at a lime stoichiometry of 0.9; for high sulfur coals, the stoichiometry is between 1.3 and 1.5. Process waste is dried as a result of a sequence of reactions resulting from simultaneous heat and mass transfer between alkali in a finely dispersed lime slurry and $SO_2$ from the gas phase. Carefully monitored water feed is used in the spray dryer to prevent total flue gas saturation. While high $SO_2$ removal requires close to adiabatic saturation (from 10 to 15 C for coal-derived flue gas), complete saturation hinders spray dryer operation due to wet solids sticking to vessel walls and inside the particulate collector. The following are the spray dryer's primary reactions.

$$Ca(OH)_2 + SO_2 \rightarrow CaSO_3 \bullet + \tfrac{1}{2}\, H_2O + \tfrac{1}{2}\, H_2O$$

$$Ca(OH)_2 + SO_3 + H_2O \rightarrow CaSO_3 \bullet + 2\, H_2O$$

$$CaSO_3 + \tfrac{1}{2}O_2 \rightarrow CaSO_4$$

At the bottom of the spray dryer, a portion of the solid dry reaction product is gathered. The separation happens at the particulate control device, where the residual solids, suspended in the flue gas, go. The filter cake on the surface of the bags may experience a significant additional removal of $SO_2$ in a process configuration where the particulate control device is a baghouse. Dry solids are disposed of from the bottom of the spray dryer and from the hopper of the particulate control device. The length of time an alkali has available for a gas-solid reaction determines how much of it can be used in a spray dryer. In a spray dryer, the residence time is typically 8 to 12 seconds. A portion of the dry solids from the particulate collector's hopper and the bottom of the spray dryer are transferred to the recycle solids slurry tank in order to maximize sorbent utilization. Alkali from earlier system passes that have partially reacted are present in the recirculated stream. The extra $SO_2$ exposure that the recycling provides for sorbents encourages higher sorbent utilization.

- **Lime Spray Drying:** The majority of sources that burn coal with low to medium sulfur use LSD. In a spray dryer, process waste is dried through a series of reactions caused by simultaneous heat and mass transfer between $SO_2$ and alkali in a finely dispersed aqueous lime slurry. According to studies, the majority of $SO_2$ capture in the spray dryer happens while the sorbent is still wet. Deliquescent additives can therefore be used to extend the amount of time the sorbent is kept moist. The purpose of  is to directly regulate $SO_2$ in the flue gas duct, situated between the particulate control device and the air pre-heater. This procedure involves injecting dry sorbent—typically hydrated lime—into the flue gas downstream of the boiler's air preheater. A separate injection of water is made from the sorbent. Within the particulate control device, fly ash, reaction products, and any unreacted sorbent are collected.
- **Furnace Sorbent Injection:** Dry sorbent is injected straight into the furnace during FSI, where temperatures range from 950 to 1,000°C. Sorbent particles break down into porous solids with a large surface area (mostly calcium hydroxide, occasionally calcium carbonate). Together with the flue gas, calcium sulfate and any leftover unreacted sorbent exit the furnace and are collected as solids in a particulate collector.
- **Circulating Fluidized Bed:** In a CFB, a humidified flue gas comes into contact with a dry sorbent (hydrated lime). Because the sorbent passes through the bed multiple times, it allows for a prolonged period of contact between the sorbent and flue gas. Good conditions for $SO_2$ mass transfer from the gas phase to the solid phase are characteristic of CFB. However, modifications to the current electrostatic precipitator might be necessary to maintain the necessary particulate emission levels because of a higher concentration of particulate matter downstream of the fluidized bed.

### *8.2.2.3. Regenerable FGD Technologies*

Regenerative FGD technologies are used sparingly both domestically in the United States and internationally. Compared to other FGD processes, these procedures have relatively high operation and maintenance (O and M) costs, and the product's sale proceeds do not make up a sizable portion of the higher process costs. The report covers four wet processes (sodium sulfite, magnesium oxide, sodium carbonate, and amine) and one dry process (activated carbon) for regenerable FGD technologies. Sulfuric acid can be produced using the concentrated stream of $SO_2$ produced by these processes.

- **Performance:** The CoalPower3 Database's design $SO_2$ removal efficiencies were examined in order to estimate the $SO_2$ reduction performance of FGD technologies. The median design efficiency for all units using wet limestone processes is approximately 90%, according to these data. On the other hand, sophisticated, modern wet scrubbers can remove $SO_2$ with an efficiency of more than 95%. According to reports, high velocity LSFO with cutting edge design options can remove more than 99.6% of $SO_2$ under test conditions. Additionally, all units using LSD have a median design efficiency of 90%, according to the data. Recent LSD installations, however, installed between 1991 and 1995, have 90–95% design $SO_2$ removal efficiencies. Analyzing the extent to which wet limestone and LSD processes have improved over time is helpful. For these processes, a consistent increase in the design $SO_2$ removal efficiency is apparent. Increased regulations on $SO_2$ control might be partially to blame for this improvement. That being said, the trends do show that the $SO_2$ removal efficiencies for the processes under consideration have increased over time.

There are numerous technical choices available to improve the efficiency of current wet FGD installations in removing $SO_2$. A perforated tray or other apparatus to enhance mass transfer, lowering the quantity of flue gas bypassed, and adding organic acid to the limestone slurry to increase its reactivity are some of the crucial alternatives. New wet FGD installations can also choose from a variety of sophisticated design, process, and sorbent options. A few of these involve buffering with organic acid, utilizing large capacity modules, and raising the flue gas velocity in the absorber. High $SO_2$ removal and/or improved operational efficiency are possible with these cutting-edge solutions.

## 8.2.3. Nitrogen Oxides (NOx) Emissions Control

A family of seven compounds is represented by NOx. Since nitrogen dioxide ($NO_2$) is the most common form of NOx produced by anthropogenic (human) activity in the atmosphere, EPA only regulates $NO_2$ as a stand-in for this family of compounds. In addition to being a significant air pollutant on its own, $NO_2$ also reacts with the atmosphere to produce acid rain and ozone ($O_3$). Roughly half of the NOx emissions come from cars and other mobile sources. Boilers in electric power plants are responsible for approximately 40% of NOx emissions coming from stationary sources. Anthropogenic sources of significant emissions include cement and steel mills, gas turbines, incinerators, gas boilers, reciprocating spark ignition, diesel engines in stationary sources, and the production of nitric acid and glass, cement, and iron and steel. Nitrogen oxides can be found in nature or in biogenic sources such as trees, bushes, grasses, lightning, forest fires, and yeasts. Different amounts of each oxide are produced by these different sources. There are three ways that NOx can form during combustion.

1. **Thermal NOx:** The molar concentrations of nitrogen and oxygen as well as the temperature at which combustion occurs determine the amount of "thermal NOx." Much lower concentrations of thermal NOx are produced during combustion at temperatures far below 1,300°C (2,370°F).
2. **Fuel NOx:** Nitrogen-containing fuels, such as coal, oxidize the nitrogen that has already been ionized within the fuel, resulting in "fuel NOx."
3. **Prompt NOx:** In fuel-rich conditions, which are present in all combustion to some degree, prompt NOx is created when airborne molecular nitrogen combines with fuel. When the fuel burns, this nitrogen oxidizes with it and turns into NOx, just like fuel NOx. The reason why different authors of articles and reports dispute the abundance of prompt NOx may be that they are considering fuels that are intrinsically either very high or very low in nitrogen, or they are considering burners that are designed to have fuel-rich regions in the flame or not.

Technology for NOx abatement and control is a rather complicated topic. All combustion sources release a lot of NOx into the flue gas, but pickling baths and facilities that produce nitric acid work to keep the NOx contained. Wet scrubbers, also known as absorbers, are able to reduce the amount of NOx released by pickling and acid plants. The liquid that absorbs the NOx can be hydrogen peroxide, alkali in water, or just water. In contrast to an upward flow of gas, the wet scrubber works by allowing liquid to flow downward due to gravity through a packing medium. Scrubbers work by switching materials between gaseous and liquid states. To achieve the desired control efficiency, a scrubber's height, packing type, liquid flow, liquid properties, gas properties, and gas flow must all work together.

The Destruction or Removal Efficiency (DRE) that each successful technology is capable of achieving is typically used to express abatement and emission control principles for combustion sources. Relative DRE is another way to express how well pollution prevention strategies work to reduce the production of NO and $NO_2$ when a preventative technology is used, compared to when it is not, the amount of NOx generation is decreased. In comparison to comparable but older systems, many modern combustion systems produce significantly less NOx thanks to design strategies that prevent it. Because of this, it might not be appropriate to take DRE—even a relative DRE—for NOx. Assessing the degree to which a new combustion system reduces NOx emissions may be best done by comparing estimated or actual NOx emissions from the system to NOx emitted from an analogous well-controlled and operated older system.

- **Method 1. Reducing Temperature:** The stoichiometric ratio—the precise ratio of chemicals that enter a reaction—can be avoided by lowering the combustion temperature. In essence, using too much fuel, air, flue gas, or steam dilutes the calories. Different variations of this method are used by combustion controls for fuels with varying nitrogen contents. The net stoichiometric ratio helps explain the control of NOx resulting from the combustion of fuels with high nitrogen content, such as coal. Lean versus rich fuel/air ratios can be used to control the NOx produced during the combustion of low nitrogen fuels (like gas and oil). In any case, by avoiding the ideal stoichiometric ratio—which results in higher temperatures and higher concentrations of thermal NOx—this method avoids it.
- **Method 2. Reducing Residence Time:** In internal combustion engines, residence time at high combustion temperatures can be decreased by timing injections or ignition. In boilers, it can also be accomplished by limiting the flame's spread to a small area where the combustion air turns into flue gas. Fuel injection, steam, additional combustion air, or flue gas recirculation happen right after this. Because of its brief residence time at its highest temperature, most nitrogen is not ionized. This has nothing to do with how long flue gas has been residing in a boiler overall.
- **Method 3. Chemical Reduction of NOx:** This method offers a chemically reducing substance to take the oxygen out of nitrogen oxides, effectively reversing the oxidization process. Examples include Fuel Reburning (FR), Selective Non-Catalytic Reduction (SNCR), which uses urea or ammonia, and Selective Catalytic Reduction (SCR), which uses ammonia. NOx is chemically reduced by non-thermal plasma, an emerging technology, in the presence of a reducing agent.
- **Method 4. Oxidation of NOx:** This method is based on the fact that NOx is more soluble at higher valences, so it purposefully increases the valence of the nitrogen ion to enable water to absorb it. One of four methods—using a catalyst, injecting hydrogen peroxide, producing ozone inside the air flow, or injecting ozone into the air flow—is used to achieve this. Non-thermal plasma can be utilized to oxidize NOx

in the absence of a reducing agent. To absorb $N_2O_5$ emissions into the atmosphere, a scrubber must be added to the process.

- **Method 5. Removal of Nitrogen From Combustion:** This is achieved by either (1) using oxygen in place of air during combustion, or (2) using ultra-low nitrogen content fuel to form less fuel NOx, in order to remove nitrogen as a reactant. Oxygen-based nitrogen elimination often results in a fairly intense flame that needs to be appropriately diluted afterwards. If air is used as the quench medium, Method 2 can rapidly lower the temperature to prevent the formation of excessive NOx, but it cannot completely remove nitrogen oxides. The air used to quench hot flue gas heats up due to this heating, which produces some thermal NOx. Since air contains 80% nitrogen, this method also involves lowering the net excess air used during combustion. Utilizing oxygen-containing fuels with ultra-low nitrogen content can almost completely remove fuel and cause NOx.
- **Method 6. Sorption, Both Adsorption and Absorption:** NOx and other pollutants, mostly sulfur, can be eliminated from flue gas by treating it with sorbents (such as carbon, ammonia, powdered limestone, aluminum oxide, or carbon). Successful attempts have been made to turn sorption products into a commodity that can be sold. This type of treatment has been used in the baghouse, flue, and combustion chamber. Despite the lack of a commercially viable product, carbon is occasionally used as an adsorbent to reduce NOx emissions. Although slurries have also been employed, the sorption process is commonly associated with the use of a dry sorbent. To remove the sorbent, this method employs either adsorption or absorption, followed by filtration and/or electrostatic precipitation.
- **Method 7. Combinations of these Methods:** A lower NOx concentration can be attained by combining many of these techniques than by using just one of them alone. For instance, fuel reburn (Method 3) and over-fire air (Method 1) can come after a fuel-rich cyclone burner (Method 1). As a result, NOx has decreased by up to 70%. The nitrogen oxide concentration is also significantly impacted by other control technologies, the main goal of which is to lower sulfur concentrations. For instance, the SOx-NOx-ROx-Box (SNRB) technology absorbs sulfur from boiler flue gas by means of a limestone sorbent. Ammonia injection and SCR utilizing catalyst fibers in the baghouse filter bags come next. A Claus process is used to recover the sulfur from the sorbent and regenerate it. This has shown that up to 90% of the NOx and 80% of the Sox can be eliminated.

## REFERENCES


1. Ebiefung, A. A., & Udo, G. (1999). An industrial pollution emission control model. *Computers & Industrial Engineering, 37*(1–2), 371–374. https://doi.org/10.1016/S0360-8352(99)00096-0.
2. Godish, T. (1997). Control of emissions from stationary sources. In *Air Quality* (3rd ed., pp. 309–340). Lewis. https://doi.org/10.1201/9780429103254.
3. Li, Z., Feng, C. J., & Ya, D. J. (2019). Air pollution and control of cargo handling equipments in ports. *E3S Web of Conferences, 93*, 02001. https://doi.org/10.1051/e3sconf/20199302001.
4. Malyshev, A. (2020). Pollution control in modern machine systems. *Sel'skohozjajstvennaja tehnika: obsluzhivanie i remont (Agricultural Machinery: Service and Repair)*, (2), 20–23. https://doi.org/10.33920/sel-10-2002-03.
5. Perry, R., & Green, D. (Eds.). (1984). *Perry's chemical engineers' handbook* (6th ed.). McGraw-Hill. https://doi.org/10.1036/0071422943.
6. Zhang, X., Wang, Q., Qin, W., & Guo, L. (2019). Sustainable policy evaluation of vehicle exhaust control—Empirical data from China's air pollution control. *Sustainability, 12*(1), 125. https://doi.org/10.3390/su12010125.
7. Zhao, Y., Zhang, J., & Nielsen, C. P. (2014). The effects of energy paths and emission controls and standards on future trends in China's emissions of primary air pollutants. *Atmospheric Chemistry and Physics Discussions, 14*(6), 7917–7963. https://doi.org/10.5194/acpd-14-7917-2014.

*Chapter* **9**

# Control of Mobile Source Emissions

## Contents

# INTRODUCTION

Any air pollution released by automobiles, aircraft, locomotives, and other movable engines and equipment is referred to as mobile source air pollution. Numerous of these pollutants harm human health in addition to degrading the environment. Environmental regulatory organizations like the U.S. have made environmental and public health regulations a priority. Policies have been established by the Environmental Protection Agency to reduce air pollution originating from mobile sources. State-level organizations are comparable to these. Because they are more numerous and more mobile than stationary sources, like power plants, mobile sources of air pollution are subject to different regulations. Through design and fuel standards, mobile sources are frequently regulated more broadly than individual emitters, such as a single vehicle. Leaded gasoline is prohibited by law in the United States, and corporate average fuel economy standards are two examples of this. The rise in the quantity of cars driven in the U.S. has complicated efforts to reduce pollution from mobile sources. Consequently, several regulatory tools have been put in place to achieve the intended emissions targets.

# 9.1. MOBILE SOURCES OF POLLUTION

Air pollution comes from a variety of mobile sources, some of which are more significant than others. As was previously mentioned, because mobile sources are more numerous and can move around, they are regulated differently than stationary sources. Different mobile sources produce different kinds and intensities of emissions and function differently. The E.P.A categorizes mobile sources as either non-road vehicles or on-road vehicles in order to distinguish between them. Regulations governing on-road and off-road vehicles frequently differ.

## 9.1.1. Milestones in Auto Emissions Control

In 1952, Haagen-Smit, a chemistry professor and researcher at the California Institute of Technology in Pasadena, established the first connection between cars and air pollution. He concluded that a significant portion of the hazy skies over Los Angeles was caused by mobile sources. He also noted that residents experienced extreme eye irritation on smoggy days and that, despite early smoke control measures, there was still an odd smell in the air similar to that of bleaching solution. In 1952, Haagen-Smit declared that ozone, the main component of smog, was produced in the atmosphere rather than being released from tailpipes or smokestacks, based on the analysis of air samples taken in five-liter flasks from refineries in the Southland. Ozone is created when sunlight drives a photochemical reaction that combines nitrogen oxides, a byproduct of combustion, with hydrocarbons from oil refineries and partially burned car exhaust.

## 9.1.2. Automobiles and Emissions

The largest mobile source of air pollution is without a doubt motor vehicles. Compared to 42% in 1970, the percentage of VOC emissions produced by on-road vehicles fell to 29% in 1999. An automobile produces pollutants primarily in three areas. Evaporative, exhaust, and refueling losses are the three main categories of emissions. One source of volatile organic compounds (VOCs) that are similar in amount to exhaust VOC emissions is gasoline evaporation.

Motor vehicle fuel systems emit evaporative volatile organic compounds under four different categories: (1) diurnal emissions; (2) running losses; (3) hot soak emissions; and (4) refueling. When the temperature rises during the day, gasoline evaporation increases, heating the fuel tank and releasing its vapor, which is what causes diurnal emissions (Figure 9.1).

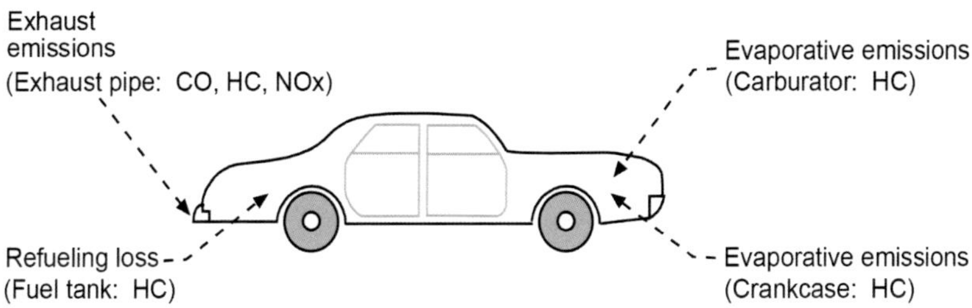


**Figure 9.1.** *Emissions and emission sources of light-duty vehicles.*

*Source: Godish, Thad, (1997). Air Quality, 3rd edition, p. 288. Lewis: New York. doi:* https://doi.org/10.1201/9780429105036.

Running losses happen when the fuel in the tank gets hot and the gasoline vapors are released from the tank while the car is moving. When the engine is switched off after running and stays hot for a while, the gasoline continues to evaporate until the engine reaches room temperature, releasing hot soak emissions. But the combustion of gas released through the exhaust pipe causes the majority of vehicle pollution. About 90% of the pollutants released by a car come from the exhaust of the engine. Unburned hydrocarbons, carbon monoxide, and various amounts of nitrogen oxides, particulates, and hazardous compounds (such as benzene, acetaldehyde, formaldehyde, and 1,3-butadiene) are the main pollutants found in the exhaust gas of a spark-ignition (SI) engine. Along with other combustion products (such as carbon dioxide and water vapor), nitric oxides and carbon monoxide are produced during the combustion process and exit the engine.

When oxygen and nitrogen atoms in the engine's combustion from the induced air break down at high temperatures and pressures, nitric oxide (NO) is formed. After recombining, these atoms produce NO, $NO_2$, and $N_2O$. They are collectively known as NOx. SI engines are generally thought to emit more than 95% of the total NOX, however in certain circumstances, $N_2O$ may account for more than 10% of the emissions.

The highest temperatures reached during the combustion process also affect the production of NO. On the other hand, other factors like flame speed, fuel to air ratio, and pre-combustion conditions also affect how much NOx is produced. Higher chamber temperatures that come with increased SI engine combustion efficiency promote the production of NOx. However, using rich air-fuel mixtures can significantly reduce NOx emissions. Reducing the amount of NOx produced can also be accomplished by postponing or retarding the spark to ignite later in the compression stroke. As a result, there is less time available to produce higher temperatures, which in turn leads to less NOx formation. In a spark-ignition engine, the air-fuel mixture is compressed as the piston rises. Lowering the engine compression ratio results in lower combustion chamber temperatures because it reduces the relationship between the initial and final volumes of the combustion chamber. As a result, NOx emissions are reduced. Additionally, this makes it easier to use unleaded fuels with lower octane ratings, which are necessary for catalytic converters to function effectively.

Similar to unburned hydrocarbons, carbon monoxide (CO) is primarily produced when carbon in fuel undergoes incomplete oxidation, the degree of which is primarily influenced by the air to fuel (A/F) ratio. When the air to fuel ratio decreases, the engine combustion reaction produces a greater amount of carbon monoxide because there isn't enough oxygen present to finish the oxidation of the air/fuel mixture. Whether the engine is operating rich, stoichiometric, or lean is the primary factor influencing CO formation. Most carbon monoxide emissions from newer SI engines are caused by excursions into fuel-rich mixtures while the engine is operating during cold starts and transient periods. By keeping combustion near stoichiometric and utilizing a catalyst control system, CO emissions from modern spark-ignition engines can be effectively managed (Figure 9.2).

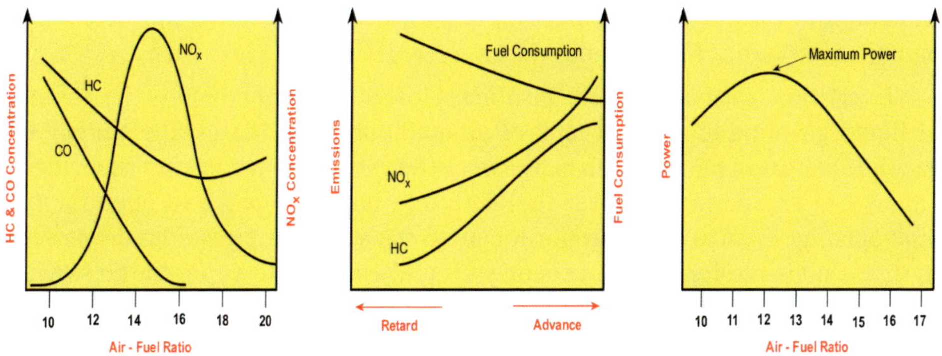


**Figure 9.2.** *Exhaust emissions related to automobile engine operating parameters.*

*Source: Godish, Thad, Air Quality, 3rd edition Lewis: New York, 1997, p. 289, 292, 297. DOI:* https://doi.org/10.1201/9780429105036.

Unburned hydrocarbons (HC) are produced by a variety of processes where fuel leaks out of the combustion chamber during flame passage. In an SI engine, between 1.5–2.0% of the gasoline fuel remains unoxidized. The way fuel escapes the combustion reaction of the engine can be influenced by numerous important processes. These include the following: absorption of fuel on carbon deposits and lubricating oil in the combustion chamber; quenching of the flame near cold cylinder walls, also referred to as quench zones; and from liquid fuel, especially during first start up. The fuel to air equivalency ratio, cylinder compression ratio, engine speed, and spark timing are the main engine parameters that influence the quantity of unburned hydrocarbons leaving a SI engine.

Compared to warm starting, cold starting an automobile results in higher trip emissions. A normal car in 2,000 emits approximately 8.7 grams of NOx and 95.1 grams of CO during a 7.5 mile drive, if the engine is already warm. However, an extra 1.4 grams of NOx and 37.8 grams of CO are produced if the engine is cold. Because of this, starting the car cold produces roughly 16% more NOx and 40% more CO for a 7.5 mile trip than starting it warm. Both the tailpipe and fuel evaporation release volatile organic compounds (VOCs) into the atmosphere. A warm engine emits about 6.3 grams of volatile organic compounds (VOCs), while evaporative emissions (during travel and cooling down) produce 2.3 grams of VOCs. 4.3 grams of VOC are produced when the car is started cold. Trip chaining, the linking of multiple trips, reduces emissions by doing away with cold starts. But even when a car is left parked all day in the summer heat, volatile organic compounds (VOCs) escape the fuel tank and control systems. Parking a car all day releases about 2.6 grams of volatile organic compounds (VOCs) into the air.

Every time a car is refueled, gasoline is lost during refueling. As the volume of liquid gasoline in the fuel tank of an automobile decreases, the amount of gasoline vapors in the fuel tank increases. When standard nozzles are used, these vapors can escape from the car fuel tank into the atmosphere through the fill pipe opening in an amount proportionate to the volume of gasoline dispensed. In order to enable further conversion of liquid gasoline to vapor in the service station tank, the gasoline removed from the storage tank would be replaced at the same time by an equivalent volume of air. When a vapor recovery nozzle is used to force gasoline vapors out of an automobile's fill pipe opening, Stage II vapor recovery systems gather them. The gathered vapors enter the nozzle's front end through a port on the outside that allows a vacuum to be introduced to aid in vapor collection, or through flexible bellows that are positioned around the nozzle spout. The vapors then enter the underground storage tank of the service station after passing through the vapor passage in the nozzle's body, the vapor hose, and the underground piping (Figure 9.3).

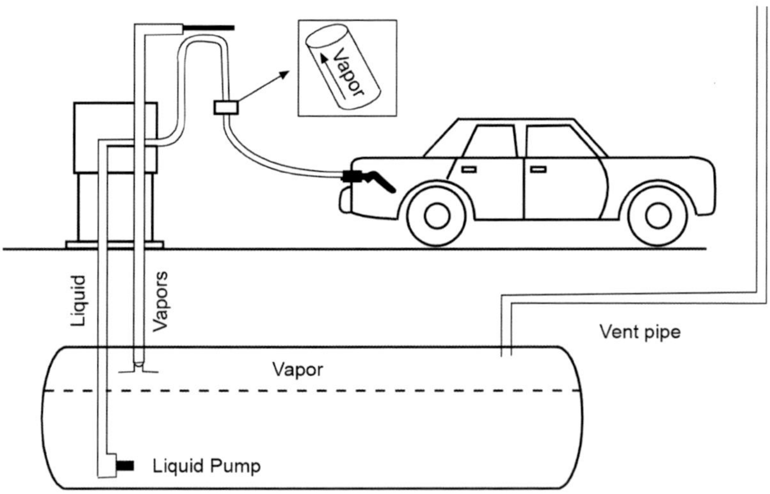


**Figure 9.3.** *Stage II gasoline vapor recovery system.*

*Source:* https://www.4cleanair.org/wp-content/uploads/APTI/452combined.pdf.

## 9.1.3. Federal Emission Standards

The federal emissions control standards for all new cars and light trucks sold in the United States are established by the Clean Air Act of 1963 and its subsequent amendments. More stringent Tier 1 emissions standards were set by the 1990

Clean Air Act Amendments, and they went into effect in 1994. Additionally, the CAAA mandated that EPA research the possibility of requiring stricter Tier 2 standards in order to meet the NAAQS. The EPA concluded in 1999 that new standards were both necessary and affordable. Beginning in 2004, all passenger vehicle classes—including light trucks and sport utility vehicles—must adhere to new nitrogen oxide average tailpipe standards of 0.07 grams per mile. The EPA released new Tier II engine and gasoline standards in December 1999. The purpose of the regulations was to lower the emissions produced by newly manufactured passenger cars and light trucks, such as sport utility vehicles, minivans, and pickup trucks. Gasoline importers and refiners in the country will have to start producing gasoline with sulfur levels limited at 300 parts per million (PPM) starting in 2004. This represents a 15% decrease from the current industry average of 347 ppm. Refiners will meet a maximum cap of 80 ppm and an average sulfur level of 30 ppm by 2006. Between 2004 and 2030, the Tier II rules are expected to reduce NOX emissions by 61% and VOC emissions by 24%.

## 9.1.4. Control of Automobile Emissions

Car emission control started out with simple devices and small engine modifications, but it has since grown to include significant design considerations and high costs. In 1961, the positive crankcase ventilation (PCV) control system was introduced, marking the beginning of control. A blowby gas of unburned fuel and exhaust combustion products is created during internal combustion and seeps past the piston ring's imperfect seal. Positive crankcase ventilation (PCV), a reasonably easy-to-use and affordable technology, has been developed to control this type of crankcase emissions. Fumigant gases are recirculated using this control method so they can burn again in the combustion chamber. Fundamental advancements in engine design, along with the addition of exhaust gas recirculation valves to lower nitrogen oxide emissions and charcoal canisters to collect hydrocarbon vapors, contributed to the 1970s emission reductions. When first-generation catalytic converters were introduced in 1975, emissions of carbon monoxide and hydrocarbons were drastically decreased. Additionally, there was a significant indirect benefit from the use of converters. Unleaded gasoline became widely available in 1975 due to the fact that lead deactivates the catalyst.

The years 1980–81 marked the next significant turning point in vehicle emission control technology. Manufacturers outfitted new cars with even more advanced emission control systems in response to stricter regulations. These systems typically included an oxygen sensor, an on-board computer, and a three-way catalyst (which helps reduce nitrogen oxides to elemental nitrogen and oxygen and converts carbon monoxide and hydrocarbons to carbon dioxide and water). This apparatus assisted in maximizing the catalytic converter's efficiency. The 1990 Clean Air Act's provisions are lowering car emissions even more. More stringent tailpipe regulations, enhanced robustness, better evaporative emission control, and computerized diagnostic systems that pinpoint defective emission controls are all features of mobile sources. One of the best methods for minimizing the production of NOx is to use an exhaust gas recirculation (EGR) system. With the help of this control mechanism, exhaust gases are circulated in precisely measured amounts back into the intake manifold, where they combine with fuel and air to initiate the combustion cycle. Because they function as a heat sink and lower peak combustion temperatures, these comparatively inert exhaust gases—which are primarily carbon dioxide and water—reduce the production of NOx.

Thermactor, a brand name for the emissions control system generally known as AIR (air injection reaction), is another device used during this time to help control emission with the catalytic converters. The AIR system's job is to introduce new air into the exhaust gas stream. This system has two advantages: (1) fresh air blasted just downstream of the exhaust ports into the exhaust gas provides oxygen ($O_2$) to allow any unburned hydrocarbons (HCs) in the exhaust gas to continue burning; and (2) more oxygen is needed by the catalytic converters to convert carbon monoxide (CO) and hydrocarbons in exhaust gases into carbon dioxide ($CO_2$) and water ($H_2O$).

Automakers will use cleaner fuels and build cleaner cars as a result of the Clean Air Act. In over forty American metropolitan areas, an inspection and maintenance (I/M) program is mandated to guarantee that cars are being maintained. It is necessary for certain areas with existing I/M programs to upgrade their emission inspection tools and practices. By employing a technique similar to IM240, enhanced I/M will produce more accurate measurements of the vehicle's emissions. This test procedure certifies compliance with federal vehicle emission standards for carbon monoxide (CO), unburned hydrocarbons (HC), and nitrogen oxides (NOx). It uses a dynamometer and driving schedules based on EPA's Federal Test Procedure (FTP75). This equipment will replace static or sitting parked emission measurements with simulated driving conditions.

### *9.1.4.1. Gross Emitters*

A significant portion of the pollution from on-road vehicles is produced by a tiny percentage of vehicles. These gross emitters include some newer cars with poorly maintained or broken emissions control systems in addition to older models. It is estimated that less than 10% of the fleet of vehicles emits roughly 50% of the VOC emissions, as depicted in the diagram. But not all criterion pollutants are grossly emittable by the same cars; for example, CO, NOx, and other pollutants may be grossly emittable by a different 10% of vehicles. Furthermore, 10–27% of the cars that fail inspections never pass the State Inspection and Maintenance exams.

### *9.1.4.2. Reformulated Gasoline and Alternative Fuels*

Cleaner fuels are also covered by the Clean Air Act of 1990. Using gasoline that is made to burn cleaner is one method of lowering the amount of air pollution produced by cars and trucks. Reformulated gasoline, or RFG, burns cleaner and is mandated by the Clean Air Act for cities with the worst pollution from smog; other cities with smog issues are free to use RFG. RFG was first implemented at the federal level in 1995, and it is presently utilized in 17 states as well as the District of Columbia. In the US, reformulated gasoline makes up about 30% of the total sold. Every oil company creates a formula that satisfies federal requirements for reducing emissions.

In comparison to conventional gasoline, the first phase of the RFG program was intended to reduce air pollution causing smog by 64,000 tons annually in the areas that use RFG; this is the equivalent of removing the emissions from over 10 million vehicles that cause smog. After Phase I RFG was superseded in 2000, Phase II RFG aims to cut smog pollutants in RFG areas by an additional 41,000 tons annually, which would remove the smog-forming emissions from roughly 16 million vehicles. A known human carcinogen, benzene, is among the harmful air pollutants whose emissions are decreased by the RFG program. In RFG areas, Phase I and Phase II RFG together reduce toxic pollutants by approximately 24,000 tons annually, which is the same as removing the toxic emissions from over 13 million cars.

Congress mandated that RFG contain 2% by weight oxygen in the Clean Air Act. The two most widely used oxygenates that raise the oxygen content of gasoline for more thorough combustion in engines are ethanol and MTBE (methyl tertiary butyl ether). MTBE has been used in U.S. gasoline at low levels since 1979 to replace lead alkalis (ethyl and tetramethyl lead) as an octane enhancer (helps prevent the engine from "knocking") and since 1992 has been used at higher concentrations in some gasoline. A chemical compound called methyl tertiary-butyl ether (MTBE) is created when isobutylene and methanol react. In 1999, the US produced over 200,000 barrels of MTBE per day. Its primary application is as a fuel additive for motor gasoline. Fuel burns more thoroughly with the aid of oxygen, which lowers harmful emissions from motor vehicles' tailpipes. In one way, sulfur and aromatics (like benzene) found in gasoline are diluted or replaced by oxygen. In a second, oxygen maximizes combustion-related oxidation. Oil companies choose which material to use in order to comply with legal requirements. For blending purposes and cost considerations, MTBE has been preferred by most refiners over other oxygenates.

MTBE is a volatile, flammable, colorless liquid that dissolves easily in water at room temperature. The most common source of gasoline contamination in water is leaking storage tanks. Inadequate disposal and minor spills are additional sources of contamination. MTBE is one of the many chemicals in gasoline that can be dangerous in water. Compared to other gasoline components, MTBE is highly soluble and moves through water more quickly and extensively. The other common oxygenate, ethanol ($CH_3CH_2OH$), is a class of chemical compounds in which a carbon atom is bonded to a hydroxyl group (-OH). Starch crops that have been reduced to simple sugars are fermented and then distilled to produce ethanol. Among the feedstocks used to make ethanol are wheat, barley, and corn. Bioethanol, another name for ethanol, can also be made from cellulosic biomass, like grasses and trees. Although it can be used at higher concentrations like E85 or E95, ethanol is most commonly blended with gasoline to create an E10 blend (10% ethanol and 90% gasoline) in certain parts of the United States. Original equipment manufacturers make flexible-fuel cars that can run on any other blend of gasoline and ethanol, including E85.

The EPA encourages and expands the use of alternative fuels that are better for the environment. There are numerous alternative fuels that can be used to address various issues related to air pollution.

- **Liquefied petroleum gas (LPG):** A fossil fuel derivative made up of 5% butanes and 95% propane. Although CO emissions are reduced, NOx emissions might be higher.
- **Natural Gas:** A fuel that can be liquefied (LNG) or compressed (CNG). CNG, which is more widely used in the transportation industry, is kept in high-pressure cylinders for storage. Compared to regular gasoline, CNG emits less CO and VOCs, and less NOx and PM than diesel fuels.
- **Methanol:** Wood alcohol derived from biomass, coal, or natural gas.
- **Ethanol:** Grain alcohol derived from woody biomass, corn, or sugarcane. Blends with ethanol may lower CO emissions, but they have very little impact on ozone.
- **Electricity:** Batteries for electric vehicles can be charged at home or at charging stations using electricity generated by power plants. While they don't emit any tailpipe emissions, power plants are the source of all emissions.

- **Hydrogen:** A fuel that burns cleanly and can be made using wind, solar, coal, gas, or oil. A car running on a fuel cell emits very little NOx and no CO or VOCs when it runs. Fuel cells produce electricity by using the reaction of hydrogen and oxygen to create water. One important factor influencing the emissions from alternative fuel vehicles is the design of the engine and vehicle.

Depending on the engine load, hybrid cars continuously switch between power sources. A small internal combustion engine and a high-energy density battery are used in a hybrid electric vehicle. The internal combustion engine's fuel, which is usually either diesel, gasoline, or compressed natural gas (CNG), determines the hybrid electric vehicle's in-use emissions.

Clean fuels are superior to regular gasoline because of a number of intrinsic qualities. These fuels generally release fewer volatile organic compounds (VOCs), and the ones they do release are less toxic and less reactive (forming ozone more slowly). Vehicles powered by electricity, natural gas, or alcohol may emit up to 90% less harmful and ozone-forming volatile organic compounds (VOCs) into the atmosphere than cars running on regular gasoline. Utilizing clean fuels may also aid in reducing the atmospheric concentration of carbon dioxide, a greenhouse gas that increases the risk of global warming. Carbon dioxide is produced when any fuel containing carbon burns. However, a fuel's overall contribution to global warming varies depending on its manufacturing process. Fuels derived generally from biomass (crops, trees, etc.) and from natural gas accumulate less carbon dioxide than fuels derived from coal or petroleum.

Utilizing clean fuel vehicles is also encouraged by giving states resources like benefits models, State Implementation Plan Credits, and the Clean Fuels Fleet program. In order to support the Energy Policy Act and Transportation Equity Act of the 21st Century programs that support alternative fuels and vehicles with the biggest environmental benefits, EPA also collaborates with the Departments of Energy and Transportation. The internal combustion engine has changed our way of life significantly over the past century, but its effects on the environment are now clearly visible. The internal combustion engine supplanted horse-drawn buggies, and it is conceivable that a new technological advancement will eventually render the automobile obsolete. But until then, carpooling, taking public transit more frequently, and switching to cleaner fuels are the most practical ways to cut down on vehicle exhaust emissions.

#### *9.1.4.3. Diesel Engines*

Unburned carbon particles, which account for the majority of diesel PM, the soluble organic fraction (SOF), which is made up of unburned hydrocarbons that have either condensed into liquid droplets or on unburned carbon particles, and sulfates, which are produced when fuel-borne sulfur in the engine's exhaust oxidizes, are the three main components of diesel PM. From 6.0 g/bhp-hr NOx and 0.6 g/bhp-hr PM in 1990 to 4.0 g/bhp-hr NOx and 0.1 g/bhp-hr PM in 1998, diesel engines have achieved significant progress in reducing engine-out emissions. The first steps toward these reductions were improvements in fuel and combustion systems. Early in the 1990s, electronic fuel systems were introduced, allowing for reduced NOx and PM levels without compromising fuel efficiency. The main technologies that allowed emission levels to be lowered to their current levels were this and rising fuel injection pressures.

Cooled exhaust gas recirculation (EGR) will be the main source of additional engine-out NOx reductions to the levels required to meet the 2004 standard of 2.5 g/bhp-hr NOx+NMHC. By using this technique, some of the exhaust is brought back to the intake manifold and drawn into the combustion chamber. The decrease in oxygen from the lower $O_2$ in the EGR and the increase in heat capacity from the $CO_2$ in the EGR could be two of the factors influencing the reduction of NOx by EGR. Nonetheless, some could contend that the shift in heat capacity is not noteworthy. According to the $O_2$ dilution theory, decreasing the $O_2$ concentration reduces the $O_2$ partial pressure, which in turn lessens the $O_2$'s tendency to oxidize $N_2$ during combustion to NO and $NO_2$ (NOx). To regulate the constituents of diesel PM, various exhaust after treatment devices have been developed. These include diesel oxidation catalysts (DOC) and various types of particulate filters, also known as traps. Although DOCs have been demonstrated to be reliable when used, they are only able to effectively regulate the part of PM that is SOF, which is only about 10–30% of total PM, especially on today's engines. As a result, the DOC by itself does not adequately address our PM concerns.

As of right now, only the PM trap can offer the degree of control required by the suggested PM standards. The PM trap has shown excellent trapping efficiency in the past, but regeneration of the collected PM has proven to be a significant difficulty. In order to collect PM, the exhaust is passed through a ceramic or metallic filter in the PM trap. Before the filter plugs, the gathered PM, primarily carbon particles but also the SOF component, needs to be burned off. "Regeneration" is the term used to describe this burning off of accumulated PM, and it can happen either:

- on a periodic basis by using base metal catalysts, or an
- active regeneration system like a microwave, fuel burner, or electrical heater; or,
- on a continuous basis by using precious metal catalysts.

Many years ago, uncatalyzed diesel particulate traps showed high efficiency in trapping particulates, but the PM standard was low enough that less expensive "in-cylinder" control methods could meet it. Furthermore, the regeneration traits lacked consistency. In order to improve regeneration, some systems used fuel burners or electrical heaters, but these complicated the system design and still didn't offer the dependability and durability needed for HD diesel applications.

## 9.1.5. Off-Road Vehicles and Equipment Emissions

Since the early 1970s, the EPA has set progressively stricter emission standards for trucks and cars used on highways. Much progress has been made in reducing emissions from highways; however, the vast array of non-road engines is now receiving attention, as they too play a major role in air pollution. The term "non-road" refers to a broad category of machinery, engines, cars, and boats. The non-road category covers a wide range of applications, such as marine vessels, aircraft, locomotives, lawn and garden equipment, recreational vehicles, outdoor power equipment, and many more. It is also sometimes referred to as off-road or off-highway. Very few of these engines were subject to any kind of emission standards prior to the mid-1990s. The Clean Air Act was amended in 1990, directing the EPA to investigate the role of non-road engines in urban air pollution and to regulate them if they were found to be a contributing factor in poor air quality. The EPA released a report in 1991 that demonstrated the significant emissions of hydrocarbons (HC), particulate matter (PM), carbon monoxide (CO), and oxides of nitrogen (NOx) from non-road equipment. Overall emissions from non-road engines were typically nearly equal to those from highway engines. When it came to particulate matter, emissions from non-road sources were considerably greater than those from highways.

### *9.1.5.1. Airplanes*

Aircraft powered by gas turbines and turbojets are another source of mobile air pollution. The amount of exhaust produced by aircraft is negligible in comparison to that of cars. Only 1% and 6%, respectively, of the nation's NOx and PM emissions come from aircraft emissions from mobile sources. These engines have a more noticeable effect in some cities with high airport traffic, and the effects on the communities surrounding large airports can be significant. Aircraft emissions are more noticeable than those from motor vehicles, despite the concentrations of nitrogen oxides, hydrocarbons, and particulates in aircraft emissions being similar. The residue left behind by jet aircraft exhaust trails constantly crisscrosses the sky. Because aircraft engines were built for high performance and dependability rather than fuel economy or other environmental considerations, they release large amounts of air pollution.

For almost 20 years, gas turbine engines used to power civil aircraft have been subject to emission regulations. Almost all commercial aircraft, including those operated by freight and scheduled airlines, have these engines. Aircraft used for general aviation or the military are exempt from the standards. Engine smoke controls and fuel venting bans were implemented in 1974 and have undergone multiple revisions since then. Gas turbine engines are subject to limits on hydrocarbon emissions per landing and takeoff cycle as of 1984. Owing to the global nature of the aviation sector, the International Civil Aviation Organization (ICAO) is crucial in establishing consistent emission standards that different countries can ratify. The ICAO's NOx and CO emission guidelines for gas turbine engines were approved in May 1997. Furthermore, it has been suggested that a second set of stricter ICAO NOx standards for gas turbine engines be put into effect in 2004.

The airline industry has made significant concessions in search of alternative strategies to lessen air pollution from air transportation as a result of public and environmental concerns about noise and air pollution near airports. Airport operating procedures have been significantly changed, and newer turbojet engines are now operating with air-fuel ratios that burn more efficiently. Strict control over arrival and departure times, for instance, has resulted in a significant reduction in the amount of time an aircraft spends idle on the ground. Auxiliary ground vehicles are also used to transport passengers to and from aircraft that are positioned remotely, and ground-handling equipment is frequently utilized to tow aircraft to and from the ramp, which minimizes the amount of time an aircraft must spend on ground taxi.

Additionally, airlines have modified the operational flight procedures used by their pilots and other staff members. For instance, the amount of time an aircraft idles for maintenance has been decreased by flight engineers. Moreover, flight patterns have evolved. Pilots today strive to reach 3,000 feet as soon as possible to minimize their impact on neighboring communities. They have also virtually abandoned the practice of spilling fuel over land and sea to improve fuel efficiency.

### *9.1.5.2. Railroad Locomotives and Marine Engines*

About 9% of NOx emissions and 4% of PM emissions from mobile sources are attributed to locomotives. Compared to any land-based, non-road diesel engine, these engines are typically larger and have longer lifespans. While HC and PM emissions from these engines will drop by 50% as a result of new emission regulations, NOx emissions will be reduced by two thirds. Three different sets of emission standards have been adopted; the standards' applicability varies according to the date on which a locomotive is manufactured. Locomotives and locomotive engines that were first produced between 1973 and 2001 are subject to the first set of standards (Tier 0), regardless of when they are produced or remanufactured. In light of the fact that locomotives typically undergo five to ten remanufacturing during the course of their 40-year or longer service lives, this special feature of the program is vital.

For locomotives and locomotive engines that were first produced between 2002 and 2004, the second set of standards (Tier 1) is applicable. Both during original manufacture and every subsequent remanufacture, these locomotives and locomotive engines must adhere to Tier 1 standards. As of 2005 and onwards, locomotives and locomotive engines are subject to the last set of standards (Tier 2). At the time of original manufacture and every subsequent remanufacture, Tier 2 locomotives and locomotive engines must adhere to the relevant standards. The sizes and types of engines used by marine vessels vary greatly. For the purposes of this discussion, gasoline-powered outboard engines and personal watercraft, stern drive and inboard gasoline engines, and marine diesel engines are the three general categories into which marine engines are classified for the purpose of setting emission standards. These marine programs cover marine diesel auxiliary engines, but gasoline-fueled auxiliary engines have to adhere to the same emission regulations as land-based engines.

## 9.2. MOBILE SOURCE PROGRAMS

Since 1990, there has been a roughly 50% decrease in mobile source emissions, or 1.5 million tons of HAPs annually. We anticipate that by 2030, these reductions will reach an 80% level due to increased fleet turnover. Furthermore, between 1990 and 2005, the amount of mobile source diesel particulate matter on and off roads dropped by roughly 27%. Between 2005 and 2030, significant additional reductions (roughly 90%) are anticipated as a result of the implementation of numerous recent mobile source regulations that target diesel engines. Tier 3 vehicle and fuel standards, released in 2014, are the EPA's most recent regulatory program that significantly reduces air toxics from mobile sources. Depending on the pollutant, these standards will cut motor vehicle emissions of air toxics by 10–30% by 2030.

The 2007 mobile source air toxics rule, which regulated gasoline's benzene content as well as vehicle emissions in cold weather and emissions from portable fuel containers, was another regulatory initiative that helped lower mobile source air toxics. A recent analysis conducted in Anchorage, Alaska, revealed that ambient benzene concentrations were lowered by more than 50% using only the fuel benzene standard. Other initiatives that are lowering mobile source air toxics include the locomotive and commercial marine rule, standards for nonroad diesel engines, controls for small spark-ignition and recreational marine engines, heavy-duty engine and vehicle standards, low-sulfur gasoline and diesel requirements, and the North American and Caribbean Emission Control Areas (ECAs), which were created to lower ship emissions.

Air toxics from mobile sources are also being reduced by non-regulatory measures. The EPA's Ports Initiative, SmartWay, Clean School Bus USA, and the National Clean Diesel Campaign are a few examples. Furthermore, the EPA developed the Diesel Emissions Reduction Program, or DERA, to implement pollution-controlling equipment in diesel fleets. Clean diesel initiatives immediately improve air quality and public health. According to EPA estimates, a community may benefit from up to $13 in additional public health benefits for every dollar invested in reducing diesel exhaust. The EPA provided $569 million between 2008 and 2013 to replace or retrofit approximately 73,000 engines in automobiles, ships, locomotives, and other equipment. Over the course of the affected engines' lifetime, the EPA projects that these initiatives will reduce emissions by 14,700 tons of PM2.5.

## 9.3. ENFORCEMENT MECHANISMS AND POLICY INSTRUMENTS

Various policy instruments are employed by federal, state, and local governments to regulate pollution originating from mobile sources. Numerous federal agencies are in charge of limiting pollution from mobile sources, or at the very least, developing regulations to that effect. This is required because a wide variety of items, including airplanes, off-road vehicles, locomotives, and on-road vehicles, are regarded as mobile sources. For instance, the Federal Aviation Administration sets limits on aircraft emissions, but the U.S. regulations pertaining to fuel efficiency of on-road vehicles are overseen by the Environmental Protection Agency and the Department of Transportation. Mandatory vehicle emissions testing programs are frequently mandated at the state level as a part of the yearly motor vehicle registration procedure.

### 9.3.1. Labeling Policies

Around the world, a large number of governments mandate that producers of specific goods include informational labels on their goods. Food nutrition and ingredient labels, Surgeon General labels on alcohol and tobacco products, and labels for common household pesticides are common examples in the United States. Similar to mobile sources of air pollution, a wide range of products may be subject to government labeling regulations; as a result, multiple federal agencies are in charge of different label-related regulation initiatives. For instance, the US Environmental Protection Agency establishes strict guidelines for the labeling of pesticides, while the US Food and Drug Administration is in charge of regulating food nutrition and ingredient labels.

Labeling regulations are primarily intended to give consumers and other product users vital information about the product. Labeling regulations are essentially intended to address the imperfect information market failure. Detailed product information may be necessary in order for consumers to allocate scarce resources, like income, in the most efficient manner. In this way, labels also aid in addressing information asymmetries that are frequently present in a wide range of market exchanges.

All new cars and light-duty trucks in the US must have labels with specific fuel economy information on them. The Corporate Average Fuel Economy (CAFE) program is administered and enforced by the National Highway Traffic Safety Administration (NHTSA), with data provided by the US Environmental Protection Agency, which computes the average fuel economy for each vehicle manufacturer. The program's two main goals are to: (1) lower emissions by mandating automakers to meet minimum fuel economy standards; and (2) give consumers fuel economy data prior to buying new cars. The labels are being redesigned by the EPA and NHTSA to give customers even more information. The Energy Independence and Security Act of 2007 mandates that the new labels include information about each vehicle's greenhouse gas emissions for the first time. The agencies are asking for feedback from the public on which of the two proposed label designs would be most beneficial to consumers.

### 9.3.2. Taxes

Taxation is another popular tool used by governments to shape market behavior. The US government has implemented numerous levies to restrict emissions from different mobile sources when it comes to air pollution from mobile sources. The Energy Tax Act of 1978 established the gas guzzler tax, which is arguably the most well-known. The act established minimum fuel economy requirements for all newly sold vehicles in the US. The tax is imposed on automakers who produce new vehicles that do not meet the mandatory 22.5% fuel economy standard. Since minivans, sport utility vehicles, and pickup trucks constituted a relatively small percentage of the US fleet at the time of the tax's establishment in 1978, they are exempt from its application. Manufacturers are required to pay a certain amount of tax, which varies depending on the average fuel efficiency of each vehicle they produce. The tax ranges from $1,000 for vehicles that achieve at least 21.5 but less than 22.5 MPG to $7,000 for vehicles that achieve less than 12.5 MPG. The gas guzzler tax does not apply to vehicles that achieve a minimum average fuel economy of 22.5 miles per gallon.

### 9.3.3. Tax Credits

In market economies, governments may also provide tax credits to promote particular kinds of conduct. For instance, the government could provide tax credits to effectively reduce the cost of each vehicle in order to encourage people to buy more fuel-efficient cars. The reasoning behind this strategy is in line with the laws of supply and demand, which state that when an item's price drops, so does the demand for it. As long as other crucial variables, like the present states of supply and demand, stay the same, this is accurate.

Currently, the US federal government uses a variety of tax credits to lower emissions from mobile sources. The Qualified Plug-In Electric Drive Motor Vehicle Tax Credit is one of the more popular tax credits. This credit can be applied to the cost of a brand-new, qualifying plug-in electric drive motor vehicle that meets certain emission standards, has a gross vehicle weight rating of up to 14,000 pounds, uses an external energy source to recharge its traction battery, and draws power from it. The credit can be anywhere between $2,500 and $7,000, based on the weight rating of the car. Customers can receive the entire $7,500 credit when they buy a new Chevrolet Volt. The Fuel Cell Motor Vehicle Tax Credit, which was initially set at $8,000 for the purchase of eligible light-duty fuel cell vehicles, is another tax credit aimed at consumers. There was a reduction in the tax credit to $4,000 on December 31, 2009.

Manufacturers of specific products may also be the target of tax credits intended to reduce pollution from mobile sources. For instance, producers of advanced biofuels or fuels made from renewable biomass other than corn kernel starch that meet the eligibility requirements can receive Advanced Biofuel Production Payments. If sufficient data is available, the net non-renewable energy content of the advanced biofuel, the number of participating producers, the amount of funds available, and the quantity and duration of production by the eligible producer will all determine whether or not these producers receive payments to support expanded production of advanced biofuels. Research from the US Department of Energy shows that biofuels burn cleaner than gasoline, resulting in fewer greenhouse gas emissions, and are fully biodegradable, unlike some fuel additives, despite the claims of many critics who claim they can actually increase greenhouse gas emissions.

### 9.3.4. Voluntary Programs

Voluntary programs are another significant tool of policy that governments can use. These initiatives aim to achieve a specific policy outcome by bringing together a variety of stakeholders. For instance, the Clean Cities initiative was started by the Department of Energy to lessen the amount of petroleum used in the transportation industry. In order to create public-private partnerships that support advanced vehicles, fuel blends, fuel economy, hybrid vehicles, idle reduction, and alternative fuels, the Clean Cities program collaborates with more than 80 volunteer organizations across the country (U.S. Environmental Protection Agency, 1999). The program's three main objectives are.

1. **Replacement:** Switch out the petroleum used in the transportation industry with renewable and alternative fuels.
2. **Reduction:** Encourage more fuel-efficient driving techniques, idle reduction, fuel-efficient cars, and cutting-edge technologies to help cut down on petroleum use.
3. **Elimination:** reducing the need for petroleum by promoting increased use of mass transit, trip-elimination strategies, and traffic-calming measures.

Since its launch in 1993, the program has saved almost 3 billion US gallons (11,000,000 $m^3$) of petroleum. The SmartWay Transport Partnership of the Environmental Protection Agency is another illustration of a voluntary program. Through enhanced fuel efficiency initiatives, this voluntary collaboration between the EPA and the ground freight sector aims to lower greenhouse gas emissions and air pollution. Benefits and services offered by EPA to partners include fleet management tools, information, technical support, public recognition, and the ability to use the SmartWay Transport Partner logo. The EPA also runs a voluntary program called Clean Construction USA, which aims to reduce the emissions of diesel exhaust from construction vehicles and equipment. The program promotes the use of cleaner fuels, emission-reducing technologies, and proper operation and maintenance.

### 9.3.5. Subsidies

Another effective tool that governments use to shape economic behavior is subsidization. There are numerous ways to provide subsidies: from direct cash payments to tax credits. For instance, the U.S. commercial service airports situated in air quality non-attainment and maintenance areas are financially supported by the Federal Aviation Administration's Voluntary Airport Low Emission Program, which aims to reduce mobile source pollution from airports. Although the funds can be utilized to lower emissions at the airport from both mobile and stationary sources, the program focuses primarily on reducing emissions from mobile sources. The initiative encourages the use of electric ground support tools, like bag tugs that move bags from an aircraft to baggage claim. Pushback tractors, which help aircraft take off from the gate, and other kinds of belt loaders are among the other electronically operated airport equipment.

The program's installation of underground fuel hydrants at airports is one of its main objectives. Fuel trucks are a significant contributor to mobile emissions, and their use would be eliminated by these. The Vision 100 Century of Aviation Reauthorization Act of 2003 created the Voluntary Airport Low Emission Program.

## 9.3.6. Command and Control: Performance Standards

In order to reduce pollution from on-road vehicles, such as cars and light-duty trucks, many states have implemented emissions-testing programs. Before any of these cars are eligible to apply for or renew a vehicle registration, they must all meet certain emissions standards. A large number of these programs are run at the county and local levels. One such program is the Clean Air Car Check, which requires all cars registered in Indiana's Lake and Porter counties to have their emissions tested. The Environmental Protection Agency declared the two counties to be non-attainment areas for ozone levels in 1977. In order to achieve and maintain specific air pollution standards, states are required to establish State Implementation Plans, and the two counties were reclassified as severe non-attainment areas by 1990. The two counties will continue their vehicle inspection and maintenance program even though they were reclassified in 2010—this time as attainment areas—because it is an essential component of Indiana's strategy to stop backsliding and keep the area in attainment.

### *9.3.6.1. Corporate Average Fuel Efficiency Standard*

In accordance with the 1975-enacted Corporate Average Fuel Economy Standard (CAFE) regulation, all US car sellers were required to meet a minimum sales-weighted average fuel efficiency of 27.5% by the year 1985. For cars that were imported and those that were made domestically, different standards had to be met. For every 1/10 of a gallon that the corporate average fuel economy fell short of the standard, there was a $5 penalty per car for not meeting the prescribed standard. The first indication of the CAFE regulation's environmental impact can be found by comparing the average fuel efficiency of domestic and foreign businesses. The effects of the CAFE regulation are greatest for domestic US manufacturers, whose corporate average fuel efficiency would decrease by 1.2 miles per gallon in the absence of the regulations. Additionally, the implementation of CAFE standards reduces fuel consumption by about 19 million US gallons (72,000 $m^3$) annually. In contrast to the CAFE guidelines, gasoline taxes apply to both new and used vehicles, so raising taxes shouldn't be expected to result in a switch to less fuel-efficient vehicles. According to

Goldberg and Pinelopi Koujianou (1998), small tax increases are insufficient to result in fuel cost savings of the same order of magnitude as CAFÉ.

## *9.3.7. Marketable Allowances*

### *9.3.7.1. Leaded Gasoline*

In order to stop engines from knocking, lead was first added to gasoline as an additive. Lead was present in almost all American gasoline during the 1970s, with an average concentration of nearly 2.4 grams per gallon. The EPA started developing plans to gradually remove lead from gasoline by the middle of the 1970s for two main reasons. Concern over lead's possible effects on human health was growing, particularly in light of its known effects on hypertension and children's cognitive development. Furthermore, the fuel standards had to be changed when catalytic converters were added to new cars made after 1975. In order to help new cars meet the hydrocarbon, carbon monoxide, and nitrogen oxide emission standards required by the 1970 Clean Air Act, catalytic converters were installed in them. Sadly, unleaded gasoline was the only fuel that the catalytic converters could run on correctly.

The EPA mandated that the average lead content of all gasoline sold be lowered from 1.7 grams per gallon after January 1, 1975 to 0.5 grams per gallon by January 1, 1979 in order to safeguard public health and guarantee that catalytic converters were functioning as intended. The average lead concentration standard goal was eventually lowered by the EPA to 0.1 gm/gal by January 1, 1986. In order to meet their lead limits every quarter, refiners who owned multiple refineries were able to trade or average their refineries according to the EPA's definition of averages. Recognizing the trading activity, the EPA allowed refiners to bank credits for use through the end of 1987. The enforcement of EPA regulations was based on mandatory reporting and ad hoc testing of fuel samples.

The EPA's initiative to phase out lead in gasoline has officially come to an end. Since 1996, automakers have been exempt from having to label all new cars with unleaded fuel only on the dashboard and near the fuel filler inlet. Furthermore, a number of record-keeping and reporting obligations for importers and refiners of gasoline have been loosened. Opponents see the lead credit trading program as a successful example of a cap and trade system that permits a pollutant to be gradually reduced. By 1987, lead credit trading had surpassed 40% of total lead use. Early in the program, an estimated 20% of refineries took part in trading; this number eventually increased to 60% of refineries.

### *9.3.7.2. Benzene in Gasoline*

The Mobile Source Air Toxics Rule was established in 2007 in an effort to reduce the harmful emissions that mobile sources' fuel combustion produces. One specific gasoline ingredient that is known to be harmful to human health is benzene. The average gasoline concentration of benzene in 2007 was 1% by volume. Beginning in 2011, the EPA required refiners and importers to start producing gasoline with an average yearly benzene content of no more than 0.62%. While refiners can petition the EPA to approve additional technologies, the EPA has listed a few that can be used to meet the new standards. Prior to 2011, refiners and importers who brought their benzene levels below 0.62% could receive credits. By offering these credits up for auction to other businesses, a marketable allowance strategy for lowering the amount of benzene in gasoline would effectively be established. The countrywide trading and banking system is not without its limitations. Even with credits, no refiner or importer could produce gasoline with benzene concentrations higher than 1.3% by volume.

## REFERENCES


1. Goldberg, P. K. (1998). The effects of the Corporate Average Fuel Efficiency Standards in the US. *The Journal of Industrial Economics, 46*(1), 1–33. https://doi.org/10.1111/1467-6451.00059. S2CID 152584179.
2. Nenes, A., Pandis, S., & Pilinis, C. (1998). ISORROPIA: A new thermodynamic equilibrium model for multiphase multicompetent inorganic aerosols. *Aquatic Geochemistry, 4*, 123–152. https://doi.org/10.1023/A:1009604003981.
3. Schroeder, J. R., Crawford, J. H., Fried, A., Walega, J., Weinheimer, A., Wisthaler, A., Müller, M., Mikoviny, T., Chen, G., Shook, M., Blake, D. R., Diskin, G., Estes, M., Thompson, A. M., Lefer, B. L., Long, R., & Mattson, E. (2016). Formaldehyde column density measurements as a suitable pathway to estimate near-surface ozone tendencies from space: $O_3$-$CH_2O$ relationship. *Journal of Geophysical Research: Atmospheres, 121*, 13088–13112. https://doi.org/10.1002/2016JD025419.
4. Simon, H., Baker, K. R., & Phillips, S. (2012). Compilation and interpretation of photochemical model performance statistics published between 2006 and 2012. *Atmospheric Environment, 61*, 124–139. https://doi.org/10.1016/j.atmosenv.2012.07.012.
5. U.S. Environmental Protection Agency. (1999). *OAQPS Economic Analysis Resource Document*. U.S. Environmental Protection Agency. Retrieved from http://www.epa.gov/ttn/ecas/econdata/Rmanual2/.
6. Valin, L. C., Fiore, A. M., Chance, K., & González Abad, G. (2016). The role of OH production in interpreting the variability of CH2O columns in the southeast U.S.: OH, VOC and CH2O columns in the southeast U.S.A. *Journal of Geophysical Research: Atmospheres, 121*, 478–493. https://doi.org/10.1002/2015JD024012.
7. Wang, Y., Li, L., Chen, C., Huang, C., Huang, H., Feng, J., Wang, S., Wang, H., Zhang, G., Zhou, M., Cheng, P., Wu, M., Sheng, G., Fu, J., Hu, Y., Russell, A. G., & Wumaer, A. (2014). Source apportionment of fine particulate matter during autumn haze episodes in Shanghai, China: Source apportionment. *Journal of Geophysical Research: Atmospheres, 119*, 1903–1914. https://doi.org/10.1002/2013JD019630.

*Chapter* **10**

# Pollution Prevention

## Contents

# INTRODUCTION

The environmental community has been focused on the measurement and management of air pollution for many years through government regulations and research projects. However, a novel approach to waste management known as "Pollution Prevention" is gradually making its way into the pages of air pollution control. For many years, experts in air pollution have focused primarily on controlling and eradicating air pollution, with little attention paid to preventing its genesis. Acknowledging this shortcoming, Congress established the foundation and structure for a successful pollution prevention program in 1990 with the passage of the Clean Air Act Amendment. The goal of pollution prevention is to minimize or stop the production of waste before it is released into the atmosphere (Theodore & McGuinn, 1992). In this new endeavor to manage air pollution, scientists, engineers, and technicians concentrate on how businesses can cut back on or completely stop producing waste at the source. In industrial or commercial settings, reducing pollution may necessitate substituting less hazardous refining process chemicals for those that pose less risk to human health or the environment. In order to minimize or completely eradicate air pollution, inventive modifications to machinery, processes, or plant operating procedures may also fall under the category of pollution prevention. Lastly, more effective maintenance practices to cut pollution or save energy at the source may be a part of pollution prevention.

Nevertheless, there is no assurance that all waste from the production process will be removed by using pollution prevention techniques like these. Rather, these approaches aim to provide industry with affordable means of preventing needless waste and pollution from entering the system at all stages of production. Strategies for reducing pollution are just one of many steps that must be taken to stop air pollution's detrimental effects on the environment and human health. As the expert in air pollution needs to understand pollution avoidance and waste management methods, this chapter provides a useful overview of the legislative background of this relatively new field and the different approaches that can be used to cut down on production waste.

## 10.1. LEGISLATIVE HISTORY

Congress and the US are mostly responsible for the air pollution problem in the US since it is mostly a plant, process, or vehicle issue. In order to reduce these pollutants, the EPA created a new management plan. With the passage of the Pollution Prevention Act (1990), the emphasis shifted from measuring and regulating industrial emissions to completely preventing their creation. Pollution Prevention is defined explicitly in the Act, even though there may not seem to be much of a difference between the two methods. In order to remove the threat to human health and the environment, the legislation defines pollution prevention as any reduction of generated waste or recycling activity carried out at the pollutant source that either (1) results in a reduction in the total volume or quantity of waste or (2) a reduction in the toxicity of the waste.

### 10.1.1. Resource Conservation and Recovery Act (1984)

In 1984, the Resource Conservation and Recovery Act (RCRA) underwent an amendment that introduced the notion of pollution prevention. Congress declared that the generation of hazardous waste was to be minimized or eliminated wherever feasible, and established "waste minimization"—as pollution prevention was originally defined—as a national policy goal. Following the RCRA, Congress released a significant report in 1986 (EPA/530-SW-033) that clarified and narrowed the definition of "waste minimization," also known as pollution prevention (Theodore & Young, 1992). More than 400 different chemicals and compounds that need to be treated or disposed of are governed by this legislation.

### 10.1.2. Superfund Amendments and Reauthorization Act (1986)

The Emergency Planning and Community Right-To-Know Act (EPCRA), also known as SARA Title III, was established by the Superfund Amendments and Reauthorization Act (SARA) of 1986. This law was created to make it easier for the public to obtain information about nearby chemical hazards and to make it easier for local and state agencies to create chemical emergency response plans. Under EPCRA, businesses falling under particular categories that produce, handle, or utilize chemicals in given amounts are required to keep track of hazardous material inventories, submit written reports, and notify anyone who spills or releases chemicals into the environment. According to EPCRA Section 313 some companies must file yearly reports called the Toxic Chemical Release Inventory (Form R), which lists any releases of "toxic chemicals." Unless a trade secret claim protects it, all information submitted in accordance with EPCRA regulations is publicly accessible. Certain companies are required to submit annual reports listing any releases of "toxic chemicals" in relation to EPCRA Section 313, and commonly known as the Toxic Chemical Release Inventory (Form R). Additionally, the EPCRA mandated the creation of State Emergency Response Commissions (SERCs), which were tasked with designating local emergency planning committees (LEPCs) and coordinating specific emergency response activities.

### 10.1.3. Pollution Prevention Act (1990)

In 1990, this historic law was passed, making the prevention of pollution a national priority. The US EPA's Office of Pollution Prevention was established by the Pollution Prevention Act (PPA), which elevated pollution prevention from "a desirable goal" to the status of national policy. The law requires manufacturing facilities to elucidate the steps taken to eliminate or reduce the quantities of any toxic chemicals they release into the environment, as well as to report information about the management of each chemical they treat or release (i.e., pollution prevention and recycling program).

Congress enacted a national policy that: under Section 6602(b) of the Pollution Prevention Act of 1990.

- Every time it is possible, pollution should be stopped or reduced at its source.
- When recycling is an option, it should be done in an environmentally responsible way to minimize pollution that cannot be avoided.
- If pollution cannot be avoided or recycled, it should, whenever possible, be handled in an environmentally responsible way.
- As a last resort, disposal or other releases into the environment should be made, but they should be done safely and with minimal impact on the environment.

According to the Pollution Prevention Act, "source reduction" and other methods that lessen or completely stop the production of pollutants are referred to as pollution prevention.

- enhanced effectiveness in the utilization of energy, water, raw materials, and other resources.
- preservation through conservation of natural resources. According to the Pollution Prevention Act, "source reduction" refers to any action taken to lessen the quantity of pollutants, hazardous materials, or other contaminants that are released into the environment—including fugitive emissions—before they are recycled, treated, or disposed of.
- minimizes the risks to the environment and public health brought on by the release of such drugs, pollutants, or contaminants.

This phrase encompasses the following: modifications to machinery or technology; adjustments to processes or procedures; reformulation or redesign of products; replacement of raw materials; and enhancements to housekeeping, maintenance, training, or inventory control. Recycling, energy recovery, treatment, and disposal are not considered forms of pollution prevention under the terms of the Pollution Prevention Act. Certain actions that are frequently labeled as "in-process recycling" might actually be considered pollution control. Many of the benefits of prevention apply to recycling when done sustainably; it can cut down on the need for treatment or disposal while also saving energy and resources. Form R Section 6607, which extends and mandates source reduction and recycling information on the EPCRA list of toxic chemicals, underwent additional requirements with the passing of PPA. Information types include: amounts of chemicals recycled, source reduction strategies, production changes from the prior year, and a list of all listed toxic chemicals entering a waste stream, along with the percentage change from the previous year (Berry & Dennison, 2000).

The Act's provisions even made it necessary to perform better at pollution prevention before a plant could continue operating or have its permit reauthorized. To further the goals of pollution prevention, local government organizations were given the legal and regulatory authority to create codes, issue licenses, or permits. In fact, one of the most important pieces of legislation in recent memory for precisely defining and outlining plans for pollution prevention practices and initiatives is the Pollution Prevention Act (Theodore & Young, 1992). The United States set a voluntary 33/50 program in February 1991 with the overall national goal of reducing the generation of 17 high-priority toxic chemicals by 50% by the year 1995. By the end of 1992, a 33% reduction was to be achieved as an interim program goal. When 750 million pounds—a decrease from 1.5 billion pounds reported in 1988—were reported for the 17 covered chemicals, the goal was achieved in 1995. There was a 60% decrease in total, or 896 million pounds, between 1988 and 1996. The program's objective was achieved, but it was discontinued in 1996.

The RCRA and the Pollution Prevention Act are undoubtedly the two most significant pieces of legislation that have an impact on federally mandated pollution prevention programs; nevertheless, a number of other regulations also give facilities legislative direction when they are creating pollution prevention programs. The Comprehensive Environmental Response, Compensation, and Liability Act (CERCLA), the Superfund Amendments and Reauthorization Act (SARA), the Hazardous and Solid Waste Amendments (HSWA), the Safe Drinking Water Act (SDWA), the Toxic Substance Control Act (TSCA), Federal Insecticide, the Fungicide and Rodenticide Act (FIFRA), and the Occupational Safety and Health Administration (OSHA) Act provide the specific details required to move each facility from pollution control to pollution prevention (Theodore & Young, 1992). Additionally, the Clean Air Act (CAA) of 1990 established MACTs with various P2 provisions under NESHAPs, which initially only regulated six specific pollutants.

## 10.2. POLLUTION PREVENTION STRATEGIES

Pollution prevention, by definition, starts at the source of the pollution. Thus, careful thought and planning are essential to the prevention of pollution. Pollution prevention is far more proactive in this sense than conventional control methods. Whatever pollutants remain after all other options for lowering or getting rid of them have been explored are recycled back into regular plant operations. Reducing the toxicity of the pollutant before it is released into the atmosphere is the most efficient control technology if the residual waste lacks significant fuel properties or recyclable qualities. Pollution disposal into wells, landfills, or the ocean ought to be reserved for extreme circumstances.

Reducing air pollution is the general pollution prevention strategy when it comes to air quality. It is best to achieve this objective by taking a series of steps in order. The importance of each step in the pollution prevention process cannot be overstated, but the order and priority of these steps also matter. Reducing sources of pollution, recycling, treating waste, and disposing of it properly are the steps of an effective pollution prevention strategy (Figure 10.1).

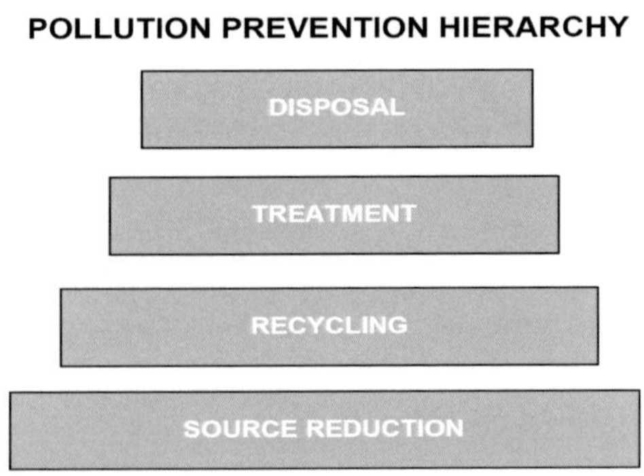


**Figure 10.1.** *Pollution prevention hierarchy.*

*Source: Theodore, Louis and McGuinn, Young C., 1992, Pollution Prevention, Van Nostrand Reinhold: New York, p. 107. ISBN: 0442006063.*

### 10.2.1. Source Reduction

Any action that lessens or stops the creation of waste at its source, usually during a production process, is referred to as source reduction (Theodore & Young, 1992). Ideally, the chance to utilize any of the numerous source reduction strategies arises prior to the generation of a pollutant. Preventing waste from arising has the benefit of completely avoiding the need to handle or dispose of it afterwards. Unlike control technology solutions that require costly retooling, source reduction options often only require minor procedural changes during processing. Substituting materials and controlling sources are two of the most effective ways to reduce sources.

When a fuel source or raw material is swapped out for one that is equally suitable for manufacturing but less volatile or toxic, this is known as material substitution. The substitution of hydrochlorofluorocarbons (HCFCs) and hydrofluorocarbons (HFCs), which are less disruptive to stratospheric ozone, for chlorofluorocarbons (CFCs) is an effective example of material substitution. The process of washing coal and other ores before heating them up and undergoing other waste-producing manufacturing processes is another excellent illustration of material substitution. When a raw material is altered or purified in this way, the quantity of byproducts that need to be decreased or removed is often decreased. Reducing pollutant emissions can also be accomplished through source control. The amount or toxicity of a pollutant can occasionally be significantly impacted by modifications made to a plant's machinery or operating conditions. Toxic spills and fugitive emissions caused by chemical mixing can also be prevented by physically separating hazardous substances from one another. Changes to plant design and operations or material handling practices can generally be an effective way to reduce the number of catastrophic public health episodes hastened by air pollution, regardless of which methods may be appropriate for a particular industry.

## 10.2.2. Recycling

Recycling methods are a crucial part of preventing pollution, but they should only be applied after all other options for reducing pollution have been explored. When a waste product can be used to power a plant's manufacturing operation as an alternative or supplement to fuel, on-site recycling of waste is particularly effective. In fact, some waste by-products have enough energy value to be used as the main fuel source for a number of industrial processes (Figure 10.2).

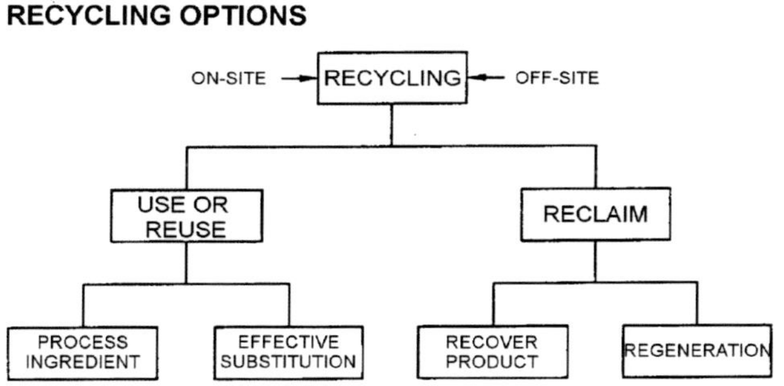


**Figure 10.2.** *Recycling options.*

*Source: Theodore, Louis and Young C. McGuinn. Pollution Prevention. Van Nostrand Reinhold: New York, 1992, p. 107.*

On-site reuse does not, however, mean that recycling is over. Waste that isn't worth energy in one industry can often be a perfect fit for another. Industries benefit economically and environmentally from exchanging waste because they can profit from each other's waste redistribution. Recycling should, nevertheless, continue at every stage, from the factory to the customer. To increase the lifespan of consumed products, undoubtedly more work needs to be done on all fronts. Reusing, transforming, and recycling raw materials for various purposes not only increases the material's lifespan and usefulness but also protects valuable resources for next generations (Theodore & Young, 1992).

### 10.2.3. Treatment

Pollutants can occasionally have a chemical makeup that makes them unsuitable for effective recycling, substitution, or reduction. One of the main obstacles facing pollution prevention specialists is dealing with these materials, which are typically highly toxic waste forms. Upon reaching this stage, chemicals need to be properly disposed of in safe places or transformed into a less hazardous state. One of the few remaining options for the safe disposal of pollutants is waste treatment, which is typically achieved through chemical or mechanical means. Incineration, chemical modification, or biological and physical treatment are some methods of handling hazardous materials. Many methods of getting rid of this kind of waste have been made possible by technology, but burning it is the best option. An ash-based method called incineration not only eliminates most of the contaminant but also changes its toxic nature. According to Theodore and Young (1992), the main objective of any treatment procedure is to lessen the quantity and toxicity of a material that cannot be removed or recycled.

### 10.2.4. Disposal

The final chance to safely protect people and the environment from the damaging effects of pollution is frequently waste disposal. Any substance that cannot be made harmless must be disposed of safely and responsibly, even though disposal is the least desirable of all the options mentioned. The Hazardous and Solid Waste Amendments (1984) delineate the distinct obligations of government agencies or manufacturers in terms of appropriately protecting the public from the disposal of hazardous waste. For instance, landfarming, deep-well injection, landfilling, and ocean dumping are the four procedures most frequently used to dispose of hazardous waste (Theodore & Young, 1992). Solid waste can be buried in landfills, injected and stored in deep wells, placed in storage containers and buried beneath the ocean, or tilled into the soil and used as fertilizer or to replenish essential nutrients. All of these approaches acknowledge that not all waste can be adequately treated, recycled, or removed from its source. However, none of them is a particularly appealing way to dispose of waste. Therefore, the issues surrounding waste disposal should strengthen our resolve to find more effective ways to stop pollution.

## 10.3. LAWS & REGULATIONS

There are still numerous fronts in the fight against air pollution: scientific, technological, and legal. On the legal front, new state and local laws governing air pollution control and stricter federal emission standards are being used to carry out the fight for clean air. As a result, laws and regulations are crucial to the management of air quality. Laws and regulations will always be one of the most crucial tools available to air pollution professionals to address a persistent issue, even though the science and technology of air pollution control should continue to advance.

### 10.3.1. Common Law

A citizen's options for stopping the spread of air pollution were limited prior to the establishment of air pollution regulations. Due to the lack of knowledge regarding the impact of air pollutants on human health and the environment, both factories and nearby residents released lethal amounts of toxins into the sky. In the past, if a citizen could demonstrate that a concrete injury had been caused, he could use common law principles and remedies to bring a lawsuit against another citizen. One of the key characteristics of common law was that it was not created by a legislative body in the way that statute law is created today. Rather, common law developed from customs and traditions in various contexts and was upheld by court rulings. It is crucial to understand that, unlike statutory laws, common law is not a set of unchanging or absolute written rules. A slow and natural process results in the orderly development of common laws as the courts consider each new set of facts in light of prior precedent. Nuisance and trespass are two common law concepts that are relevant to air pollution control.

#### *10.3.1.1. Nuisance*

Pollution was first classified as a nuisance in 1,611 when an English court upheld an injunction and granted damages to a plaintiff after it was demonstrated that the defendant had built a hog sty so close to the plaintiff's home that it contaminated the air.

Under common law, the right to breathe clean air was a property right that could be safeguarded through a nuisance action rather than a personal environmental right. Throughout the Industrial Revolution, a community's vibrancy and financial prosperity were greatly enhanced by its factories. A judge would only find an unreasonable interference with the public's right to common areas if the pollutant's effects affected every member of the community. The aristocracy's property rights were always valued more highly than the annoyance of a single person. As a result, courts rarely granted relief against complaints of nuisance in the form of a permanent injunction.

Tort law, a type of legislation that allowed a remedy for actual bodily harm or damage done to an individual, covered all civil suits. A nuisance claim under tort law presumed that the plaintiff had interfered unreasonable with the other party's use of the plaintiff's property. A deliberate or careless act could cause nuisance, but it was the injured party's responsibility to prove to a judge or jury that the act constituted "unreasonable" interference. Common law established that the injured party was only entitled to the exact amount of damages if they won. Common law was rarely favorable to an harmed party because the effects of air pollution on human health and the environment were frequently hard to measure and even harder to prove.

A substantial and unreasonable interference with another person's use of their property is the definition of a common law nuisance, unchanged from more than a century ago. Courts continue to weigh the degree of harm a defendant's actions cause in relation to the defendant when determining whether or not a nuisance has been established. When granting an injunction, for instance, to shut down or restrict the defendant's conduct, the court will take into account a number of factors, including the defendant's operating hours, the extent of the harm caused, who arrived first (noticing the nuisance), laches (the length of time the plaintiff tolerated the interference before filing a lawsuit), how both parties used their property, how other members of the community behaved, how an injunction would affect people's ability to go about their daily lives, and the availability of a method for controlling pollution.

In places without an air pollution control standard, the nuisance theory can be applied with good results. In an odor case, for instance, the most persuasive evidence comes from witnesses who can testify about how odors from a source make them sick, make their food taste strange, keep them awake at night, etc. Nuisance acts fall into two categories: private and public. Only the plaintiff is harmed by a private nuisance. The community is harmed by a public nuisance (i.e., casino, or roadblock on a highway). Therefore, in our case of the smells, if only one person complains about the smells, he or she can, typically without the assistance of environmental agencies, file a private nuisance action in court. An environmental agency will typically file a public nuisance action on behalf of a community that files complaints regarding the smells.

### *10.3.1.2. Trespass*

Another common law tool that a citizen could use to pursue legal action was the idea of trespass. It was said that trespassing happened when a landowner was not allowed to use his land exclusively. Many claimed that smoke carrying particulates that blew onto a plot of land and hurt people violated the tort law's trespass clause. Because the requirements for "unreasonable interference" and proving actual damages were generally waived, a plaintiff was generally more likely to succeed in proving trespass rather than nuisance. Unfortunately, environmental claims faced significant opposition from the political and social milieu of the newly industrialized British Empire. Therefore, neither the trespass argument nor the nuisance claims helped to address the bigger issue of air pollution.

## 10.3.2. Statutory Law

Ultimately, one of the main causes of the rise of statutory law in America was the common law's incapacity to safeguard the larger public interest. The colonists from America, having escaped the injustices of their native England, were well aware of the influence that social class and wealth had on common law cases. They made the decision to give legal authority to an elected legislative body that would enact laws to fairly regulate public behavior, as opposed to establishing a government based on the same abuse from which they had fled. The administrative branch was established to carry out the enforcement of statutory law, while the judicial branch was established to interpret it in cooperation with the legislative body.

### *10.3.2.1. Police Power*

The foundation of state and local authority to regulate air pollution is police power. In a nutshell, police power is the authority granted to the states by the U.S. Constitution to enact laws for the prevention, detection, and sentencing of criminal activity as well as the maintenance of public peace, order, and morality. While local governments are granted home-rule power by their state constitution, or if the state grants them the authority to pass laws through enabling statutes, local governments do not have the police power to enact laws. Legislation pertaining to pollution control can be deemed constitutional as long as it is based on the need to safeguard public health. Nonetheless, some laws pertaining to pollution control also address other topics, like leisure and artistic pursuits. It is probably safe to say that when a statute's scope expands beyond issues related to public health, its constitutionality becomes less strong.

### 10.3.3. Administrative Law

Practically speaking, though, Congress was unable to anticipate every administrative rule or regulation that would be required to restrict public behavior. In this sense, organizations like the US EPA were established with congressional power to draft the regulations required to correctly carry out statutory law. A comprehensive corpus of administrative law was established, consisting of guidelines and directives commonly denoted as administrative/regulatory law. Regulatory law is always being contested, reinterpreted, and modified to conform to congressional intent, even though it carries the full weight and authority of the US Congress. Administrative agencies have no inherent or residual authority, but can act (pass laws, etc.) only from authority that is "delegated" to them from state or federal statutes. These laws are referred to as "enabling" laws or acts. Usually, enabling legislation will outline the protocols that an administrative agency needs to adhere to. The agency is required to follow these protocols. The agency must adhere to the Administrative Procedure Act's guidelines if the enabling legislation is silent on agency procedures (APA). State APAs are available for state agencies, and federal APAs are available for federal agencies.

Historically, administrative agencies have been established to address specialized legal matters by completing the details of a broad statutory framework. This practice dates back to the 1930s. This has been applied to the field of air pollution control with great success. Lawmakers create agencies, which then employ experts and technical staff to draft laws and regulations in particular fields. This strategy has the following advantages.

1. More objectivity and predictability.
2. In the case of an emission regulation violation, the legislature has essentially declared that any emission that exceeds the standard is illegal, so no harm needs to be demonstrated.

Depending on which governmental body passed the law, the laws that are passed are either statutes, regulations, or ordinances. Laws passed by legislative bodies are referred to as "statutes" (i.e., state environmental acts and federal CAA). Laws passed by local governments are referred to as "ordinances." Laws passed by state or federal governmental bodies are referred to as "regulations." The "Code of Federal Regulations" (CFR) contains all agency regulations at the federal level. Every day, the "Federal Register" publishes the federal regulations, both proposed and final. State agencies operate in a comparable manner.

## 10.3.4. Constitutional Law

The federal and state constitutions define the authorities of governments as well as the rights of individuals. The federal government is a limited power government that can only take action using the authorities specified in the Constitution. These broadly construed powers include the authority to enact federal laws safeguarding the welfare and health of the public. The Constitution also lists the rights of individuals, and there is a long history of judicial interpretation and definition of these rights. Inspectors must take into account a number of US Constitutional amendments in the course of their work. Governmental action is restricted by these constitutional protections.

### *10.3.4.1. Right of Entry: The Fourth Amendment*

This Amendment provides a guarantee against arbitrary search and seizure for all citizens, including corporations. When entry is refused, a search warrant must be obtained. A violation of the law must be shown to have "probable cause" in order to obtain a search warrant. In Camara v Municipal Ct. of San Francisco (1967), the U.S. Supreme Court said that only "reasonable cause" is needed for administrative searches; (this standard requires less proof than is necessary for "probable cause").

The majority of air pollution control statutes contain a clause that permits entry into a facility at reasonable hours for the purpose of looking into suspected pollution activities, which is why an inspector may find themselves in a search and seizure situation. In most cases, inspectors can begin an investigation without first obtaining a search warrant. The argument that air pollution regulations are in place to safeguard the public's health typically serves as justification for warrantless searches. If a warrant is not needed, the reasonableness of the search—also known as "reasonable cause"—is what determines its propriety. A certain level of propriety must be used and the search must be reasonably related to controlling air pollution (i.e., the inspection must be conducted during reasonable hours by individuals who can properly identify themselves to the facility's owner or manager). Thus, when there is "reasonable cause" (i.e., visible emissions or a complaint) that a statute (CAA) is being violated at the facility, or when the plant was inspected due to an existing scheme or plan for entries (to prevent malevolent motivated inspections), a warrantless search will be justified.

In Marshall v. Barlow (1978), the U.S Supreme Court held that the Occupational Safety and Health Administration (OSHA) warrantless search statute was unconstitutional. This meant that in order to verify if plant property was in compliance with OSHA regulations, OSHA inspectors had to get warrants before entering it. There isn't currently a rule like this for air pollution inspections. This may be as a result of the restricted scope of warrantless administrative searches permitted by Section 114(a)(2) of the CAA "Right of Entry" Statute. Or, the public's right to an investigation of the source appears stronger than in an in-plant scenario since the public is directly impacted by an air pollution violation. Administrative searches, in any case, bring up a dilemma in which one must balance a person's right to privacy with the necessity for a local, state, or federal program to safeguard its residents from risks to their health or welfare. Since there is no requirement for a search warrant when the owner or operator has granted his consent, warrants are typically only sought after entry is denied. In the event that an EPA inspector is denied entry, the agency will typically seek a search warrant rather than attempting to invoke its statutory authority. This prevents a constitutional conflict.

### *10.3.4.2. Prohibition Against Self-Incrimination: The Fifth Amendment*

This Amendment ensures that people cannot be forced to testify against their own interests. This defense is applicable in criminal prosecutions. The Fifth Amendment does not apply if the evidence is only used in civil cases involving fines or injunctions. Additionally, only individuals are covered by the Fifth Amendment; partnerships or corporations are not. A requirement found in the majority of air pollution control statutes requires sources to provide the enforcement agency with emission inventories and other data. When air pollution enforcement actions are conducted as civil lawsuits rather than criminal prosecutions, the Fifth Amendment has not proven to be a barrier to the use of self-reported data. The person in charge at the location where evidence is to be taken should be informed of his rights to remain silent, to have an attorney, and that any evidence taken could be used against him in a later criminal action in the event that the evidence or samples taken are used in a criminal action.

### *10.3.4.3. Due Process: The Fourteenth Amendment*

Equal protection under the law and due process for all citizens are guaranteed by this amendment. It declares that a person's life, liberty, or property cannot be taken away from them by the government without due process. Due process comes in two flavors: substantive and procedural. The right to a hearing or other procedures (i.e., notice) prior to the initiation of governmental action is known as "procedural due process." One of the best ways to ensure procedural due process, for instance, is to give the public an opportunity to voice their opinions before regulations are adopted or before significant agency action. "Substantive due process," the second kind of due process, examines government actions.

When a government action infringes upon a fundamental right, such as the freedom of speech and religion, the right to privacy, etc., the term "strict scrutiny" is applied. When a less important right is at risk, such as when people's health is safeguarded by legislation against air pollution, "minimum scrutiny" is applied. An environmental law must pass the "minimum scrutiny" test in order for the enforcing agency to apply it reasonably and for any standards adopted to have a reasonable relationship to the interest (public health, the quality of air resources, etc.) that the law is intended to protect.).

#### *10.3.4.4. Equal Protection: The Fourteenth Amendment*

The Equal Protection Clause of the Fourteenth Amendment prohibits the creation of laws that allow discrimination against individuals who belong to the same classification or that establish arbitrary systems of classification. This amendment examines state action with the same rigor as due process. According to the equal protection "minimum scrutiny" test, individuals or subjects may be categorized for legislative purposes if the classifications are reasonable and have a logical connection to the regulation's goal. For instance, smaller tanks might be exempt from a VOC control regulation that targets hydrocarbon storage tanks of a specific size. This would make sense because storage tank size affects evaporative loss and is a legitimate factor in volatile organic compound regulation. A rule would be arbitrary, though, if it said that any hydrocarbon storage tank that had a painting of a bumblebee on its side would not be subject to it. Controlling air pollution would not be reasonably impacted by such a rule.

## 10.4. COMPLIANCE AND ENFORCEMENT

The goal of an agency's enforcement operations is to execute fieldwork aimed at bringing sources into compliance with regulations as soon as feasible and keeping them that way. Agencies differ greatly in their enforcement tactics, and the policies that determine their tactics are just as individualistic as the agencies themselves. Different enforcement mechanisms are presented in this section. Inspectors are limited in what mechanisms they can select from, but hopefully this will broaden their enforcement horizons.

### 10.4.1. Enforcement Systems

Five enforcement systems are implemented with control strategies woven into them: (a) construction and operation permits; (b) compliance monitoring; (c) enforcement of compliance plans; (d) surveillance and complaint handling; and (e) filing of an enforcement action (response).

#### *10.4.1.1. Permits to Construct*

In the event that emissions from a source: (a) violate applicable sections of the control strategy; (b) prevent attainment or maintenance of one or more of the National Ambient Air Quality Standards (NAAQS); or (c) significantly deteriorate the air quality that is currently better than that required by the NAAQS, the permit's goal is to prevent the construction of a new stationary source or the modification of an existing source. The permit systems help with enforcement and can also help the applicant in the following ways: (a) allowing for engineering review before construction, so any necessary changes can be made for less money than after construction begins (not applicable to registration); (b) stopping construction before it begins if the new source does not comply in all respects; (c) requiring, if necessary, that the permit documents highlight parameters which are important to the proper functioning of the control equipment; (d) ensuring that the required emission monitors will be installed; (e) requiring, in some cases, that the permit document include an operations and maintenance program; (f) denying operating permits if inspection or tests reveal noncompliance, so the source cannot legally operate until it is in compliance; (g) giving.

#### *10.4.1.2. Operating Permits*

The 1990 Clean Air Act Amendments imposed strict guidelines on industrial operations regarding the discharge of pollutants into the atmosphere. One of the CAA Amendments' Titles V directed the U.S. EPA to develop a unified permit program comprising the following components: (1) standard application with predetermined approval criteria; (2) periodic monitoring and reporting obligations; (3) annual renewal fee payment; (4) minimum training requirements for site personnel; and (5) establish US EPA authority to issue permits and ensure compliance.

One effective enforcement mechanism is to make sources submit applications for the renewal of their operating permit on a regular basis. The objective is to arrange a timetabled examination of all sources and, based on the outcome of the assessment procedure, to either reissue or refuse an operating permit. Operating licenses support law enforcement in the following ways.

i. Subjecting sources to periodic review.
ii. Allowing an operating permit to be denied, which would make it illegal for a source to operate.
iii. Delivering regular updates to the original permit documents about changes in ownership, procedures, materials, etc.
iv. Updating emission inventory

v. Providing a systematic check of compliance status by:
   a. Observing visible emissions
   b. examining the instrumentation of control devices and emission monitors
   c. Reviewing recorded emissions data
   d. checking control devices for proper maintenance and operation practices.

The permit system's objectives are to monitor data regarding air pollution sources, develop suitable program components, and assess potential future management approaches. Sources of air pollution must apply for permits in order to build and run a facility in order for government regulations to be effective. Control agencies must adhere to the permitting requirements in order to accurately maintain up-to-date records regarding the kind, location, and amount of emissions coming from each source of air pollution. Ensuring that each facility precisely identifies each source of its air pollution is the aim of the permit system. The permit contains various ancillary provisions, a schedule of compliance, monitoring and reporting requirements, and emissions criteria for each source. The purpose of a U.S. EPA permit to operate is that it reduces pertinent information about each facility to one document and creates a contractual agreement between government regulatory agencies and private industry for continued operation.

### *10.4.1.3. Compliance Monitoring and Inspections*

Regulatory bodies use compliance monitoring as a tool to assess whether environmental laws and regulations are being followed. Continuous self-monitoring of a regulated facility and compliance inspections are the two methods of compliance monitoring that are employed. The majority of state regulatory bodies mandate that specific facilities report their compliance status and conduct self-monitoring. Because there are more industrial sites than there are state inspection teams, and because state inspection teams have fewer resources, proper compliance requires full participation from every facility. A typical monitoring regulation will outline the obligation of a facility to track and report precise emissions data as well as the emission thresholds for each criterion pollutant present in the facility. For a site to remain operational, full adherence to the terms of the self-monitoring agreement is often required. An additional way to guarantee adherence to legal requirements is by conducting either planned or spontaneous facility inspections. Such an inspection's goals are to assess an operational facility's compliance status, obtain proof of wrongdoing, and establish a conspicuous presence of enforcement officers at regulated facilities. Sometimes, the mere "threat" of an inspection is sufficient to discourage violations at a facility since there are rarely enough inspection teams to perform frequent, routine site visits.

Inspections may be conducted as part of a regular plant visit, in response to a complaint or possible infraction, or as a follow-up to an earlier complaint. Although it is ideal for all regulated facilities to undergo routine inspections, no agency has the staff or funding to do so for every facility under its purview. Regulatory bodies are required to set inspection priorities as a result. A triage list that categorizes all facilities that pose the biggest risk to public health and the environment can be created, not too dissimilar from other businesses that go through times when resources are scarce. In urgent or urgent situations, the list might consist only of the people with the worst compliance histories. On the other hand, regulatory bodies can benefit tremendously from the creation of such a classification when they are establishing priorities and inspection schedules to ensure industrial compliance. An inspection for compliance monitoring has four main goals:

- Examine the building and note any particular environmental issues that may arise. Using this data, EPA will be able to assess if the facility complies.
- Give the EPA accurate information regarding the compliance status of a site or facility and guarantee the accuracy of self-reported data.
- Gather and store documentation of any particular issues that seem to be infractions.
- The inspection itself demonstrates a regulatory awareness that serves as a stop sign for the facility's future infractions.

### *10.4.1.4. Conducting an Inspection*

There are five steps in conducting an inspection:

1. Review rules and source records. Identify the essential details about the facility, such as names, titles, maps, process flow diagrams, and any potential safety hazards. Examine the facility's history of compliance and enforcement with regard to applicable federal, state, and local laws and regulations (i.e., self-reporting data and reports). Examine the facility's most current permits to determine the permitted uses, special circumstances, exemptions, and waivers, as well as the acceptable operating parameters.
2. Prepare an inspection plan. Understand the goal of the examination and the duties required to achieve it. Develop a plan for scheduling the inspection.

3. Enter the facility and hold an opening conference: This meeting is intended to inform the plant official of the inspection's goal, the authority that will carry it out, and the protocols that must be adhered to.
4. Conduct an inspection, obtain samples, and hold a closing conference. Arranging a post-inspection meeting with relevant facility officials will give a last chance to collect data, address queries, and sign confidentiality agreements. It is important to let facility officials know that they have the right to simultaneous measurements or a duplicate of any sample collected for laboratory analysis.
5. Prepare a report. The inspection report fulfills two functions: it gives the agency quick access to facility data and forms a significant portion of the supporting documentation for any further enforcement actions. This report needs to be precise, pertinent, thorough, coordinated, objective, neat, and readable.

### *10.4.1.5. Compliance Monitoring Strategy*

The Clean Air Act Stationary Source Compliance Monitoring Strategy (CMS) was updated by the USEPA in April 2001. The 1991 CMS is superseded by the updated policy. The following are the main modifications to the policy.

- Title V major sources and a small subset of synthetic minor sources have received special attention.
- It has been suggested that minimum frequencies be used to assess whether or not the facilities covered by this policy are in compliance. It is possible to create and discuss alternatives with the Regions to give States and local governments the ability to handle significant local compliance issues.
- The policy states clearly that there are several tools available that should be used to assess compliance, including self-certifications and conventional stack tests. It also acknowledges that, considering the abundance of self-reported data, including annual Title V compliance certifications, deviation reports, and semiannual monitoring reports based on periodic monitoring and compliance assurance monitoring, on-site visits might not be required to assess a facility's compliance status. However, it has been suggested that a minimum frequency of on-site visits be made in order to guarantee a compliance presence in the field.
- The 1987 Clean Air Act Compliance/Enforcement Guidance Manual's current levels of inspection are replaced by three categories

of compliance monitoring. The three new categories for compliance monitoring are investigations, partial compliance evaluations, and full compliance evaluations.

- CMS plans can now be submitted once every two years instead of having to be submitted annually.

The primary objective of the new 2001 compliance monitoring strategy is to give States and local governments the flexibility to address local air pollution and compliance issues while also ensuring national consistency in the development of stationary source air compliance monitoring programs.

The following procedure for putting Regions, States, and Local compliance monitoring strategy plans into action is outlined in the revised 2001 strategy:

- Every two years, states and localities submit a CMS plan to the Regions for review and approval. Every two years, regions also draft a plan for discussion with their respective States and locals.
- The annual regional response to the OECA Memorandum of Agreement (MOA) includes a summary of the plans.
- State, local, and regional compliance monitoring records are kept, and facility-specific compliance data is entered into the national air compliance database (AIRS/AFS, or its successor) by various entities.
- Every year, states, municipalities, and regions evaluate the outcomes of the compliance monitoring efforts and, if required, create an annual update to the biennial plan. Significant detours are talked about as they happen.
- Periodically, regions carry out comprehensive assessments of the entire State/local compliance-monitoring program. The Regional programs are also evaluated in a similar manner by Headquarters.

### *10.4.1.6. Compliance Monitoring Categories*

Three types of compliance monitoring are defined by the CMS in 2001: investigations, partial compliance evaluations, and full compliance evaluations. They are known as evaluations of full compliance.

#### *10.4.1.6.1. Full Compliance Evaluations*

A thorough assessment of a facility's compliance status is known as a full compliance evaluation. It covers every regulated emission unit and every regulated pollutant. It also discusses the state of each emission unit's compliance at the moment and the facility's ability to keep each emission unit compliant in the future. It ought to contain the following items:

- a review of all required reports and, if needed, the supporting documentation. This covers all tracked information submitted to the regulatory body (i.e., reports on CEM and continuous parameter monitoring, malfunctions, and excess emissions). A review of the semiannual and periodic monitoring reports, the Title V self-certifications, and any other reports mandated by the permit are also included.
- an evaluation of the operating conditions of the process and control devices, as necessary. Depending on the availability of continuous emission and periodic monitoring data, compliance certifications, and deviation reports, an on-site visit may not be required to make this assessment. Certain source categories, such as gas-fired compressor stations, boilers in large office and apartment buildings, peaking stations, and gas turbines, might not require an on-site visit to evaluate compliance.
- An observation of visible emissions when required. An examination of the operating logs and facility records.
- an evaluation of the feed rates, raw material compositions, and process rates, among other process parameters.
- An evaluation of the performance parameters of control equipment (i.e., water flow rates, temperature, pressure drop, and power levels of electrostatic precipitators).
- When no other method exists to ascertain compliance with the emission limits, a stack test is used. States and local governments should evaluate the following factors when deciding whether to conduct a stack test: the size of the emission unit; the amount of time that has passed since the last stack test; the results of that test and the margin of compliance; the state of the control equipment; and the availability and outcomes of related monitoring data. States and localities should perform a stack test whenever they think it is appropriate, in addition to when there is no other way to assess compliance.

Every two years at the very least, all Title V major sources—aside from those categorized as mega-sites—should undergo a Full Compliance Evaluation (very large sites). A Full Compliance Evaluation for mega-sites ought to be carried out, at the very least, every three years. For synthetic minor sources that emit or have the potential to emit at or above 80% of the Title V major source threshold, a Full Compliance Evaluation ought to be carried out, at the very least, once every five years. At all Title V major sources, an on-site visit should be carried out, at the very least, once every five years to verify record reviews, ensure compliance in the field, observe modifications or new construction, and identify any significant permit deviations. It is possible to conduct a Full Compliance Evaluation in stages by starting with a Partial Compliance Evaluation.

### *10.4.1.6.2. Partial Compliance Evaluations*

A documented compliance assessment that concentrates on a subset of regulated pollutants, legal requirements, or emission units at a specific facility is called a partial compliance evaluation. A superficial examination of individual reports should not be substituted for a more thorough Partial Compliance Evaluation. It can be carried out in order to evaluate a single aspect of a facility, or it can be combined with other studies over a period of one year (or three years at mega-sites) in order to fulfill the requirements of a Full Compliance Evaluation. If a chemical manufacturing facility's main area of concern is compliance with the HON MACT requirements, this kind of evaluation could be used to effectively assess that compliance. The results of the MACT evaluation could be combined with later evaluations that concentrated on the balance of other CAA requirements, should the regulatory agency determine later in the year that a Full Compliance Evaluation was required.

### *10.4.1.6.3. Investigations*

An investigation differs from the other two categories in that it usually involves a more thorough evaluation of a specific issue, is more resource-intensive, and is restricted to a specific area of a facility. It is typically based on data gleaned from Full Compliance Evaluations, industry, regulatory, or statutory targeted uses, and analysis of data not contained in EPA data systems. It works best when dealing with problems that are challenging to assess during a standard Full Compliance Evaluation due to scheduling conflicts, the kind of preliminary fieldwork needed, or the degree of analytical know-how required to assess compliance. These compliance monitoring activities include, for example, the extensive PSD/NSR and NSPS reviews of the pulp, utility, and petroleum refining industries carried out by the EPA. These investigations were started as a result of analyzes of data on industry growth that was made available to the public and a comparison of that data with records kept by the regulatory bodies regarding the quantity of PSD/NSR permits granted over the same period of time. The analyzes showed that a large number of facilities were not able to secure the required permits. The facilities saw considerable financial gains as a result of not controlling pollutant emissions as needed.

### 10.4.1.7. Compliance Plan Enforcement

Examining the progress made toward particular compliance plan, administrative order, court order, or Section 113(d) delayed compliance order milestones is the aim of a compliance plan inspection. Administrative orders are often formalized into negotiated compliance plans by many agencies. Plans for compliance can also be created for cases that are filed in court or before a hearing board. After these plans are agreed upon or implemented, an on-site inspection is required to confirm their accuracy. The planned tasks and their completion dates, which are included in the plan as identifiable benchmarks showing progress toward the final compliance status, are the most crucial components of the compliance plan. There should always be a penalty included in the compliance plan for missing the different deadlines for completion. A major facility's construction or modification would involve the following milestones in a compliance plan:

- To produce process and emission data as well as cost estimates, engineering studies, pilot projects, and source testing are used.
- Approval of funds by management (Board of Directors).
- Completion of final design, bids taken and best proposal selected.
- Approval of APC Agency. Construction permit secured.
- Order placed for equipment.
- Delivery of equipment.
- Installation and system start-up.
- Source test for compliance.
- Operating permit secured.

Enforcement actions are significantly aided by a closely watched compliance plan because:

- It is possible to identify time slippage and take action to quicken the compliance program.
- If there is no action or a significant departure from the timeline, a significant amount of time is saved in creating legal compulsion.
- Penalties for noncompliance are apparent to the source.

### 10.4.1.8. Surveillance and Complaint Response

A methodical program of searching the inspector's designated district for observable violations is how surveillance is carried out. From outside the source boundary, observations are made either randomly or in accordance with a schedule based on the source class or time. When circumstances and agency personnel allow, surveillance should be conducted around-the-clock and should involve surprise on-site inspections. One of the inspector's main responsibilities is to respond to citizen complaints. A complaint may pertain to a specific infraction, like visible emission, or it may, and frequently does, have to do with nuisance.

### *10.4.1.9. Initiation of an Enforcement Action*

When the inspector gathers evidence and establishes facts to support the occurrence of a violation of the terms and conditions of an order or compliance plan, a permit to construct or operate, or the laws and regulations, enforcement action must be taken. Enforcing compliance is meant to get all sources up to speed as quickly as feasible. The alternatives for enforcement include filing a lawsuit first and having a conversation afterwards, or voluntary compliance upon specific notice. The majority of agencies function within a range that varies based on various factors such as the nature of the infraction, agency capabilities, statute details, and the efficiency and accessibility of legal aid, including local court procedures. Other methods of enforcement consist of:

- notification of the infraction along with corrective administrative orders. Used for relatively new requirements and for first-time offenders.
- A formal administrative abatement order and a mutually agreed-upon compliance schedule are the results of an administrative conference or hearing.
- Like traffic tickets, citations are paid for without a court appearance penalties that are administratively enforced in accordance with a schedule of fines.
- Legal actions involving civil or criminal charges that result in fines and other penalties, as well as court orders with detailed compliance schedules.
- Court ordered injunctions to stop the violating practice.
- Administratively revoking a building permit or an operating permit, which renders additional construction or operation illegal.

Field staff are responsible for implementing an agency's enforcement policy, which includes decisions about notification procedures and enforcement actions. Inspectors working in the field will only be able to respond consistently to similar situations if there is a well-defined enforcement policy or a set of enforcement procedures in place.

### *10.4.1.10. Steps in an Enforcement Action*

There are five steps to an enforcement action:

- An inspection report contains the violation's location and documentation.
- Agency decision regarding the kind and extent of enforcement action to be taken.
- prepared and submitted enforcement documents. A formal complaint, compliance order, or notice of noncompliance may need to be written entirely or in part by the inspector.
- Settlement talks started. The inspector is probably going to be consulted, if not actively involved in the team that negotiates a settlement with the violator.
- Trial or hearing held. The inspector might find his notes and other documents subject to discovery in a litigated case, and he might even be called as a government witness.

### ***10.4.1.11. The Goals of Enforcement***

In essence, enforcement actions are what an agency does when there is an environmental law violation. The circumstances and gravity of the infraction determine how severe the enforcement action will be, which is primarily a subjective decision. When choosing an appropriate enforcement action, the US EPA and state agencies typically consider the following factors or objectives:

- As soon as you can, remedy the infraction. A violation has frequently resulted in an environmental issue or threat. The Agency wants to get rid of that threat or issue as soon as possible.
- prevent the same party or other parties from breaking the law in the future.
- When determining fines for comparable infractions, continue to be impartial and consistent.
- Seek criminal sanctions against the most severe violators.
- Utilize enforcement resources as effectively as possible to meet environmental and public health objectives.

## 10.4.2. Penalty Assessment Criteria

In judicial or administrative proceedings as well as citizen suits, courts and administrative law judges must take the same penalty assessment factors into account for both stationary and mobile source violations, with a few minor exceptions. The following elements must be taken into account when determining the penalty: the size of the company, the penalty's economic impact, the company's history of compliance, the length of the infraction, prior penalties for the same offense, the financial gain the company receives from noncompliance, and the gravity of the infraction.

### *10.4.2.1. Compliance Certification and Enhanced Monitoring*

Section 114(a)(3), as revised by the 1990 CAA Amendments, requires major sources to submit annual compliance certificates, reports and enhanced monitoring. Information about a source's noncompliance status ought to be included in annual compliance certifications. Congress meant for the source to bear the responsibility for compliance assurance. They are hoping that by providing this information, they will be able to claim it as proof of liability and admissions.

The EPA established the "Compliance Assurance Monitoring" (CAM) Rule (62 Fed. Reg.) in response to Congress's mandate for "enhanced monitoring" of significant sources. Reg. 54900 (1997). This Rule aims to ensure ongoing adherence to CAA emission limitations by regularly checking the control equipment at a facility (i.e., keep an eye on particular performance indicators of a control device, such as pressure, temperature, and airflow.) or any other procedures important to reaching compliance. The data generated for CAM monitoring needs to be guaranteed to "present valid and sufficient information on the actual conditions being monitored," according to performance specifications. Title V major sources are subject to the CAM Rule. These sources are in charge of creating the monitoring strategy that they think will meet CAM requirements. The CAM plan must be reviewed and approved by the Title V permitting authority. The CAM Rule has the effect of making it easier to obtain proof of liability.

### *10.4.2.2. The Any Credible Evidence Rule*

Congress granted the EPA the right to use "any credible evidence" to prove a violation in place of applicable test methods under Section 113(e)(1) of the 1990 Amendments. This eliminated the requirement for stringent proof of violations based on reference test methods approved by the EPA, using "any credible evidence." To put these provisions into effect, the EPA created regulations (See: 62 Fed. Reg. 8314 (1997). The Any Credible Evidence Rule is the name given to these regulations. Any reliable evidence may be used for enforcement under this Rule. The EPA mandated that a violation of an emission standard must be demonstrated through the use of EPA reference test procedures prior to the Any Credible Evidence Rule. According to this Rule, any reliable source of evidence, including data from continuous emission monitoring or other emissions records, may be used to demonstrate noncompliance. This Rule's benefits extend beyond EPA enforcement; it also greatly facilitates citizen suits by simplifying the process of obtaining evidence of noncompliance through the use of freedom of information acts.

### *10.4.2.3. Compliance Audits*

A business ought to think about carrying out an internal environmental compliance audit in light of the CAA's increased criminal liability. While the business will have an obligation to report CAA violations upon discovery, voluntary disclosure and collaboration with the EPA will lessen the likelihood of an enforcement action. The Environmental Protection Agency (EPA) published a notice in 1995 stating that if a company found, voluntarily reported, and quickly corrected violations as part of an audit program, the EPA would seek lower civil penalties and not pursue criminal action. Additionally, audit privilege laws have been enacted in many states. The audit privilege laws generally provide immunity (or reduction) from penalties for violations found during a compliance audit, or they provide a qualified evidentiary privilege. The majority of privilege laws mandate that the business takes appropriate corrective action.

### *10.4.2.4. Enforcement Responsibility*

According to Section 101(a)(3) of the Clean Air Act, state and local governments bear primary responsibility for preventing and controlling air pollution at its source. All governmental levels monitor and enforce laws and regulations pertaining to air pollution, but most compliance and enforcement actions are typically handled by state and local agencies. The U.S. usually, EPA doesn't get involved until a state shows that it is unwilling to uphold the terms of its own control programs. State Implementation Plans (SIPs) are thus the most crucial instrument and the primary source of state authority to impose pollution control programs. The Code of Federal Regulations, Title 40, Part 51 (40 CFR 51), gives the states particular legal authority to carry out enforcement plans (US CFR, 40 CRF 51). States have a variety of options for ensuring compliance. For instance, they can check that the right control technology is installed to meet compliance deadlines by looking over new construction plans. States are also permitted to use emissions testing and site inspections to obtain data regarding the sources of pollution. Additionally, according to public law, a source needs to keep thorough records and submit reports on a regular basis that precisely measure plant emissions. Lastly, laws have given the states wide investigative authority to look into public complaints about air pollution and find out about infractions.

All State Implementation Plans (SIPs) must have a comprehensive enforcement component, and each state must make sure that its staff is adequately trained to look into compliance issues, even though the U.S. EPA is the one who approves SIPs. States may occasionally assign local air pollution agencies enforcement authority. For instance, in California, all air pollution control in and around the Los Angeles area is under the jurisdiction of the South Coast Air Quality Management District. The EPA is authorized by Section 105 of the Clean Air Act to provide funding to air pollution control agencies so they can carry out air pollution control initiatives. This is one of the main sources of funding that municipal and state governments use to run their air pollution programs.

## REFERENCES


1. Berry, J. F., & Dennison, M. S. (2000). *The Environmental Law and Compliance Handbook*. McGraw-Hill. ISBN 0071340947, 9780071340946.
2. Godish, T., Davis, W. T., & Fu, J. S. (2003). Regulation and public policy. In *Air Quality* (5th ed., pp. 48). CRC Press. https://doi.org/10.1201/9780429103254.
3. Kunz, C., Schmedemann, D., Erlinder, C., & Downs, M. (2008). *The Process of Legal Research*. Little, Brown and Company. ISBN 9780735569775.
4. Sullivan, T. F. P. (2003). Environmental law handbook (17th ed.). *Management of Environmental Quality, 14*(5), 621. https://doi.org/10.1108/meq.2003.14.5.621.1.
5. Theodore, L., & McGuinn, Y. C. (1992). *Pollution Prevention*. Van Nostrand Reinhold. ISBN 0442006063.
6. U.S. Environmental Protection Agency. (1992). *Air pollution Orientation Course* (pp. 10–15). U.S. Environmental Protection Agency.

# INDEX

**D**



**E**

Kolibri 360 GmbH
Ferdinand-Jühlke-Straße 7
99095 Erfurt, Deutschland
produktsicherheit@kolibri360.de